Automated Methods in Cryptographic Fault Analysis

Jakub Breier • Xiaolu Hou • Shivam Bhasin
Editors

Automated Methods in Cryptographic Fault Analysis

Foreword by Joan Daemen – Co-Designer of AES & SHA-3

Springer

Editors
Jakub Breier
Underwriters Laboratories
Singapore, Singapore

Xiaolu Hou
Acronis
Singapore, Singapore

Shivam Bhasin
Temasek Laboratories
Nanyang Technological University
Singapore, Singapore

ISBN 978-3-030-11332-2 ISBN 978-3-030-11333-9 (eBook)
https://doi.org/10.1007/978-3-030-11333-9

Library of Congress Control Number: 2019933913

This Springer imprint is published by the registered company Springer Nature Switzerland AG.
The registered company address is: Gewerbestrasse 11, 6330 Cham, Switzerland

The greatest of all faults is to be conscious of none.

–Thomas Carlyle

Foreword

Building cryptographic schemes that offer resistance against determined adversaries has never been easy, and the number of failed attempts largely outweighs the successful ones. However, thanks to open research, we have seen huge progress in the last 50 years. In particular, at conceptual and mathematical level, we are able to build complex cryptosystems that offer security with high assurance against adversaries that have only access to the system's input and output and not its secret or private keys. This picture becomes somewhat less rosy if the adversary can get access to side-channel information such as the power consumption or electromagnetic emanations of the—ultimately physical—devices performing the actual cryptographic computations. It becomes outright worrisome if an adversary can disturb these devices, e.g., through the power supply, ambient temperature, radiation, etc., to cause it to make faults.

We are witnessing the transition of our world into a cyber-age with ubiquitous smartphones, sensors, cameras, and all appliances thinkable connected to the Internet. The evidently huge security and privacy risks require these devices to support cryptographic schemes. Clearly, in many realistic attack scenarios, these devices are effectively exposed to physical measurements and disturbance. Therefore, there is a need to design, evaluate, and build these systems so that they offer resistance against real-world adversaries.

The vast proliferation of embedded devices, and, more importantly, the increase of their diversity, poses a humongous task for designers, evaluation labs, and implementers. While side-channel attacks still pose a challenge, evaluation of the related security is nowadays largely routine, thanks to code running in constant time and automated evaluation tools such as t-tests. For fault attacks, however, in the last decades, we have witnessed an exponential increase in attack vectors and even in ways to exploit the behavior of devices under disturbance, and the only hope we have is automation. And, that is exactly the subject of this book. It treats automation in analysis, design, and deployment of fault countermeasures in cryptosystems in a sequence of chapters that have been written by the top researchers that are active in the domain today. It will be of great use to hardware and/or software designers and implementers, evaluators, and academics active in the field of applied cryptography, and it can even be the material for a specialized course on automating fault analysis in different stages.

Nijmegen, The Netherlands
November 2018

Joan Daemen

Preface

Today's era of digital communication necessitates the need of security and reliability. Confidentiality and integrity, guaranteed by cryptography, have become imperative to our everyday lives, protecting our digital identities and limiting the attack vectors that can come from anywhere in the world at any time. Implementation attacks were shown to be capable of breaking these guarantees easily, extending the security evaluations of encryption algorithms from the realm of cryptanalysis to realms of software and electrical engineering. Evaluation standards, such as Common Criteria, consider both side-channel attacks and fault injection attacks a highly potent threat and require implementations of countermeasures for security-critical applications. However, identifying possible vulnerabilities and applying remedies requires vast amount of expertise in multiple domains and often results in a complex, time-consuming process. Moreover, it is highly influenced by a subjective perception of a person conducting these tasks.

In this book, our objective is to introduce the reader to various methods and tools that bring automation to fault analysis. More specifically, we provide ways to automate both the evaluation and the implementation of protecting measures. We believe that each part of this book brings state-of-the-art understanding of given topic and provides explanations and guidance to both academic and industrial practitioners. Each of the four parts focuses on different stages in analysis and countermeasure deployment, and these can be read in arbitrary order. Here, we summarize the covered material in more detail:

Part I Automated Fault Analysis of Symmetric Block Ciphers
This part deals with identifying potential vulnerabilities to fault injection attacks in unprotected cipher design specifications, software implementations, and hardware implementations.

Part II Automated Design and Deployment of Fault Countermeasures
After identification of vulnerabilities, this part offers automated methods of protecting software and hardware implementations.

Part III Automated Analysis of Fault Countermeasures
Whether the applied protection was implemented correctly and effectively can be evaluated by using one of the methods from this part. Again, both software, and hardware-level approaches are considered.

Part IV Automated Fault Attack Experiments
To be able to find a fault experimentally, instead of performing a manual device profiling and fault characterization, one can use methods detailed in this part, focusing on electromagnetic and laser fault injection.

The book therefore aims to follow a logical sequence that is required for securing cryptographic applications: first, analyze the existing implementation; second, deploy protection; third, evaluate the protection theoretically; and finally, conduct experimental evaluation to support the obtained analysis findings.

We believe the covered topics are of significance to the ever-growing community of hardware security practitioners that either develop new or protect the existing encryption techniques.

Singapore, Singapore — Jakub Breier
Singapore, Singapore — Xiaolu Hou
Singapore, Singapore — Shivam Bhasin
November 2018

Acknowledgments

First of all, the authors would like to express their gratitude to all the contributors for delivering high-quality chapters and making this book possible. They would also like to thank Joan Daemen for writing an inspiring foreword. Furthermore, they would like to thank Springer editorial staff who took care of the manuscript from the first proposal until the print, more specifically to Charles Glaser and Cynthya Pushparaj.

Jakub would like to thank his parents for their constant support and encouragement.

Xiaolu would like to thank her mother and grandfather for their love and inspiration.

Shivam would like to thank to his family who has always been a constant source of love and support.

Contents

Contributors

Lejla Batina Digital Security Group, Radboud University, Nijmegen, The Netherlands

Bernd Becker University of Freiburg, Freiburg, Germany

Shivam Bhasin Temasek Laboratories, Nanyang Technological University, Singapore, Singapore

Jakub Breier Underwriters Laboratories, Singapore, Singapore

Anupam Chattopadhyay School of Computer Science and Engineering, Nanyang Technological University, Singapore, Singapore

Samuel Chef Temasek Laboratories, Nanyang Technological University, Singapore, Singapore

Kais Chibani Secure-IC S.A.S., Cesson-Sévigné, France

Pallab Dasgupta Department of Computer Science and Engineering, Indian Institute of Technology, Kharagpur, India

Adrien Facon Secure-IC S.A.S., Cesson-Sévigné, France

LTCI, CNRS, Télécom ParisTech, Université Paris-Saclay, Paris, France

École Normale Supérieure, département d'informatique, CNRS, PSL Research University, Paris, France

Chee Lip Gan Temasek Laboratories, Nanyang Technological University, Singapore, Singapore

Mael Gay University of Stuttgart, Stuttgart, Germany

Dawu Gu Department of Computer Science and Engineering, Shanghai Jiao Tong University, Shanghai, China

Sylvain Guilley Secure-IC S.A.S., Cesson-Sévigné, France

LTCI, CNRS, Télécom ParisTech, Université Paris-Saclay, Paris, France

École Normale Supérieure, département d'informatique, CNRS, PSL Research University, Paris, France

Shize Guo Institute of North Electronic Equipment, Beijing, China

Wei He Shield Laboratory, Huawei International Pte. Ltd., Singapore, Singapore

Xiaolu Hou Acronis, Singapore, Singapore

Dirmanto Jap Temasek Laboratories, Nanyang Technological University, Singapore, Singapore

Mustafa Khairallah School of Physical and Mathematical Sciences, Nanyang Technological University, Singapore, Singapore

Yang Liu School of Computer Science and Engineering, Nanyang Technological University, Singapore, Singapore

Antun Maldini Faculty of Electrical Engineering and Computing, University of Zagreb, Zagreb, Croatia

Damien Marion Secure-IC S.A.S., Cesson-Sévigné, France

Yves Mathieu LTCI, CNRS, Télécom ParisTech, Université Paris-Saclay, Paris, France

Debdeep Mukhopadhyay Department of Computer Science and Engineering, Indian Institute of Technology, Kharagpur, India

Hock Guan Ong Smart Memories Pte Ltd, Singapore, Singapore

Sikhar Patranabis Department of Computer Science and Engineering, Indian Institute of Technology, Kharagpur, India

Tobias Paxian University of Freiburg, Freiburg Germany

Stjepan Picek Cyber Security Group, Delft University of Technology, Delft, The Netherlands

Ilia Polian University of Stuttgart, Stuttgart, Germany

Sayandeep Saha Department of Computer Science and Engineering, Indian Institute of Technology, Kharagpur, India

Niels Samwel Digital Security Group, Radboud University, Nijmegen, The Netherlands

Matthias Sauer University of Freiburg, Freiburg, Germany

Laurent Sauvage Secure-IC S.A.S., Cesson-Sévigné, France

LTCI, CNRS, Télécom ParisTech, Université Paris-Saclay, Paris, France

Youssef Souissi Secure-IC S.A.S., Cesson-Sévigné, France

Francois-Xavier Standaert UCL Crypto Group, Louvain-la-Neuve, Belgium

Sofiane Takarabt Secure-IC S.A.S., Cesson-Sévigné, France

Tao Wang Department of Information Engineering, Ordnance Engineering College, Hebei, China

Bolin Yang College of Information Science and Electronic Engineering, Zhejiang University, Hangzhou, China

Fan Zhang College of Information Science and Electronic Engineering, Zhejiang University, Hangzhou, China

Institute of Cyberspace Research, Zhejiang University, Hangzhou, China

State Key Laboratory of Cryptology, Beijing, China

Xinjie Zhao Institute of North Electronic Equipment, Beijing, China

Acronyms

ADFA	Algebraic Differential Fault Analysis
AES	Advanced Encryption Standard
AFA	Algebraic Fault Analysis
ALU	Arithmetic Logic Unit
ANF	Algebraic Normal Form
ASIC	Application-Specific Integrated Circuit
BFS	Breadth-First Search
BGA	Ball Grid Array
BRAM	Block Random Access Memory
CDG	Cipher Dependency Graph
CED	Concurrent Error Detection
CFA	Collision Fault Analysis
CLB	Configurable Logic Block
CNF	Conjunctive Normal Form
CPA	Correlation Power Analysis
CPLD	Complex Programmable Logic Device
CTF	Capture the Flag
DATAC	DFA Automation Tool for Assembly Code
DBA	Dynamic Binary Analyzer
DC	Differential Characteristic
DCA	Differential Computation Analysis
DCM	Digital Clock Manager
DDT	Difference Distribution Table
DES	Data Encryption Standard
DFA	Differential Fault Analysis
DFARPA	Differential Fault Attack Resistant Physical Design Automation
DFG	Data Flow Graph
DFIA	Differential Fault Intensity Analysis
DPL	Dual-Rail Precharge Logic
DRAM	Dynamic Random Access Memory
DSP	Digital Signal Processor

DUT	Device Under Test
ECC	Elliptic Curve Cryptography/Error-Correcting Code
EDA	Electronic Design Automation
EDC	Error-Detecting Code
EHM	Empirical Hardness Model
EM	Electromagnetic
EMFI	Electromagnetic Fault Injection
FA	Fault Attack
FF	Flip-Flop
FIA	Fault Injection Attack
FPGA	Field-Programmable Gate Array
GA	Genetic Algorithm
GCC	GNU Compiler Collection
GDB	GNU Debugger
HDL	Hardware Description Language
HW	Hamming Weight
IC	Integrated Circuit
IFA	Ineffective Fault Analysis
IO	Input-Output
IoT	Internet Of Things
IP	Intellectual Property
IR	Intermediate Representation/Infrared Radiation
ISA	Instruction Set Architecture
ISS	Instruction Set Simulator
LFI	Laser Fault Injection
LRA	Linear Regression Analysis
LUT	Look-Up Table
MAC	Message Authentication Code
MDS	Maximum Distance Separable
MitM	Meet-in-the-Middle
ML	Machine Learning
NUEVA	Non-Uniform Error Value Analysis
PAVT	Pre-silicon Assisted Verification Tool
PC	Program Counter
PCB	Printed Circuit Board
PD	Phase Detector
PHD	Pseudo-Hadamard Transform
PIN	Personal Identification Number
PKC	Public Key Cryptography
PLL	Phase-Locked Loop
PoI	Points-of-Interest
PRP	Pseudo-Random Permutation
PUF	Physical Unclonable Function
RAM	Random Access Memory
RISC	Reduced Instruction Set Computer

RO	Ring Oscillator
ROC	Receiver Operating Characteristic
RSA	Rivest-Shamir-Adleman
RTL	Register Transfer Level
SEA	Safe-Error Analysis
SCA	Side-Channel Attack
SHA	Secure Hash Algorithm
SKC	Secret Key Cryptography
SoC	System on Chip
SPN	Substitution-Permutation Network
SR	Success Rate
SRAM	Static Random Access Memory
SSA	Static Single Assignment
VLSI	Very Large-Scale Integration
WBC	White-Box Cryptography
WRO	Watchdog Ring Oscillator
XOR	Exclusive OR

Chapter 1
Introduction to Fault Analysis in Cryptography

Jakub Breier and Xiaolu Hou

1.1 Cryptography Background

Cryptography is ubiquitous for the modern connected world. People communicate and exchange information through electronic devices, such as mobile phones, laptops, and tablets, and these exchanges are often private in nature, and thus require creating a secure channel over a nonsecure network. This requirement had brought cryptography from the military world to every day's usage, shifting the computations from highly specialized secure computers to a wide range of microcontrollers, often with low computational power and restricted memory size.

Security of ciphers is normally determined by the length of the key, which, according to Kerckhoffs's principle, should be the only secret component of the encryption system [20]. Key length recommendations are provided by standardization bodies, such as National Institute of Standards and Technology (NIST) for USA, Agence nationale de la sécurité des systèmes d'information (ANSSI) for France, and Bundesamt für Sicherheit in der Informationstechnik (BSI) for Germany.

We can divide ciphers according to the way how encryption and decryption are performed, to *symmetric* and *public key* (or *asymmetric*) cryptography. While the symmetric ciphers use the same secret key for encryption and decryption, in case

This research was conducted when author "X. Hou" was with Nanyang Technological University.

J. Breier (✉)
Underwriters Laboratories, Singapore, Singapore
e-mail: jbreier@jbreier.com

X. Hou
Acronis, Singapore, Singapore
e-mail: ho0001lu@e.ntu.edu.sg

J. Breier et al. (eds.), *Automated Methods in Cryptographic Fault Analysis*,
https://doi.org/10.1007/978-3-030-11333-9_1

of asymmetric ciphers, there is a key pair, consisting of a private key and a public key. Depending on use case (digital signature, and encryption), one key is used for encryption, and the other for decryption. When it comes to key lengths, according to latest BSI report [27], it should be at least 128 bits for the symmetric cryptosystems, 2000 bits for RSA modulus, and 250 bits for elliptic curve cryptography.

Another division of ciphers is according to data size that can be encrypted in one run of the algorithm. *Block ciphers* encrypt the data by chunks, providing a convenient way when dealing with long messages of a fixed size. On the other hand, *stream ciphers* generate a key stream that can be used to encrypt the message bit by bit.

Cryptanalysis, as a counterpart to cryptography, tries to find methods to break the security of the cryptographic algorithms in a more efficient manner compared to exhaustive search over the key space. For established algorithms, finding a "weak" spot in the cipher is normally hard. For example, the best known cryptanalysis of AES-128 only reduces the brute-force complexity from 2^{128} to $2^{126.1}$, which is still infeasible to compute on current machines. On the other hand, a well-aimed laser fault attack can recover the secret key of AES with just one faulty and correct ciphertext pair [10].

Since the rest of the book is aimed mostly at symmetric block ciphers, we provide more details on these below.

Symmetric Block Cipher

A symmetric block cipher is an algorithm operating on blocks of data of a predefined size. It specifies two processes, encryption and decryption. The encryption takes a plaintext and a secret key as inputs and produces a ciphertext as an output. Similarly, for a ciphertext and a secret key as inputs for decryption, it outputs the plaintext. Block cipher normally consists of several rounds where each round consists of a small number of operations. Those operations scramble the input by using various transformations and adding key-dependent data. A key used in a round is referred to as *round key*. Round keys are derived from the secret key, which is called master key, by a key scheduling algorithm that works as an invertible transformation. Therefore, by getting information about a certain round key, it is possible to get the master key by using an inverse key scheduling algorithm. This is important in the context of fault attacks that normally try to recover the last round key.

1.2 Fault Injection Attacks

Fault injection attacks (FIAs) exploit a possibility to introduce a fault during the execution of a cryptographic algorithm for the purpose of getting information about the secret key. This attack method falls within *active semi-invasive* physical attacks on cryptosystems, complementary to side-channel attacks which are passive and mostly noninvasive.[1] The attack has two stages: *fault injection stage* that disturbs the device by various techniques and causes the fault (see Sect. 1.4), and *fault analysis stage* that utilizes different methods to analyze the fault propagation and recover the secret information (see Sect. 1.3).

FIAs were first introduced by Boneh et al. [8] to attack the calculation of exponentiation using Chinese Remainder Theorem in RSA. They showed that the private key can be recovered with just one pair of faulty and correct outputs. Afterwards, numerous fault analysis methods were developed for attacking various cryptography primitives. In practice, the attack is normally mounted on a real-world device, in an implementation that is either hardware—or software-based. There are many different ways to attack such implementations—one can corrupt the instruction opcodes resulting in instruction change, skip the instructions completely, flip the bits in processed constant values or register addresses, or change the values in the registers and memories directly [4].

1.3 Fault Attack Methods

When a fault is injected into a cryptographic circuit, it either produces a faulty output or an error—this happens in case it is correctly identified as an unwanted behavior and the device stops the faulty ciphertext from being outputted. Both of these states can be exploited by the attacker using various fault analysis methods. In this section, we will briefly discuss the most significant methods developed to date.

Differential Fault Analysis (DFA) Differential fault analysis (DFA) is among the most popular fault analysis methods for analyzing symmetric block ciphers. The concept of DFA was introduced by Biham and Shamir in [6] in ’97. The authors showed that it is possible to break Data Encryption Standard by analyzing 50–200 faulty ciphertexts with a bit-flip fault model.

DFA attack consists of injecting a fault into the intermediate state of the cipher, normally during one of the last encryption rounds. The fault then propagates, resulting into a faulty ciphertext. Difference between the original and the faulty ciphertext is then analyzed, giving the attacker information about the secret key. DFA exploits the properties of the non-linear operation that is used in the cryptosystem. The whole concept is similar to a classical differential cryptanalysis [5] of a reduced-round cipher.

Most of the block cryptosystems were shown to be vulnerable against DFA, since there are no efficient cipher designs that could prevent this analysis up to date. For further reading, one can find DFA on AES [39], DES [34], PRESENT [2], SIMON & SPECK [40], etc.

[1] In some cases, side-channel attacks can be semi-invasive—device decapsulation might be required when the signal is too low, e.g., in case of low-power smartcards.

Algebraic Fault Analysis [11] Algebraic fault analysis is similar to DFA. It also exploits the differences between faulted ciphertexts and original ciphertexts. The difference is that DFA relies on manual analysis, while AFA expresses the cryptographic algorithm in the form of algebraic equations and feeds the resulting system of equations to a SATisfiability (SAT) solver.

Collision Fault Analysis [7] In this fault attack, the attacker invokes a fault in the beginning of the algorithm and then tries to find a plaintext, which encrypts into the same ciphertext as the faulty ciphertext by using the same key. Such a collision can give the attacker information on the secret key. Due to the diffusion nature of cryptographic algorithms, to find a collision, the attacker injects a fault in the early rounds so that the fault propagation will enable collisions.

Ineffective Fault Analysis [7] The goal is to find a fault that does not change the intermediate result, therefore leading into a correct ciphertext. In fact, the attacker gains information from both scenarios after a fault injection—when faults do not locally modify intermediate values, and also when they do. Such attack only requires the knowledge whether the output is correct or not, ultimately leading to the information on the secret key.

Safe-Error Analysis [41] Similar to ineffective fault analysis, safe-error analysis also exploits a situation when the ciphertext does not change after the fault injection. However, safe-error analysis requires the change of the intermediate result. It utilizes a state when the data is changed but it is not used.

Fault Sensitivity Analysis [25] Exploits the side-channel information, such as sensitivity of a device to faults and uses this information to retrieve the secret key. It does not use the values of faulty ciphertexts.

Linear Fault Analysis [21] Linear fault analysis examines linear characteristics for some consecutive rounds of a block cipher. It is a combination of linear cryptanalysis and fault analysis.

1.4 Fault Injection Techniques

To physically inject a fault, the attacker normally needs an additional device that can tamper with the device implementing a cryptographic circuit. This fault injection device uses one of the fault injection techniques to momentarily disturb the integrated circuit without destroying it. There are various fault injection techniques in practice, with different levels of precision, ranging from devices that cost several hundreds of dollars to hundreds of thousands of dollars. Various levels of expertise to control these devices are also required. Figure 1.1 depicts examples of three different fault injection setups: voltage glitch, electromagnetic pulse injection, and laser fault injection. In this part, we will detail the most popular techniques.

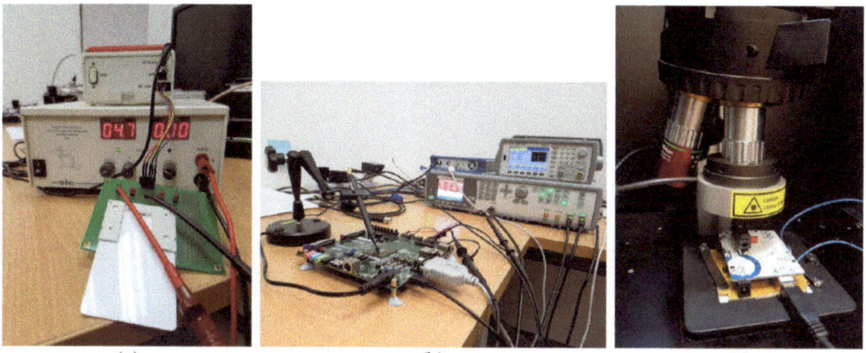

(a) (b) (c)

Fig. 1.1 Different fault injection techniques in practice, on various targets. (**a**) Voltage glitch on a smart card. (**b**) Electromagnetic pulse injection on an FPGA. (**c**) LASER fault injection on a microcontroller

Clock/Power Glitch If the device uses an external clock, then a sudden variation in the clock can result in a fault in the execution. The technique was used in [12], for example. Similarly, a glitch in the supply voltage can also result in a fault [4]. Glitch is an efficient and commonly employed method for fault injection in practice. Such technique is the most accessible way of perturbing the encryption device since the cost of the fault injection device is low, as well as the expertise required to do the attack. However, the precision of the injected fault also stays low, mostly resulting into skipping an instruction or sequences of instructions.

Optical Injection (LASER/X-Ray/Focused Ion Beam) Induction of faults by optical sources is another commonly used method. Often, this type of attack is done with a focused laser beam at a particular location of the chip, e.g. [36]. In this technique, the chip has to be depackaged to be accessible by the beam. Also, the correct location for injecting the fault has to be found by inspection, which requires specific expertise and time. While the laser setup falls into less affordable techniques, it was shown that even with a camera flash and a specific lens, the light beam can be targeted in a relatively precise manner [14]. Compared to basic optical techniques, X-ray offers an advantage of keeping the device package untouched. However, this is balanced by the cost and expertise required to mount such attack [1]. Focused ion beam falls within the most expensive injection techniques, normally used for testing and validation of integrated circuits [18]. A method for automated profiling of FPGA to laser fault injection attack is provided in Chap. 14.

Electromagnetic (EM) Emanation The transistors on the chip can be effectively forced to produce erroneous output by a high energy EM pulse for a short period of time [29]. It has a few advantages over other techniques. For example, effect of the clock/voltage glitch is generally global to the chip, whereas EM emanation can be made local to a particular location by suitably placing the probe [35]. Moreover, it does not require depackaging of the chip as in the case of optical fault, and the injection devices are cheaper.

Temperature Variation Forcing a chip to work outside its operating temperature can cause faulty operations [22]. While such technique is of extremely low cost, the resulting faults are very hard to control. Therefore, there are no practical works up to date that would show breaking the cryptographic algorithms by varying the temperature.

Other Techniques for fault injection include, for example, *rowhammer* [23]. Rowhammer is a known side effect in dynamic random-access memory (DRAM). Repeatedly accessing a row in a DRAM can lead to flipping bits in the adjacent rows of the memory. This behavior can be achieved by a malicious software (malware), running on the target machine. However, memory chip manufacturers are aware of this behavior, and newer chips utilize error-correcting mechanisms to avoid faults.

Another source of a fault can be a *Hardware Trojan*, as shown in [9]. Hardware Trojan consists of a trigger and a payload. Trigger is a condition that activates a dormant Trojan to start a malicious behavior—to deliver the payload. Hardware Trojan can be used for many adversarial purposes, the one implemented in [9] precisely flips the bits in a cryptographic circuit to achieve a desired fault propagation.

1.5 Countermeasures

Since the first reported attacks, various ways have been developed, protecting the implementation of ciphers. When selecting a countermeasure, one needs to decide what degree of protection to implement, taking into account the data value and protection cost. There is no universal countermeasure, each method has its advantages and limitations. In general, countermeasures against fault attacks focus on techniques that allow two different ways to protect the underlying implementation: *detection* and *prevention*.

Detection Countermeasures

Fault detection countermeasures aim at detecting an anomaly online to raise an alarm. Once the alarm is raised, it can be decided what will happen next—a standard emergency procedure is to prevent the device from outputting the ciphertext to thwart the attacker from analyzing it. However, as it was shown in [12], the attacker might get information about the secret even without the faulty ciphertext analysis, simply by getting the knowledge that a fault happened. Therefore, in case there is a need for strong security, it might be of interest to erase the secret information and render the device unusable. Of course, the risk of false positives has to be considered before implementing such protection.

The initial fault countermeasures used detection principles from information theory and concurrent error detection like parity code [19] or non-linear codes, etc. Concurrent error detection adds redundancy to the sensitive data processed, which allows the detection of data modification under given fault model. Such countermeasures are easy to implement and incur acceptable overhead. However, the protection is limited to specific fault models and the countermeasure can be bypassed [16]. In Chaps. 10 and 11, we show how to automatically analyze code-based countermeasures for software and hardware, respectively.

Operation-level redundancy can also be used for fault detection, e.g., duplicated computation-and-compare (for detection) or triple-modular redundancy for correction. In fact, industrial encryption schemes often involve a simple duplication, since the overheads stay within reasonable limits, and precise fault injection into both redundant circuits requires a significant effort on the attacker's side. Although, it has been demonstrated several times that such an attack is practically achievable [37, 38].

Fault detection can also be done at circuit level, by monitoring physical conditions that can be exploited for fault injection. This essentially involves design of physical sensors which often work in plug and play configuration staying algorithm independent [15]. Automated deployment of such sensors is detailed in Chap. 9.

Use of randomization has also been proposed to boost the security of fault detection countermeasures [26]. Implementation of such randomized scheme can be found in [3], where authors utilize a protocol-level detection technique against faults.

Prevention Countermeasures

Fault prevention focuses either on preventing the attacker from accessing the device or preventing her from getting a meaningful information from the faulty output.

To prevent the attacker from opening the device, chip shielding and protection packages can be used [4]. In case the attacker tries to access the chip, it would be destroyed together with the package. Infection is an algorithmic-level prevention against fault attacks [32]. It causes deeper diffusion (or pollution) of faulty value upon detection such that the faulty value is no longer usable by the attacker. Some infection approaches simply replace the faulty value by a random number. Software-level prevention can be achieved by idempotent instruction sequences [30] that normally offer fault tolerance for single fault injections. More sophisticated approaches combine redundancy with bit slicing, such as intra-instruction redundancy [33] and internal redundancy countermeasure [24]. On the protocol level, there are key refreshing techniques that are designed to prevent fault and side-channel attacks [28]. The protocol updates the session key every time with a fixed master key. Since fault attacks normally require several correct and faulty ciphertext pairs, the attack is prevented. When it comes to protecting the public key encryption schemes, it was shown that they can profit from strong arithmetic structure to prevent faults [13].

1.6 Chapter Summary

In this chapter, we presented the necessary background that will help the reader understand the latter parts of this book. For more details on fault injection attacks, techniques, and countermeasures, we advise the reader to reach to some of the books in this area that provide a comprehensive coverage of each of the topics [17, 31].

References

1. S. Anceau, P. Bleuet, J. Clédière, L. Maingault, J.-L. Rainard, R. Tucoulou, Nanofocused x-ray beam to reprogram secure circuits, in *International Conference on Cryptographic Hardware and Embedded Systems* (Springer, Berlin, 2017), pp. 175–188
2. N. Bagheri, R. Ebrahimpour, N. Ghaedi, New differential fault analysis on present. EURASIP J. Adv. Signal Process. **2013**(1), 145 (2013)
3. A. Baksi, S. Bhasin, J. Breier, M. Khairallah, T. Peyrin, Protecting block ciphers against differential fault attacks without re-keying, in *2018 IEEE International Symposium on Hardware Oriented Security and Trust (HOST)* (IEEE, Piscataway, 2018), pp.191–194
4. H. Bar-El, H. Choukri, D. Naccache, M. Tunstall, C. Whelan. The sorcerer's apprentice guide to fault attacks. Proc. IEEE **94**(2), 370–382 (2006)
5. E. Biham, A. Shamir, Differential cryptanalysis of DES-like cryptosystems, in *Advances in Cryptology-CRYPTO*, vol. 90 (Springer, Berlin, 1991), pp. 2–21
6. E. Biham, A. Shamir, Differential fault analysis of secret key cryptosystems, in *Advances in Cryptology – CRYPTO '97*, ed. by B.S. Kaliski. Lecture Notes in Computer Science, vol. 1294. (Springer, Berlin, 1997), pp. 513–525
7. J. Blömer, J.-P. Seifert, Fault based cryptanalysis of the advanced encryption standard (AES), in *International Conference on Financial Cryptography* (Springer, Berlin, 2003), pp. 162–181
8. D. Boneh, R.A. DeMillo, R.J. Lipton, On the importance of checking cryptographic protocols for faults (extended abstract), in *Advances in Cryptology – EUROCRYPT '97, Proceedings of the International Conference on the Theory and Application of Cryptographic Techniques* (Konstanz, 1997), pp. 37–51
9. J. Breier, W. He, Multiple fault attack on present with a hardware Trojan implementation in FPGA, in *2015 International Workshop on Secure Internet of Things (SIoT)* (IEEE, Piscataway, 2015), pp. 58–64
10. J. Breier, D. Jap, C.-N. Chen, Laser profiling for the back-side fault attacks: with a practical laser skip instruction attack on AES, in *Proceedings of the 1st ACM Workshop on Cyber-Physical System Security, CPSS '15* (ACM, New York, 2015), pp. 99–103
11. N.T Courtois, K. Jackson, D. Ware, Fault-algebraic attacks on inner rounds of DES, in *e-Smart'10 Proceedings: The Future of Digital Security Technologies* (Strategies Telecom and Multimedia, Montreuil, 2010)
12. C. Dobraunig, M. Eichlseder, T. Korak, S. Mangard, F. Mendel, R. Primas, Exploiting ineffective fault inductions on symmetric cryptography. Technical report, Cryptology ePrint Archive, Report 2018/071, 2018. https://eprint.iacr.org/2018/071
13. P.-A. Fouque, R. Lercier, D. Réal, F. Valette, Fault attack on elliptic curve Montgomery ladder implementation, in *2008 5th Workshop on Fault Diagnosis and Tolerance in Cryptography* (IEEE, Piscataway, 2008), pp. 92–98
14. O. Guillen, M. Gruber, F. De Santis, Low-cost setup for localized semi-invasive optical fault injection attacks—how low can we go? in *International Workshop on Constructive Side-Channel Analysis and Secure Design* (Springer, Berlin, 2017), pp. 207–222

15. W. He, J. Breier, S. Bhasin, Cheap and cheerful: a low-cost digital sensor for detecting laser fault injection attacks, in *International Conference on Security, Privacy, and Applied Cryptography Engineering* (Springer, Berlin, 2016), pp. 27–46
16. W. He, J. Breier, S. Bhasin, A. Chattopadhyay, Bypassing parity protected cryptography using laser fault injection in cyber-physical system, in *Proceedings of the 2nd ACM International Workshop on Cyber-Physical System Security* (ACM, New York, 2016), pp. 15–21
17. M. Joye, M. Tunstall. *Fault Analysis in Cryptography*, vol. 147 (Springer, Berlin, 2012)
18. J. Karlsson, P. Liden, P. Dahlgren, R. Johansson, U. Gunneflo, Using heavy-ion radiation to validate fault-handling mechanisms. IEEE Micro **14**(1), 8–23 (1994)
19. R. Karri, G. Kuznetsov, M. Goessel, Parity-based concurrent error detection of substitution-permutation network block ciphers, in *Proceedings of the Cryptographic Hardware and Embedded Systems* (IEEE, Piscataway, 2003), pp. 113–124
20. A. Kerckhoffs, A. kerckhoffs, La cryptographie militaire. J. Sci. Mil. **9**, 38 (1883)
21. C.H. Kim, Improved differential fault analysis on AES key schedule. IEEE Trans. Inf. Forensics Secur. **7**(1), 41–50 (2012)
22. C.H. Kim, J.-J. Quisquater, Faults, injection methods, and fault attacks. IEEE Des. Test Comput. **24**(6), 544–545 (2007)
23. Y. Kim, R. Daly, J. Kim, C. Fallin, J.H. Lee, D. Lee, C. Wilkerson, K. Lai, O. Mutlu, Flipping bits in memory without accessing them: an experimental study of dram disturbance errors, in *2014 ACM/IEEE 41st International Symposium on Computer Architecture (ISCA)*, vol. 42 (IEEE Press, New York, 2014), pp. 361–372
24. B. Lac, A. Canteaut, J. Fournier, R. Sirdey, Thwarting fault attacks using the internal redundancy countermeasure (IRC), in *International Symposium on Circuits and Systems (ISCAS) 2018* (Florence, 2018)
25. Y. Li, K. Sakiyama, S. Gomisawa, T. Fukunaga, J. Takahashi, K. Ohta, Fault sensitivity analysis, in *International Workshop on Cryptographic Hardware and Embedded Systems* (Springer, Berlin, 2010), pp. 320–334
26. V. Lomné, T. Roche, A. Thillard. On the need of randomness in fault attack countermeasures-application to AES, in *Proceedings of Fault Diagnosis and Tolerance in Cryptography* (IEEE, Piscataway, 2012), pp. 85–94
27. M. Margraf, Kryptographische verfahren: empfehlungen und schlüssellangen, in *Technische Richtlinie TR-02102, Bundesamt fur Sicherheit in der Informationstechnik*, 2008
28. M. Medwed, F.-X. Standaert, J. Großschädl, F. Regazzoni, Fresh re-keying: security against side-channel and fault attacks for low-cost devices, in *International Conference on Cryptology in Africa* (Springer, Berlin, 2010), pp. 279–296
29. N. Moro, A. Dehbaoui, K. Heydemann, B. Robisson, E. Encrenaz, Electromagnetic fault injection: towards a fault model on a 32-bit microcontroller, in *2013 Workshop on Fault Diagnosis and Tolerance in Cryptography* (IEEE, Piscataway, 2013), pp. 77–88
30. N. Moro, K. Heydemann, E. Encrenaz, B. Robisson, Formal verification of a software countermeasure against instruction skip attacks. J. Cryptogr. Eng. **4**(3), 145–156 (2014)
31. S. Patranabis, D. Mukhopadhyay, *Fault Tolerant Architectures for Cryptography and Hardware Security* (Springer, Berlin, 2018)
32. S. Patranabis, A. Chakraborty, D. Mukhopadhyay, Fault tolerant infective countermeasure for AES. J. Hardw. Syst. Secur. **1**(1), 3–17 (2017)
33. C. Patrick, B. Yuce, N.F. Ghalaty, P. Schaumont, Lightweight fault attack resistance in software using intra-instruction redundancy, in *International Conference on Selected Areas in Cryptography* (Springer, Berlin, 2016), pp 231–244
34. M. Rivain, Differential fault analysis on DES middle rounds, in ed. by C. Clavier, K. Gaj *Proceedings of the 11th International Workshop on Cryptographic Hardware and Embedded Systems (CHES 2009)* (Springer, Berlin), pp. 457–469
35. L. Riviere, Z. Najm, P. Rauzy, J.-L. Danger, J. Bringer, L. Sauvage, High precision fault injections on the instruction cache of ARMv7-M architectures. arXiv preprint arXiv:1510.01537 (2015)

36. B. Selmke, S. Brummer, J. Heyszl, G. Sigl, Precise laser fault injections into 90 nm and 45 nm SRAM-cells, in *International Conference on Smart Card Research and Advanced Applications* (Springer, Berlin, 2015), pp. 193–205
37. B. Selmke, J. Heyszl, G. Sigl, Attack on a DFA protected AES by simultaneous laser fault injections, in *2016 Workshop on Fault Diagnosis and Tolerance in Cryptography (FDTC)* (IEEE, Piscataway, 2016), pp. 36–46
38. E. Trichina, R. Korkikyan, Multi fault laser attacks on protected CRT-RSA, in *2010 Workshop on Fault Diagnosis and Tolerance in Cryptography* (IEEE, Santa Barbara, 2010), pp. 75–86
39. M. Tunstall, D. Mukhopadhyay, S. Ali, Differential fault analysis of the advanced encryption standard using a single fault, in *IFIP International Workshop on Information Security Theory and Practices* (Springer, Berlin, 2011), pp. 224–233
40. H. Tupsamudre, S. Bisht, D. Mukhopadhyay, Differential fault analysis on the families of SIMON and SPECK ciphers, in *2014 Workshop on Fault Diagnosis and Tolerance in Cryptography* (IEEE, Piscataway, 2014), pp. 40–48
41. S.-M. Yen, M. Joye, Checking before output may not be enough against fault-based cryptanalysis. IEEE Trans. Comput. **49**(9), 967–970 (2000)

Part I
Automated Fault Analysis of Symmetric Block Ciphers

Chapter 2
ExpFault: An Automated Framework for Block Cipher Fault Analysis

Sayandeep Saha, Debdeep Mukhopadhyay, and Pallab Dasgupta

2.1 Introduction

Support for cryptographic computation has become indispensable for modern embedded computing devices. Block ciphers, being one of the prime constituents of cryptographic protocols, are deployed with most of the modern devices. The wide variation of resource and performance requirements in computing platforms has led to the development of a broad range of block cipher designs. For example, several lightweight block cipher algorithms like PRESENT [6], LED [12], and SKINNY [4] have been deployed in recent years targeting low-resource embedded devices. However, given the fact that performance requirements are getting stringent day-by-day, there is an increasing trend of designing precisely engineered, application -specific cipher algorithms.[1] Further, the practice of designing proprietary ciphers is increasingly prevalent among defense and civil organizations. In essence, numerous commercially usable block ciphers are available today, and this number is expected to increase significantly in the coming years.

The continual increase in block cipher usage makes fault attack a necessary evil. The threat has become more severe with the advent of small embedded devices, where ensuring security is essential but challenging due to resource-constraints.[2]

[1]In fact, recently there is a call from National Institute of Standards and Technology (NIST) for standardizing lightweight cipher designs (available online at https://csrc.nist.gov/Projects/Lightweight-Cryptography).

[2]Fault attack countermeasures, in general, are resource-hungry.

S. Saha (✉) · D. Mukhopadhyay · P. Dasgupta
Department of Computer Science and Engineering, Indian Institute of Technology, Kharagpur, India
e-mail: sahasayandeep@iitkgp.ac.in; debdeep@iitkgp.ac.in; pallab@iitkgp.ac.in

J. Breier et al. (eds.), *Automated Methods in Cryptographic Fault Analysis*,
https://doi.org/10.1007/978-3-030-11333-9_2

The common trend in cipher design is to evaluate the security of a cipher against classical attacks like differential and linear cryptanalysis before it is deployed. However, evaluation of cipher algorithms against fault attacks has been largely undermined so far. Given the fact that fault attacks exploit several algorithm-level features, and are very close to classical differential cryptanalysis attacks by nature, an algorithmic evaluation is essential before implementation and deployment. Usually, countermeasures are deployed at various levels of abstractions (algorithm-level, implementation-level or system- level) to defend against fault attacks. On the other hand, countermeasures have overheads which have to be optimized carefully in order to provide proper security bounds within specified resource-constraints. Such precisely engineered countermeasures can only be devised if the complete attack space of a given cipher is well-understood. While finding a single attack instance for a system is sufficient from the perspective of an attacker, certifying a system for fault attack resilience demands comprehensive knowledge of the entire attack space. Unfortunately, most of the fault analysis efforts till date are manual, and they primarily target a couple of well-known block ciphers like AES [8] and PRESENT. Consequently, most of the existing block ciphers are yet to be analyzed extensively, to the extent that their entire attack spaces can be understood. Given a large number of available ciphers and their diverse structures, exploration of the complete fault attack space seems to be quite an arduous task with traditional manual fault analysis approaches. The example of the AES algorithm highlights the nature of this problem, where it took the research community 9 years of extensive research to discover the optimal attack.

The automation of fault attacks promises to solve the attack space characterization problem within feasible time. In this chapter, we address the problem of attack space exploration in the context of differential fault analysis (DFA) attacks [5, 10, 22]. DFA is the most widely explored and complex class of fault attacks so far and is particularly interesting given their (relatively) low data/fault complexity. Also, they form the mathematical basis for many new classes of fault attacks. It is well-established that even a single properly placed malicious fault is able to compromise the security of mathematically strong crypto-primitives in certain cases. Interestingly, not every possible fault in a cipher is exploitable in a practical sense to cause an attack, and determining the exploitability of a fault instance is nontrivial.[3] Characterization of the exploitable fault space of a block cipher is a difficult task given the formidable size and diversity of the fault space even for a single block cipher. Typically, faults in a cipher (let us focus on the block ciphers only) are specified by multiple attributes (e.g., the location, width and value of the fault, fault model, plaintext and the mathematical structure of the cipher), which eventually lead to a fault space of formidable size, accounting the need for automation in this context.

[3]This claim is certainly not restricted to DFAs only and is valid for other classes of fault attacks as well.

The very first step towards building such automation is to devise a core engine which can determine the exploitability status of an individual fault instance on a given block cipher. However, the problem of exploitable fault space exploration demands certain properties to be fulfilled by any such automated tool. Any automated methodology, which targets individual fault instances for exploitability evaluation should be sufficiently fast and scalable in order to practically explore huge fault space. On the other hand, the automation should be applicable to most of the available block ciphers and fault models. Another point of concern is the interpretability of an attack instance. Interpretability of an attack is extremely important to get necessary insights which may eventually lead to improved cipher and countermeasure designs. Designing automation which fulfills all the abovementioned properties is nontrivial, but indeed an interesting research problem.

The aim of this chapter is to introduce some of the recent advances in automated fault analysis. Research in this area is still in its infancy, and only a few references can be found. Here, we shall mainly focus on the work presented recently at CHES 2018, which simultaneously meets the goals specified above. The framework, known as *ExpFault* [20] is able to perform automated fault analysis on any given block cipher and fault model. Although ExpFault at its current form focuses on DFA, it can be extended for other classes of fault attacks as well. The compelling feature of this framework is that it can automatically figure out the attack algorithm along with an estimation of attack complexity.

In order to explain the key concepts behind ExpFault, three attack examples corresponding to AES [8] and PRESENT [6] block ciphers have been used in this chapter. We shall explain how ExpFault discovers these attacks in steps. Finally, we analyze a recently proposed cipher GIFT [1], and automatically figure out several attacks corresponding to different fault models. To provide a glimpse of the capacity of ExpFault, here we summarize some interesting results for GIFT. ExpFault found that the 2^{128} bit key-space of GIFT can be reduced to a size of $2^{7.06}$ by means of two nibble faults injected consecutively at the 25th and 23rd round of the cipher on average. The overall computational complexity of the attack is, however, $2^{17.53}$. Moreover, the attack is found to be optimal from an information theoretic perspective.[4]

The chapter is organized as follows. We start by describing two well-known ciphers—AES and PRESENT, for which we provide three attack examples using the proposed framework (Sect. 2.2). Next, we present a brief overview of some early efforts in automated fault analysis followed by a summary of the main features of ExpFault (Sect. 2.3). The next three sections present the detailed technical concepts behind the framework along with examples. In Sect. 2.4, the cipher and the fault models are formalized. Detailed description of the framework is presented next in Sect. 2.5. Proof-of-concept evaluations of three known attacks on AES and PRESENT are used as examples while describing the scheme. Section 2.6 presents DFA results on the GIFT block cipher. Finally, we conclude the chapter in Sect. 2.7.

[4] We refer to [20] for the proof of optimality.

The appendix at the end of this chapter further explains the implementation aspects of ExpFault, and certain other intricacies related to the attacks on GIFT. A detailed discussion over two related work is also provided in the appendix.

2.2 Preliminaries

In this section, we introduce some basic terminology encountered frequently in this chapter. Some of them will be formally defined according to our cipher model in the following section. We also provide a brief description of the two ciphers AES and PRESENT in this section which are to be used as examples throughout this chapter.

2.2.1 *Basic Terminology*

Block ciphers are the realizations of pseudo-random permutations (PRP). In general, block ciphers are constructed by repeating a *round* multiple times (perhaps with slight modifications in some iterations). Each round is a sequence of *sub-operations*. In this chapter, the input of each sub-operation is called an *intermediate state* (also known simply as a *state*). With the injection of a fault, states assume faulty values which differ from the correct values assumed by them in the absence of the fault. We use the term *state differential* to represent the XOR-difference between the correct and faulty computation of a state. Each state differential consists of word variables known as *state differential variables*, where the word size typically depends on the cipher under consideration.

2.2.2 *AES*

The AES block cipher is the current worldwide standard for symmetric key cryptography. The widely used version AES-128 uses a block size of 128 bits and a master key of the same size, all of which are processed as 1-byte chunks. The encryption is realized by iterating a round function 10 times. The round function of AES consists of four sub-operations, namely `SubBytes`, `ShiftRows`, `Mixcolumns`, and `AddRoundKey`. The `SubBytes` consists of 16 identical 8×8 nonlinear S-Boxes. The `ShiftRows` sub-operation is a permutation realized at the byte level, whereas the `Mixcolumns` is a linear transformation by means of a maximum distance separable (MDS) matrix. The last sub-operation in a round is the `AddRoundKey`, which performs a bitwise XOR operations between the state and 128-bit round keys generated by means of a key schedule for each round, from the master key. It is worth mentioning that the round function in the last round of AES does not include the `Mixcolumns` sub-operation.

AES is the most widely analyzed cipher in the context of fault attacks, especially for DFA [5]. Most of the DFA attempts on AES till date target the last three rounds of the cipher. The most optimal attack on AES is due to Tunstall et al. [22], which requires only a single byte fault injection at the beginning of the eighth round of the cipher resulting in a keyspace of size 2^8. The computational complexity of this attack is 2^{32}. In [19], Saha et al. have shown that the same attack can still be realized even if multiple bytes at the beginning of eighth round get faulty. The only constraint is that the faulty bytes must remain within the same diagonal in the matrix representation of the AES state. Further, in [9], Derbez et al. proposed impossible differential fault attack (IDFA) and meet-in-the-middle (MitM) attack, both of which target the beginning of the seventh round of AES. Finally, Kim et al. proposed integral fault attacks on AES [16]. Being well explored, AES is a major mean for experimentally validating our framework in this chapter. More specifically, we shall show that our framework in its current state can detect the standard DFA attempts on AES including the IDFA attacks.

2.2.3 *PRESENT*

The PRESENT [6] is a widely known block cipher of the lightweight genre. The PRESENT-80 version of the cipher utilizes an 80 bit master key with a 64 bit block size. Round keys of 64 bits are generated from the 80 bit key state for 31 iterations having the same round structure. The constituent sub-operations for the round function are `AddKey`, `sBoxLayer`, and `pLayer`, of which the `sBoxLayer` is a nonlinear layer consisting 16 identical 4×4 bijective S-Boxes. The linear diffusion layer of PRESENT is constructed with a simple bit-permutation operation which is significantly different and simpler than that of the MDS based diffusion functions of AES.

Just like AES, PRESENT has gone through several fault analysis attempts mostly targeting the 28, 29th rounds of the cipher as well as the key schedule [18, 23–25]. In this chapter, we shall use the attack proposed by Jeong et al. [14] on the 28th round of the cipher to explain various parts of the framework. This attack requires two instances of 16 bit faults injected at the beginning of the 28th round. The computational complexity of the attack is $O(2^{32})$.

2.3 Automated Fault Analysis: A Brief Discussion

2.3.1 *Some Early Automation Efforts*

Most of the efforts in fault attack automation are fairly recent. Perhaps the most popular among them is the algebraic fault analysis (AFA) framework [24]. The main idea of AFA is to encode the cipher and a fault instance to a low-degree system of multivariate polynomial equations, which is then solved with SAT solvers by converting it to a Boolean formula in conjunctive normal form (CNF). A detailed discussion comparing AFA with ExpFault will be presented in the appendix of this chapter. However, as a justification of not following the path of AFA during the ExpFault development, here we want to point out that analyzing a single fault instance in AFA involves solving a SAT problem, which often requires prohibitively long time. This makes AFA not a very convenient choice in the exploitable fault space characterization context. Moreover, the attacks reported by a SAT solver are often difficult to interpret. As a result, they do not provide any clue by which one may improve the design and implementation of the cipher. Also, certain things like attach complexity analysis are extremely cumbersome and computationally impractical with AFA. However, the algebraic representation of AFA automatically handles all the available information at once, which become important in certain applications.[5] Recently, Barthe et al. [2] have proposed a framework for synthesizing fault attacks automatically given a software implementation using concepts of program synthesis. However, their framework mainly targets for public key cryptosystems.

The most relevant work in the present context is due to Khanna et al. [15], who proposed a framework called XFC based on principles somewhat similar to that of ExpFault. The key component of XFC is the characterization of the fault propagation path by means of coloring, where each color represents a variable. The coloring based static analysis eventually provides a scalable way for the calculation of the attack complexity as well. Albeit being scalable, the usability of the XFC scheme is found to be limited to a specific class of DFAs. More specifically, it fails to detect distinguishers, which typically exploit the constraints on the values that certain fault difference variables may assume. Impossible differential fault analysis (IDFA) attacks are prominent examples of such cases. Further, XFC scheme lacks proper automation in its attack complexity analysis algorithm and makes certain simplifying assumptions, which fails to capture the most generic scenario.

2.3.2 *ExpFault: An Overview*

As pointed out in the last subsection, the AFA approach encodes every necessary mathematical property within a cipher and fault. However, such extensive information content often becomes a double-edged sword by making AFA computationally expensive. One should recall that the three prime criteria for automatic fault space exploration are fault characterization of individual faults, genericness, and scalability. The AFA at its basic form can only achieve the genericness. In order to achieve these three goals simultaneously, a simple strategy has been adopted in ExpFault,

[5]In the next chapter, we shall present another framework which fully utilizes the aforementioned advantage of AFA while making it fairly scalable.

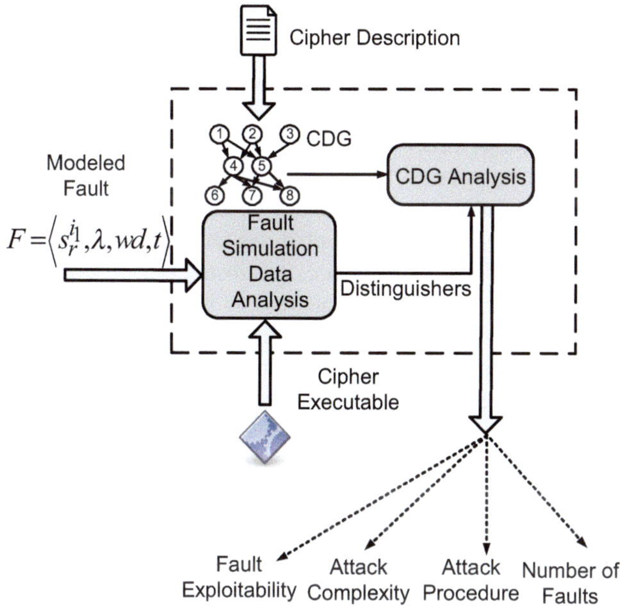

Fig. 2.1 The ExpFault framework (the abbreviation CDG stands for cipher dependency graph, which will be defined in the subsequent sections)

which just estimates the attack complexity instead of doing the attack explicitly to recover the secret. In the light of this simple strategy, and a rigorous formalization of the cipher description and DFA, ExpFault evaluates fault exploitability in three steps, the first among which is the identification of a set of potential wrong key distinguishers. The generic distinguisher identification step is realized by analyzing fault simulation data with assistance from standard data-mining strategies. The goodness of each DFA distinguisher is also evaluated by means of a metric based on Shannon entropy. The next step to distinguisher identification is the evaluation of attack complexity. We propose a graph based abstraction of the cipher to realize this step, which works by automatically identifying a divide-and-conquer strategy for evaluating distinguishers on different key guesses. The choice of the divide-and-conquer strategy has a major role in determining the attack complexity. Finally, we figure out the overall attack complexity by calculating the size of the keyspace after a single fault injection and estimate the number of fault injections required to reasonably figure out the key. A schematic of the whole framework is presented in Fig. 2.1.

One implicit but extremely crucial feature of ExpFault are the abstractions that are applied at various steps of the framework. In other words, unlike considering all information regarding the cipher and fault like AFA, ExpFault only considers the relevant parts. Especially, the internal mathematical structures of the nonlinear sub-operations are abstracted. Further, the fault and plaintext values are not considered explicitly during the analysis. Both of these abstractions are consistent with classical fault manual analysis approaches. By ignoring these details we make the fault space much compact so that it can be explored in a scalable manner. Further, the abstraction makes attack complexity estimation and attack interpretation feasible. Even with certain information abstracted, ExpFault automatically finds the average attack complexity, required number of injections, and the attack algorithm. In general, these outputs provide sufficient information about the fault space of the cipher. However, for certain applications like cipher design this is not sufficient. For those applications we have to consider a rather exact, abstraction-free representation. The next chapter of this book is devoted for addressing this issue.

The ExpFault framework mines distinguishers from fault simulation data. This data analysis approach of distinguisher identification shows enough potential to be extended for other genres of fault attacks, viz. integral fault attacks [16] and differential fault intensity analysis (DFIA) attacks [11]. A unified framework for automated fault analysis will be the ultimate goal which has been initiated in this work by means of ExpFault. One should notice that it is not straightforward to extend equation based approaches like AFA for attacks like DFIA which are mainly statistical in nature. Data analysis thus seems to be a better alternative for such cases.

From the next section onward, we shall describe the ExpFault framework in detail. Designing a cipher-oblivious analysis framework requires a generalization of the fault attack itself. In the next section we provide a theory generalizing the fault attacks, which is the backbone of ExpFault. The rest of the framework will be built upon this theory in the subsequent sections.

2.4 A Formalization of the Differential Fault Analysis

In this section, we construct a formal notion of the cipher representation as well as the differential fault analysis, which perfectly suits our purpose in this chapter. We begin with a general view of the DFA attacks and eventually present the formal framework.

2.4.1 DFA on Block Ciphers: A Generic View

The general concept of DFA remains the same for most of the ciphers, except some manual cipher-specific tricks, which make the automation a challenging task. DFAs broadly follow three major steps:

1. **Distinguisher Identification:** The key step of DFA is to identify wrong key distinguishers. Distinguishers are constraints defined over the state differential variables, which make the probability distribution of the state differential statistically biased (i.e., the distributions deviate from uniform distribution.).

According to the well-known wrong key assumption, a distinguisher attains a uniform distribution with a wrong key guess and a non-uniform one with a correct key guess, which eventually helps to reduce the candidate keyspace. In the context of DFA, however, distinguishers are mostly described in the form of constraints rather than statistical distributions.

2. **Divide-and-Conquer:** The step following the distinguisher identification is the evaluation of the same with different key guesses to filter out the wrong keys in a computationally efficient manner. Not every distinguisher is efficiently computable and the computational efficiency lies in two facts: (1) whether it can be partitioned into independent subparts and (2) whether each subpart is efficiently computable, that is with a reasonable number of exhaustive key guesses.
3. **Estimating the Number of Possible Key Candidates:** The sole idea of DFA is to reduce the complexity of the exhaustive key search by means of the distinguisher. However, the reduction of the search space typically depends upon the distinguisher used. If the distinguisher is unable to sufficiently reduce the search space complexity, more faults should be injected. Thus, the quality of a distinguisher must be quantified to achieve successful and practical attacks.

Automation of the above-mentioned steps demands a mathematical specification of the cipher and the faults, to begin with. The following subsections present a formalization of the cipher and the differential fault attacks in this context. To maintain clarity, a list of notations used is provided in Table 2.1.

2.4.2 Representing a Block Cipher

A block cipher is a mapping $\mathcal{F}_k : \mathcal{P} \rightarrow \mathcal{C}$, where $\mathcal{P}$ and $\mathcal{C}$ denote the plaintext and ciphertext space, respectively. The mapping is typically specified by a key $k \in \mathcal{K}$. Structurally, they can be represented as a tuple of invertible functions as:

$$\mathcal{F}_k = \langle o_1^1, o_1^2, \ldots, o_1^d, o_2^1, o_2^2, \ldots, o_2^d, \ldots . o_R^1, o_R^2, \ldots, o_R^d \rangle \tag{2.1}$$

Typically, for a given $p \in \mathcal{P}$ and a fixed $k \in \mathcal{K}$, there exists a unique $c \in \mathcal{C}$ such that $c = o_R^d(o_R^{d-1}(\ldots(o_1^2(o_1^1(p)) \ldots)$. Here, each o_j^i represents the i-th sub-operation in the j-th round of an R round cipher. Further, each o_j^i can be represented as:

$$o_j^i(x_1, x_2, \ldots . x_l) = \bigoplus_{h_1=1}^{h_1=l} a_{h_1} \cdot x_{h_1}, \quad \text{if } o_j^i \text{ is linear} \tag{2.2}$$

$$o_j^i(x_1, x_2, \ldots . x_l) = \bigoplus_{h_1=1}^{h_1=2^l} a_{h_1} \cdot \prod_{h_2 \in I} x_{h_2}, \quad \text{if } o_j^i \text{ is nonlinear} \tag{2.3}$$

Table 2.1 List of notations used

Notation	Meaning
$\|\cdot\|$	Size of a set
$\mathcal{F}_k$	Block cipher
$\mathcal{P}, \mathcal{C}, \mathcal{K}$	Plaintext, ciphertext, and keyspace
R	Total number of iterative rounds in the block cipher
d	Total number of sub-operations in each round
o^i_j	The i-th sub-operation in the j-th round
$\mathcal{E}_k$	Data-centric view of the cipher
s^i_j	The state at the input of the i-th sub-operation in the j-th round
λ, m	Block size; word-size (size of each word in the cipher state in bits)
$l = \frac{\lambda}{m}$	Word count
F	A fault instance
X	Fault affected register
r, wd, t, f	Fault round, width, location and value
δ^i_j	State differential at the input of the i-th sub-operation in the j-th round
w^{ij}_z	A state differential variable (discrete random variable) corresponding to an m-bit word of the state differential δ^i_j
Δ_k	Set of state differentials of the cipher
$p^{w^{ij}}_z$	Probability distribution of w^{ij}_z
$H(\cdot)$	Entropy
$T(.)$	Dataset for state differentials for each w^{ij}_z.
$\{\mathcal{D}^i_j\}$	Set of distinguishers formed with state differentials
$\mathcal{T}$	The enumeration algorithm for the key set using distinguisher
$Comp(\mathcal{T})$	The complexity of the distinguisher enumeration algorithm
$\mathcal{R}$	Remaining keyspace
IS, VS	Itemset and variable set
MKS, VG	Maximal independent key set, and variable group

Here, $I \subseteq \{1, 2, \ldots .l\}$, and each a_{h_1} is a constant. The data width of the function inputs is a notable factor in this description. Given the block width of a cipher is λ bits, it is processed as m-bit words, where $m = \frac{\lambda}{l}$. We call m as the *word size* of the cipher. It is worth mentioning that the data width of each sub-operation might not be the same for a given cipher. In such cases, we assume the data width of the input of nonlinear sub-operations as the *word size*.

In an alternative data-centric view, the cipher $\mathcal{F}_k$ is represented as a sequence of states as follows:

$$\mathcal{E}_k = \langle s^1_1, s^2_1, \ldots, s^d_1, s^1_2, s^2_2, \ldots, s^d_2, \ldots .s^1_R, s^2_R, \ldots, s^d_R, c\rangle \tag{2.4}$$

where each s^i_j represents the input of the i-th sub-operation in the j-th round of a R round cipher. Intuitively this representation presents an execution trace of the cipher on a plaintext p and a key k. Each s^i_j actually refers to an *internal state* (or simply *state*) of the cipher. The $\mathcal{E}_k$ is also referred to as the *execution trace* of the cipher. The state sequence begins with the plaintext $p = s_1^1$. Each state s_j^i is a vector of length l of m-bit words. The values assumed by the state vectors are subject to change with the variation of the plaintext and the key. Intuitively, the data-centric specification formally represents the simulation data from the cipher.

2.4.3 *Formalization of the DFA*

The formalization of DFA requires a precise specification of the injected faults. In general, it is assumed that injected faults are localized and transient so that they can affect at least one bit from a chunk of contiguous bits within a state, at some specific round. If a fault affects some part of the input state of the sub-function o_j^i, the output of o_j^i will differ from its expected value. We provide a formal representation of a fault as a tuple $F = \langle s_r^{i_1}, \lambda, wd, t\rangle$, which is similar to that of [24]. Here, $s_r^{i_1}$ represents the state, where the fault is injected. It is apparent that $r < R$. The λ parameter denotes the data width of the state, wd is the width of the fault, and t is the fault location within the state. Let us denote any $s_j^i = \langle V_1, V_2, \ldots . V_l\rangle$, where each V_z ($z \in \{1, 2, \ldots, l\}$) is an m-bit variable. The localized fault, depending on the scenario, will affect one or more of these variables. In general, this is determined by the width of the fault wd. To simplify the matter we assume that wd is either $\leq m$ or it is a multiple of m. As a result, one or more of the V_zs can be affected by the fault. For simplicity, it is further assumed that only consecutive V_zs can be affected by the fault and the location of that is indicated by the fault location parameter t, in the fault model. The width of the fault wd is often used to represent the fault models. In this work, we only consider standard fault models (the bit ($wd = 1$), nibble ($wd = 4$), and byte ($wd = 8$) fault models), although the framework is not limited to them.

The reader should notice that the fault model here does not take the value of the fault and the plaintext into account. In the most general case, every plaintext and fault value should give rise to a distinct pattern of fault propagation (while all other parameters like location and width remain unchanged.). Although the attack complexity, in certain ciphers, may vary over different fault and plaintext values, classically fault attacks ignore this variation. Instead, they infer a general constraint for deriving the key by considering all these fault patterns together. This is indeed an abstraction, and it makes the attack complexity evaluation problem scalable. The cost of this abstraction is that we always obtain an average case estimation of the attack complexity.

Once the cipher and the fault model are determined, we can now formally describe the DFA attack on a cipher. In order to construct a general model for DFA, we first need to formally define the state differentials and the state differential variables, already introduced in Sect. 2.2.1. Let us consider the execution trace $\mathcal{E}_k$ of a cipher, as described in (2.4). In order to capture the effect of an injected fault, we define a *faulty execution trace* $\mathcal{E}'_k = \langle s_1^1, s_1^2, \ldots, s_1^d, \ldots, s_r^{'i_1}, s_r^{'i_1+1}, \ldots, s_r^{'d}, \ldots, s_R^{'d}, c' \rangle$. Here each $s_j^{'i}$ denotes the faulty input of the i-th sub-operation in the j-th round ($r \leq j \leq R$) starting from the injection point of the fault at round r. Before the r-th round the states remain the same. A *state differential* is defined as $\delta_j^i = s_j^i \bigoplus s'^i_j = \langle V_1^{ij} \oplus V_1^{'ij}, V_2^{ij} \oplus V_2^{'ij}, \ldots, V_l^{ij} \oplus V_l^{'ij} \rangle = \langle w_1^{ij}, w_2^{ij}, \ldots, w_l^{ij} \rangle, \;\; r \leq j < R$, where V_z^{ij} denotes the z-th m-bit correct state variable, and $V_z^{'ij}$ denotes the corresponding faulty state variable. For each such word, $\oplus$ denote the bitwise XOR operation. Each w_z^{ij} denotes a *state differential variable*. Finally, we define another formal structure called *differential execution trace* Δ_k as $\Delta_k = \langle \delta_r^{i_1}, \delta_r^{i_1+1}, \ldots, \delta_r^d, \ldots, \delta_R^1, \delta_R^2, \ldots, \delta_R^d \rangle$. Each of the state differentials δ_j^i in Δ_k may potentially form a distinguisher.

Given a cipher $\mathcal{F}_k$ and a fault F in it, the DFA can be formally described as

$$\mathcal{A} = \langle \{\mathcal{D}_j^i\}, \mathcal{T}, \mathcal{R} \rangle \tag{2.5}$$

where

- $\{\mathcal{D}_j^i\}$ denotes a set of distinguishers defined over the state differential variables of some state differential δ_j^i.
- $\mathcal{T}$ is the exhaustive enumeration algorithm for the key set $\mathcal{K}$ via distinguisher evaluation. A proper divide-and-conquer strategy is essential for this enumeration algorithm, which enables the evaluation of the distinguishers in parts. The time complexity of the enumeration algorithm is one of the determining factors of the overall DFA complexity, which is $O(2^n)$, with $n \leq \log_2(|\mathcal{K}|)$. For practical cases $n \ll \log_2(|\mathcal{K}|)$, whereas $n = \log_2(|\mathcal{K}|)$ implies no gain from the perspective of an attacker.
- $\mathcal{R}$ is the remaining key search space after the injection of a single instance of the fault F. The evaluation of the distinguishers over the complete key set $\mathcal{K}$ partitions the set into two non-overlapping subsets $\mathcal{K}_w$ and $\mathcal{K}_{cr}$; the first one being the set of wrong keys and the second one being the set of candidate keys one of which is the correct key. Evidently, $\mathcal{R} = \mathcal{K}_{cr}$ and $|\mathcal{R}| \ll |\mathcal{K}|$ for an efficient fault attack. One should note that it is sufficient to consider the search space reduction for one single fault instance, as the reduction for multiple fault instances can be easily calculated from that. $\mathcal{R}$ is often represented as the solution set of a system of equations or inequations, involving the keys and distinguisher variables.

2.5 A Framework for Exploitable Fault Characterization

In this section, we describe the proposed automated framework in detail. The following subsections will provide generic algorithms for computing each of the components described in (2.5). The input to the framework is a mathematical description (linear layers as matrices and the S-Boxes as tables) and an executable model (software/hardware implementation) of the target block cipher along with an enumeration of the fault space under consideration. The output is the exploitable fault space.

2.5.1 *Automatic Identification of Distinguishers*

In the last section, we have abstractly defined distinguishers as state differentials having certain mathematical or statistical properties. However, a metric is required which can identify the state differentials having such special properties and also quantify the goodness of distinguishers. We define such a metric based on the *entropy of state differentials*. Here the state differential variables are considered as random variables.

Definition 2.1 (Entropy of a State Differential) The entropy of a state differential $\delta^i_j = \langle w^{ij}_1, w^{ij}_2, \ldots, w^{ij}_l \rangle$, where each w^{ij}_z is a discrete random variable with probability distribution $p^{w^{ij}}_z$, is defined as $H(\delta^i_j) = H(w^{ij}_1, w^{ij}_2, \ldots, w^{ij}_l)$, that is the joint entropy of the random variables in the state differential.

Definition 2.2 (Maximum Entropy of a State Differential) The maximum entropy of a state differential $\delta^i_j = \langle w^{ij}_1, w^{ij}_2, \ldots, w^{ij}_l \rangle$, is defined as $H_{\max}(\delta^i_j) = \sum_{z=1}^{l} H_{\max}(w^{ij}_z) = \sum_{z=1}^{l}\left(-\sum_{q=0}^{2^m-1} p^{w^{ij}_z}_q \log_2(p^{w^{ij}_z}_q)\right)$, where each w^{ij}_z is independent and uniformly distributed within the range $[0, 2^m - 1]$, given m is the bit width of variable w^{ij}_z.

Note that the maximum entropy defined here assumes the uniformity and independence of the associated random variables within a specific range $[0, 2^m - 1]$, where m is the bit length of each variable. In case, the variable is not uniform within this complete range the entropy will be less than the maximum entropy. Correlations among the variables will also cause entropy reduction. Next, we define the *distinguishing criteria*—the decision criterion for determining the distinguishing capability of state differentials.

Definition 2.3 (Distinguisher Criteria) A state differential δ^i_j is called a distinguisher if the entropy $H(\delta^i_j)$ is less than the maximum entropy of the state differential.

The main idea of our dynamic distinguisher identification scheme is to learn the distinguishers from the fault simulation data, acquired from the executable cipher model by varying the plaintexts, keys, and the fault values. Let us consider the differential execution trace Δ_k corresponding to a fault F. The values assumed by the variables associated with Δ_k vary with the change of plaintext, key, and the fault value. Such variations result in a fault simulation dataset which is analyzed to

Algorithm 1: Procedure *RngChk*

Input : The dataset for a state δ^i_j as $T_{\delta^i_j} = \langle T_{w^{ij}_1},\ T_{w^{ij}_2}, \ldots,\ T_{w^{ij}_l}\rangle$
Output : $\langle\{Rng_{w^{ij}_z}\}^l_{z=1},\ H_{Ind}(\delta^i_j)\rangle$

```
H_Ind(δ^i_j) := 0;
for each T_{w^{ij}_z} ∈ T_{δ^i_j} [6] do
    Store all distinct values assumed by w^{ij}_z in Rng_{w^{ij}_z}          ▷ [7];
    Calculate the probability distribution of w^{ij}_z as p'^{w^{ij}_z}      ▷ [8];
    Calculate the Entropy of w^{ij}_z as H_Ind(w^{ij}_z) using p'^{w^{ij}_z};
return: ⟨{Rng_{w^{ij}_z}}^l_{z=1}, H_Ind(δ^i_j)⟩
```

identify distinguishers. Let us denote the datasets corresponding to each state differential δ^i_j as $T_{\delta^i_j}$. Each $T_{\delta^i_j}$ is a table, containing l, m-bit variables w^{ij}_z $(1 \le z \le l)$ and data values, corresponding to each of them. For convenience, we further denote each column of a $T_{\delta^i_j}$ as $T_{w^{ij}_z}$. Corresponding to each fault according to our formalization, we have many such tables corresponding to each state differential in Δ_k. Typically, a subset of the possible state differentials actually qualifies as potential distinguishers. We denote $T_{\Delta_k} = \langle T_{\delta^1_r}, T_{\delta^2_r}, \ldots, T_{\delta^d_r}, \ldots, T_{\delta^1_R}, T_{\delta^2_R}, \ldots, T_{\delta^d_R}\rangle$ as the set of the tables for the state differentials. Our data-based framework tests each δ^i_j separately and decides whether it constructs a distinguisher. Two distinct cases can be identified in the course of the distinguisher identification which we outline next.

2.5.1.1 Case 1: The Variables Are Independent, But Not Uniform Within the Complete Range

In this typical case, the probability distributions of individual state differential variables change, while they still remain independent. Decrease in individual entropies of the variables due to their non-uniformity over the complete range $[0, 2^m - 1]$ (note that uniformity may still hold over some sub-range of $[0, 2^m - 1]$) causes a drop in the total state differential entropy. The situation is described in Algorithm 1, where the changed probability distributions are denoted as $p'^{w^{ij}}_z$ $(z = 1, 2, \ldots, l)$, and the joint state differential entropy as $H_{Ind}(\delta^i_j) = \sum^l_{z=1}(-\sum^{2^m-1}_{q=0} p'^{w^{ij}_q}_z \log_2(p'^{w^{ij}_q}_z))$. Each column of the table $T_{\delta^i_j}$ (denoted as $T_{w^{ij}_z}$), corresponding to each variable w^{ij}_z

[6] $z = 1, 2, \ldots, l$.
[7] Values of w^{ij}_z belongs to the set $\{0, 1, \ldots 2^m - 1\}$.
[8] $p'^{w^{ij}_z}_q := \frac{\#q}{|T_{w^{ij}_z}|}$, where $\#q$ denote the frequency of $q \in \{0, 1, \ldots 2^m - 1\}$ in $T_{w^{ij}_z}$.

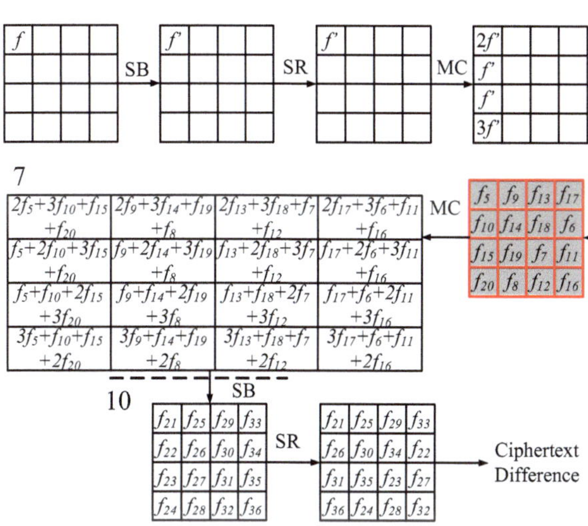

Fig. 2.2 Fault propagation in impossible differential fault attack on AES and formation of the IDFA property (marked in red). None of the variables in this state differential can assume the value 0 for the correct key guess [20]

is treated separately for missing values (if any) within the range of $[0, 2^m - 1]$. As a concrete example, if a state differential poses an impossible differential property, none of the w_z^{ij}s can assume value 0, and as a result, the value 0 will be missing in the table $T_{w_z^{ij}}$ for any z. Information regarding the values which are not missing are important in the context of the distinguisher and hence preserved for each w_z^{ij} in the set $Rng_{w_z^{ij}}$. Typical examples of Case 1 include the IDFA attack on AES and the attack on PRESENT described in [14].

Example 2.1 (IDFA Attack Distinguisher on AES) IDFA attacks exploit a typical cipher property that, depending on the fault, the variables of a state differential corresponding to some internal state of a cipher may not attain certain values within their domains. Such a property is used in IDFA attacks to distinguish correct key guesses from wrong ones. For the IDFA attack on AES, a byte fault is injected at the beginning of the seventh round of the cipher resulting in some state differentials none of whose variables can assume the value 0, with the correct key guess. The situation is elaborated in Fig. 2.2, where each large square represents an intermediate state differential of AES, with the fault injected at the beginning of the seventh round in the zeroth byte location. Each small square in the figure represents a state differential variable of size one byte. The shaded states in Fig. 2.2 denote the existence of an impossible differential, with all bytes being active (fault difference cannot be 0). It is convenient to use the last among them as a distinguisher due to its proximity to the ciphertext (marked red in Fig. 2.2).

It is apparent that the distinguisher identification framework of ours identify this impossible differential property as an instance of Case 1. The `RngChk` function detects the absence of 0 in each of the variables, and as a result, the entropy becomes $H_{Ind}(\delta_9^3) = \sum_{z=1}^{16}(-\sum_{q=0}^{2^8-1} p_z^{\prime w_q^{ij}} \log_2(p_z^{\prime w_q^{ij}})) = \sum_{z=1}^{16}(-0 - \sum_{q=1}^{2^8-1} \frac{1}{255} \log_2(\frac{1}{255})) = \sum_{z=1}^{16}(-255 \times \frac{1}{255} \log_2(\frac{1}{255})) = 127.90$, which makes the state differential qualify as a distinguisher. One should note that the state differentials δ_9^1 and δ_9^2 also possess the impossible differential property and are detected by the `RngChk` routine.

Example 2.2 (A Distinguisher on PRESENT) In this example, a fault is injected at the beginning of the 28th round of the PRESENT cipher. The width of the fault is of 16 bits. The distinguisher identification algorithm, in this case, identifies the input state of the S-Box of the 30th round (δ_{30}^1) as the best distinguisher (i.e., with lowest entropy). The `RngChk` function identifies that each of the 4 bit variables in this state differential can assume only two values among 2^4 possible values (although the two values assumed may change depending on the fault locations), and as a result, the entropy becomes $H_{Ind}(\delta_{30}^2) = \sum_{z=1}^{16}(-2 \times \frac{1}{2} \log_2(\frac{1}{2})) = 16$. This example establishes that the distinguisher identification algorithm works fine with multiple nibble/byte fault models.

2.5.1.2 Case 2: The Variables Are Not Independent

The second case of the distinguisher identification problem deals with the scenarios where correlations exist between some of the variables within a state differential, which eventually cause the reduction of state differential entropy. Typical examples exist for the ciphers with MDS matrices. Detection of the associations/correlations among the variables is crucial for calculating the entropy $H_{Assn}(\delta_j^i) = H(w_1^{ij}, w_2^{ij}, \ldots, w_l^{ij})$ in this case. We utilize well-known association rule mining (itemset mining) strategies for this purpose.

Frequent Itemset and Association Rule Mining

Association rule/itemset mining is a widely explored, classical problem in the domain of data mining, which refers to the discovery of association relationships or correlations among a set of items. Formally, given a large number of variables (attributes) $(var_1, var_2, \ldots, var_n)$, and a table/database of values they assume within their respective domains, an *item* is defined as $var_q = val$, where val lies in the domain of var_q. The simplest case occurs while dealing with discrete-valued variables having small ranges, where each item can be defined precisely. If $I = \{i_1, i_2, \ldots, i_a\}$ is a set of all items constructed from a table of discrete valued variables, then any $I_s \subset I$ is called an *itemset*. The prime task of an association rule mining algorithm is to figure out associations (if any) of the form $A \Rightarrow B$, where both A and B are propositional logic formulae over the items.

In the present context, we are mainly interested in itemsets and the variables associated with them. The number of all possible itemsets is exponential with the size of I, and most of them are not interesting for practical purpose. This fact leads to the

Table 2.2 Frequent itemset mining: toy example

TID	v_1	v_2	v_3	v_4	v_5
1	1	5	7	8	11
2	2	4	6	9	13
3	1	5	7	10	2
4	2	4	6	11	4
5	3	9	8	6	5
6	1	10	11	9	8

finding of itemsets occurring frequently in a table, which is known as *frequent itemset mining*. The frequent itemset mining task is governed by a statistical parameter *support*, which represents the frequency of occurrence of an itemset in the database. Formally support of an itemset I_s in a table/database DB is defined as $supp(I_s) = |I_s(t_i)|/|DB|$, where $I_s(t_i) = \{t_i | t_i \ \mathtt{is\ an\ entry\ in}\ DB\ \mathtt{and}\ t_i\ \mathtt{contains}\ I_s\}$. An itemset is called a frequent itemset if its support is greater than or equal to some predefined minimum support value. Further, an itemset is called a *maximal frequent itemset* if none of its immediate supersets is frequent.

To illustrate the above-mentioned concepts precisely, let us consider the toy database presented in Table 2.2. There are five discrete valued variables in this table having value ranges from 1 to 13. We set the support as $\frac{2}{6} = 0.33$. It can be easily figured out from Table 2.2 that there are two itemsets of size 3, beyond this support threshold—namely, $(v_1 = 1, v_2 = 5, v_3 = 7)$ and $(v_1 = 2, v_2 = 4, v_3 = 6)$. It is worth to note that no superset of these itemsets is frequent (that is, these are the maximal frequent itemsets), and all subsets of these are frequent. Further, it is interesting to note that for variable v_4 and v_5 all the itemsets are of cardinality 1. Intuitively, this implies that the variables v_4 and v_5 are statistically uncorrelated. Note that setting the proper support is imperative, as otherwise, one may obtain a large number of itemsets of little practical interest.

Finding Itemsets Within State Differentials

In the context of distinguisher identification, we are mainly interested in the maximal frequent itemsets within some reasonable support. The key idea is to figure out the variables within a state differential, which are strongly correlated. For this purpose, we utilize the well-known *Apriori* association rule mining framework. The complete procedure is described in Algorithm 2. The algorithm takes a $T_{\delta^i_j}$ as input, which is then fed to the `Apriori` function after some basic preprocessing. From each of the itemsets generated by the miner, we separate out the variables and create sets called *Variable Sets*. Variables within the same variable set are dependent, whereas they are assumed to be independent across different variable sets. Multiple itemsets exist corresponding to each *Variable Set* and a table is formed which stores each *Variable Set*, along with its corresponding itemsets. This table contains complete information regarding the distinguisher of our interest and is represented

Algorithm 2: Procedure *Miner*

Input : $T_{\delta_j^i} = \langle T_{w_1^{ij}}, T_{w_2^{ij}}, \ldots, T_{w_l^{ij}} \rangle$

Output : $\langle VS_{\delta_j^i}, \{IS^v_{\delta_j^i}\}_{v=1}^{|VS_{\delta_j^i}|}, H_{Assn}(\delta_j^i) \rangle$

1 $\langle VS_{\delta_j^i}, \{IS^v_{\delta_j^i}\}_{v=1}^{|VS_{\delta_j^i}|} \rangle := \texttt{Apriori}(T_{\delta_j^i})$;

2 $H_{Assn}(\delta_j^i) := 0$;

3 **for** ***each*** $v \in VS_{\delta_j^i}$ **do**

4 $\quad tot := \texttt{VarCount}(v) \times m$ $\triangleright$ [9];

5 $\quad p'^v_q := \frac{1}{|IS^v_{\delta_j^i}|}, \forall q \in IS^v_{\delta_j^i}$ $\triangleright$ [10];

6 $\quad p'^v_q := 0, \forall q \notin IS^v_{\delta_j^i}$;

7 $\quad H_{Assn}(v) := -\sum_{q=0}^{2^{tot}-1} p'^v_q \log_2(p'^v_q)$ $\triangleright$ [11];

8 $\quad H_{Assn}(\delta_j^i) := H_{Assn}(\delta_j^i) + H_{Assn}(v)$;

9 **return:** $\langle VS_{\delta_j^i}, \{IS^v_{\delta_j^i}\}_{v=1}^{|VS_{\delta_j^i}|}, H_{Assn}(\delta_j^i) \rangle$.

here as a pair $(VS_{\delta_j^i}, \{IS^v_{\delta_j^i}\}_{v=1}^{|VS_{\delta_j^i}|})$, where $VS_{\delta_j^i}$ denote the set of all variable sets and $\{IS^v_{\delta_j^i}\}_{v=1}^{|VS_{\delta_j^i}|}$ denote the set of itemsets corresponding to each variable set v. Next, the state differential entropy is calculated using this table, which involves the calculation of the joint distribution followed by the joint entropy of each variable set $v \in VS_{\delta_j^i}$ (line 6–8 in Algorithm 2). Using the independence assumption of the variable sets, these entropies can be summed up giving the total entropy of the state as $H_{Assn}(\delta_j^i)$.

Example 2.3 (A Distinguisher for AES with a Byte Fault Injected at the Beginning of Eighth Round) This example elaborates the Case 2 of the distinguisher identification problem. In this attack a byte fault is injected at the zeroth byte of AES state at the beginning of eighth round. The propagation of the fault is illustrated in Fig. 2.3. Let us consider the state differential δ^1_{10}, which is the output of the ninth round MixColumn. This state differential achieves the smallest entropy value and is eventually selected as the potential distinguisher for the attack. We now elaborate the entropy calculation for this state differential. The state differential $\delta^1_{10} = \langle w_1^{110}, w_2^{110}, \ldots, w_l^{110} \rangle$ contains 16 state differential variables, each with

[9] `VarCount` returns the number of variables in a variable set.

[10] Calculate the probability distribution of each variable set.

[11] Calculate the entropy of variable sets.

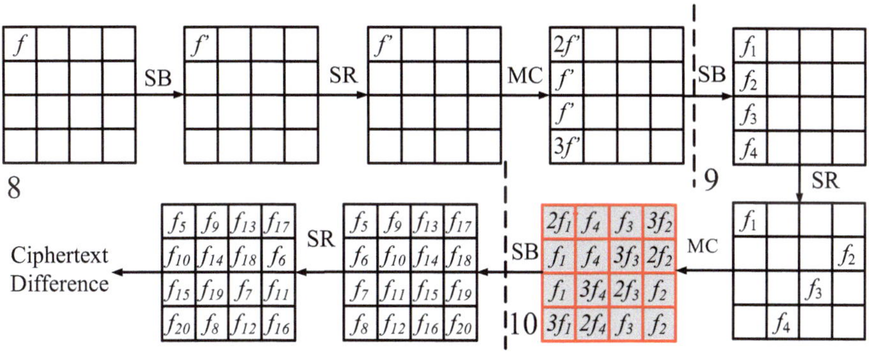

Fig. 2.3 Fault propagation in AES with the fault injected at the beginning of eighth round. Distinguisher is formed at the input of the tenth round S-Box (marked in red) [20]

bit-width $m=8$. The maximum entropy here is $H_{\max}(\delta_{10}^1)=128$. However, the function `Miner` reveals variable associations. More specifically, there are four variable sets $(w_1^{110}, w_2^{110}, w_3^{110}, w_4^{110})$, $(w_5^{110}, w_6^{110}, w_7^{110}, w_8^{110})$, $(w_9^{110}, w_{10}^{110}, w_{11}^{110}, w_{12}^{110})$, and $(w_{13}^{110}, w_{14}^{110}, w_{15}^{110}, w_{16}^{110})$ (variable numbering was done column-wise maintaining the convention in AES), each having 255 itemsets for them. The joint entropy of each variable set v becomes $H_{Assn}(v) = \sum_{q=1}^{255} \frac{1}{255} \log_2(255) = 7.99$, which finally results in the state differential entropy of $H_{Assn}(\delta_{10}^1) = 4 \times 7.99 = 31.96$.

Complete Distinguisher Identification Flow

The complete distinguisher identification algorithm takes the dataset $T_{\Delta_k} = \langle T_{\delta_r^1}, T_{\delta_r^2}, \ldots, T_{\delta_R^d} \rangle$ as input, and outputs a set $Dist = \{\langle \mathcal{D}_j^i, H_j^i \rangle\}$, where $\mathcal{D}_j^i$ is a distinguisher corresponding to the state δ_j^i (only if δ_j^i satisfies the distinguishing criterion), and H_j^i is the entropy of this distinguisher. The entropy H_j^i is typically the minimum of $H_{Ind}(\delta_j^i)$, $H_{Assn}(\delta_j^i)$ (returned by `RngChk` and `Miner`, respectively), and $H_{\max}(\delta_j^i)$ (calculated according to Definition 2.2). Indeed, $H_j^i < H_{\max}(\delta_j^i)$ is the essential criterion for a state differential to qualify as a distinguisher. It is worth to note that $\mathcal{D}_j^i$ contains the complete description of a distinguisher, obtained by combining the outputs of `RngChk` and `Miner`, given by $\mathcal{D}_j^i := \langle \{w_z^{ij}\}_{z=1}^l, \{Rng_{w_z^{ij}}\}_{z=1}^l, VS_{\delta_j^i}, \{IS_{\delta_j^i}^v\}_{v=1}^{|VS_{\delta_j^i}|} \rangle$. The pseudocode for this algorithm is rather straightforward and is thus omitted here.

Determining the Proper Distinguisher

The distinguisher identification step usually returns a set of potential distinguishers with their respective entropies specifying their qualities. In general, the distinguisher having the lowest entropy is the best for obvious reasons. However, the evaluation complexity of a given distinguisher plays a crucial role in its selection for a practical attack, as will be shown in the next subsection. After the completion of the first phase of the algorithm, we simply retain all the discovered distinguishers. This is because their usefulness is difficult to decide at this point. However, some of the distinguishers can be instantly eliminated based on some simple rules. It is mandatory to have an S-Box between any distinguisher and the ciphertext. In an even strict sense, if one intends to extract round keys from a specific round with a given distinguisher, he/she must have an S-Box between the distinguisher and the key addition step. Otherwise, the difference equations for key extraction cannot be constructed. Based on this rule, one can clearly eliminate some of the distinguishers, if possible. A concrete example of such a situation is discussed in the next section in the context of IDFA attack on AES.

2.5.2 *Enabling Divide-and-Conquer in Distinguisher Enumeration Algorithm $\mathcal{T}$*

Injection of a fault results in a set of distinguishers with different entropy values, as shown in the previous subsection. However, only a few of them are practically utilizable for attack, as the usability of a distinguisher depends on the complexity of evaluating it exhaustively. In DFA, the distinguishers are evaluated in the form of a system of difference equations (or inequations) and the solution space of the system results in a reduced set of candidate keys containing the correct key. Given this system, the practicality of a DFA attack depends on two factors:

1. *Distinguisher Evaluation Complexity:* The complexity of exhaustively enumerating the system for all possible key guesses.
2. *Offline Complexity:* Size of the remaining keyspace after distinguisher evaluation, which has to be searched exhaustively.

Following the notation described in Eq. (2.5), we denote the *Distinguisher Evaluation Complexity* with $Comp(\mathcal{T})$, where $\mathcal{T}$ is the distinguisher enumeration algorithm and $Comp(\cdot)$ denotes the complexity of an algorithm. The *Offline Complexity*, on the other hand, is denoted with $|\mathcal{R}|$, where $\mathcal{R}$ is the remaining keyspace after distinguisher evaluation. The $Comp(\mathcal{T})$ and the $|\mathcal{R}|$ can be estimated once the systems of equations for the distinguishers are in place. The *Attack Complexity* of a DFA can be determined as:

$$Comp(\mathcal{A}) = max(Comp(\mathcal{T}), |\mathcal{R}|) \tag{2.6}$$

One should note that the definition of attack complexity at this point assumes that only one fault instance has been injected. The attack complexity indeed depends on the number of faults injected. However, for most of the cases the required number of injections for making an attack practical can be determined from the value of $Comp(\mathcal{A})$ with a single fault instance and so we define it in terms of a single injection.

Knowing the systems of difference equations a priori is not a very practical assumption for an automated tool, as it depends upon the distinguisher(s) chosen. The most critical factor associated with distinguisher evaluation is to choose a proper divide-and-conquer strategy for enumerating the solution space of the difference equation system. Instead of guessing the complete key at once (which has a prohibitively large complexity), such a strategy allows guessing small key parts exhaustively and as a result the correct key can be recovered in parts with a practical time complexity. In this work, we construct such equation systems automatically in an abstract form, which is suitable for the purpose of attack complexity evaluation. Further, this abstract description can be extended to concrete fault difference equations, if required. To automatically determine the divide-and-conquer strategy we propose a graph based abstraction of the cipher called *cipher dependency graph* (CDG). Let us represent each state s^i_j as $s^i_j = \langle b^{ij}_1, b^{ij}_2, \ldots, b^{ij}_\lambda \rangle$, where each b^{ij}_z corresponds to a *bit variable*.[12] Given this representation of the states, we define the CDG for a block cipher as follows:

Definition 2.4 (Cipher Dependency Graph) A cipher dependency graph (CDG) for a block cipher is a directed acyclic graph (DAG) $\mathcal{G}\langle\mathbb{V}, \mathbb{E}\rangle$, where every node $v \in \mathbb{V}$ corresponds to a bit variable b^{ij}_z $(1 \leq z \leq \lambda)$ at the input of round j and sub-operation i of the cipher. A directed edge $e \in \mathbb{E}$ represents the dependency between two bit variables belonging to two consecutive states s^i_j and s^{i+1}_j (or s^1_{j+1}) imposed by the sub-operation o^{i+1}_j at the abstraction level of bits, considering the bit variable of s^i_j as input, and that of s^{i+1}_j as the output, respectively.

Certain simplifying assumptions were made, while constructing the CDGs. Some basic CDG building blocks are illustrated in Fig. 2.4. For the S-Boxes, we assume that each output variable is dependent on all the S-Box inputs (Fig. 2.4a). The key addition operations are represented by structures shown in Fig. 2.4b. Permutation layers are often straightforward and thus not shown here. However, some linear operations like MDS matrices need special care (more specifically the linear layers which involves XOR operations). Figure 2.4c, d represents one such scenario for 8 bit variables, which are shown in groups for convenience. Figure 2.4c represents how an individual 8 bit variable is treated in the CDG for MDS, and Fig. 2.4d represents the complete graph structure for one complete column of the MDS layer output. The MDS structures are also complete graphs (of 32 vertices in this example). It is worth to mention that the graph $\mathcal{G}$ is completely cipher-specific, and

[12]This is in contrast to the last subsection, where they (the states) were represented as vectors of variables of size m bits.

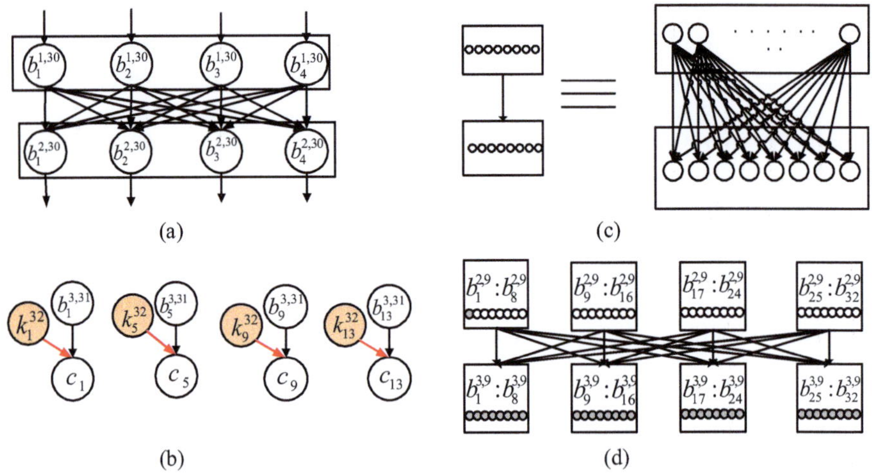

Fig. 2.4 Example: subgraphs corresponding to different sub-operations of a cipher [20]

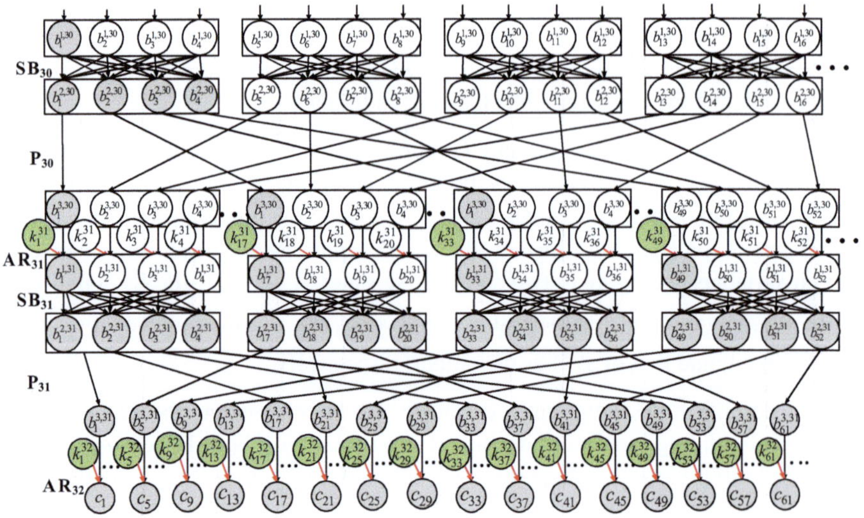

Fig. 2.5 Example: finding key parts for the distinguisher evaluations in PRESENT [20]

thus one needs to construct it only once while doing the exploitable fault analysis for a specific cipher. A CDG corresponding to a fault attack test case on PRESENT is illustrated in Fig. 2.5. For ease of understanding, only the sub-graph relevant to an attack example is shown.[13] Interestingly, the CDG is already divided into clearly identifiable levels.

---

[13]We have described the distinguisher corresponding to this attack in Example 2.2.

The CDG graph of a cipher abstracts several mathematical details of the S-Boxes and certain linear layers like MDS matrices. However, such abstractions are again consistent with the assumptions of classical fault analysis. Moreover, the aim of the ExpFault tool is to construct the attack structure and to estimate the complexity which can be achieved in a scalable way only by means of such abstractions.

Construction of a Divide and Conquer Strategy

The next step to the CDG construction is the identification of independent key parts to be guessed. For a given distinguisher, we initiate a series of breadth first searches (BFS) up to the ciphertexts nodes of the CDG. Each BFS search begins with a bit variable at the state, where the distinguisher has been constructed. The search typically figures out all the mutually dependent bit variables starting from the start node, in the form of the BFS tree (refer to Fig. 2.5 for example). Once the BFS tree is obtained, one can figure out the key nodes attached to it in $O(1)$ complexity.

Example 2.4 For the sake of illustration, let us refer to Fig. 2.5 once again. The distinguisher under consideration is the one described in Example 2.2, which is being constructed at the input of the 30th round S-Box operation (the first layer of nodes shown in Fig. 2.5.) In the figure, the colored circles represent the associated state bits one must compute to calculate the first bit in the distinguisher. The key bits one need to guess to calculate the shaded state bits are shown in red, while the associated state bits are represented in grey. All the colored variables here are the part of a BFS tree. Further, from the BFS tree of Fig. 2.5, the key variables to be guessed can be extracted which are 20 in number for the first bit. In summary, to calculate the first bit of the distinguisher, it is sufficient to guess these 20 bits together and no other key bit is required to be guessed. This provides the divide-and-conquer we require.

Optimizations

Certain intricacies are there to be taken care of while collecting the independent key parts for a distinguisher. Interestingly, not all key variables obtained by the BFS search are necessary. To illustrate this, we refer to Fig. 2.6a, which corresponds to the partial CDG for the IDFA attack on AES.[14] For convenience, the word level representation of the CDG is also provided along with (Fig. 2.6b). It is easy to observe from the word level representation that the key bits corresponding to k_9 are not required to be guessed for distinguisher evaluation. The reason behind this

[14]The IDFA distinguisher was described in Example 2.1.

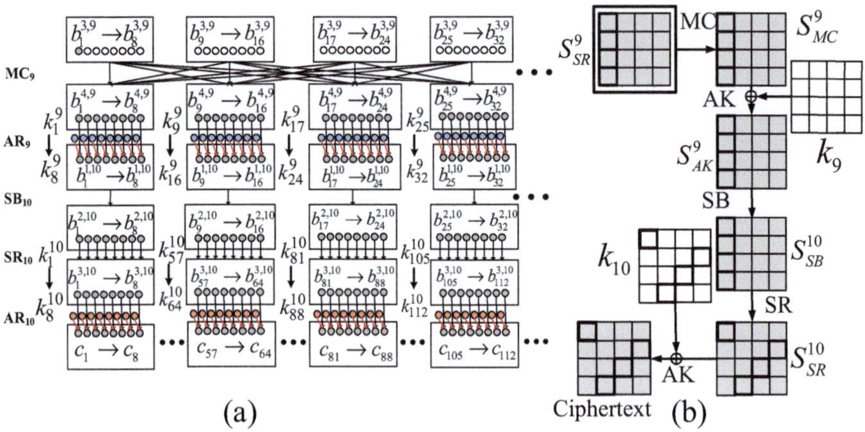

Fig. 2.6 Example: finding key parts for the distinguisher evaluations in AES [20]

fact is that there is no nonlinear layer between the key variables in k_9 and the distinguisher in S^9_{SR}. As a result, these key variables get cancelled out with the calculation of the differential. However, these key variables will still be detected by the BFS based search. Fortunately, we can easily enhance the proposed mechanism to encompass such scenarios. The idea is to keep the track of the nonlinear layers (S-Boxes) encountered, at each level of the CDG during the BFS traversal. This can be easily done by maintaining counters within the nodes of the CDG. While collecting the key variables, if it is found that the level corresponding to the key variables is not preceded by any S-Box level, the keys can be discarded. Referring to Fig. 2.6a, the key nodes in blue color thus can be discarded. The first bit of this distinguisher can be evaluated by guessing just 32 key bits of k_{10}.

Calculation of the Distinguisher Evaluation Complexity

The BFS based key part finding algorithm actually returns sets of key bits, corresponding to each bit of the distinguisher state. However, in order to calculate the quantities $Comp(\mathcal{T})$ and $|\mathcal{R}|$ we need to exploit some more structural properties of the cipher, already present in the CDG. As for most of the time, we are dealing with m bit distinguisher variables, it is trivial to combine the key bit sets corresponding to each m bit variable. One should also consider combining the key bit sets corresponding to the variable sets (if any). While evaluating any of these variables/variable sets, the corresponding keys must be guessed simultaneously. At this point, certain other things are to be taken care of. Let us consider a distinguisher $\delta^i_j = \langle w^{ij}_1, w^{ij}_2, \ldots, w^{ij}_l \rangle$. Corresponding to each w^{ij}_z, there exists a set of key bit variables. An obvious way is to view the relationships as a bipartite graph, as shown in Fig. 2.7. Without loss of generality, we just consider variables and not the variable sets in this discussion, although the same logic applies to the later one. Let us denote the key set corresponding to each variable w^{ij}_z as *Key Set*

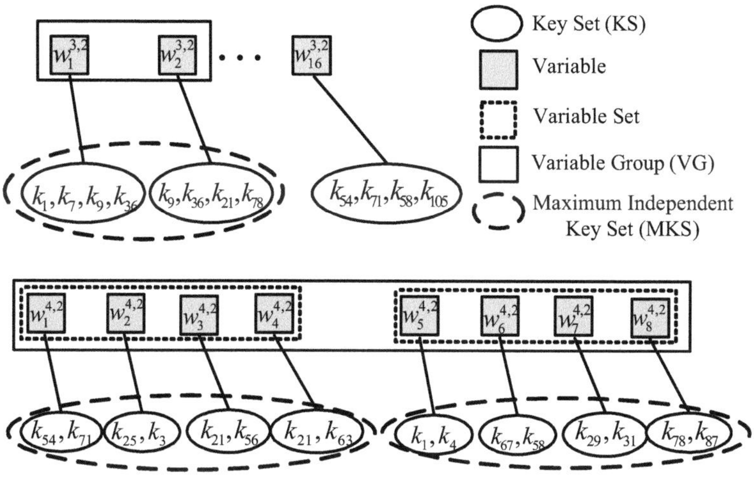

Fig. 2.7 Illustration of the relationships between key and distinguisher variables [20]

(KS_z). The key sets, however, may have overlaps. As a concrete example, one may consider the PRESENT case study depicted in Fig. 2.5. All four consecutive nibbles in the distinguisher at round 30 (shown in the diagram as layer SB_{30}) depend upon the same 16 round key bits from the last round. Such overlaps are extremely important from the divide-and-conquer point of view. This is because the overlaps indicate that all the difference equations that can be constructed involving these key bits and the associated variables w_z^{ij} will share the key variables. As a result they must be evaluated simultaneously. Putting it in a more simplified manner, if there are overlaps, computation related to all the overlapping variables must be performed simultaneously. To deal with such cases, we define *maximum independent key sets* (MKS), which are non-overlapping subsets of key variables, constructed by taking the union of overlapping KS_zs. Each MKS_h also imposes a grouping on the corresponding w_z^{ij}s attached to its component KSs. We call such groupings as *Variable Groups* (VG).[15] Intuitively, each $\langle VG_h, MKS_h \rangle$ tuple refers to a set of independent equations to be solved for the key extraction. In our graph based representation, we informally refer them as *independently computable chunks/subparts*.

Calculation of $Comp(\mathcal{T})$ becomes trivial after the above-mentioned grouping. Let us consider an MKS as MKS_h and the corresponding variable group as VG_h (note that variable groups may also include variable sets as its elements.). Each VG_h can be evaluated independently. Let us assume that we have M such VG_hs along with their corresponding MKS_hs. The time complexity of computing each of them is given as $Comp(\mathcal{T}_h) = 2^{|MKS_h|}, \quad 1 \leq h \leq M$. It is quite obvious that

[15]Note that we have used the term "group" to differentiate it from the variable sets. From this point onwards, we shall use *variable set* and *variable group* to identify these two separate entities. Variable sets can be members of variable groups.

such a search can be performed (and should be) in a parallel manner. As a result, the overall complexity of the distinguisher enumeration algorithm $\mathcal{T}$ becomes:

$$Comp(\mathcal{T}) = max_h(Comp(\mathcal{T}_1), Comp(\mathcal{T}_2), \ldots, Comp(\mathcal{T}_M)) \quad (2.7)$$

Example 2.5 (IDFA Attack on AES) In this example, we figure out the evaluation complexity of the IDFA distinguisher. It is observed that each byte of the distinguisher depends on 32 key bits from the tenth round. Further, four consecutive state differential variables are found to depend on the same 32 key bits. Following our notation size of each VG_h here is four state differential variables and the associated MKS_h contains 32 bits of key. Clearly, the distinguisher evaluation complexity $Comp(\mathcal{T})$ becomes 2^{32} in this case.

Example 2.6 (AES with Eighth Round Fault Injection) In this case, the distinguisher consists of four variable sets, containing four variables each. Each 8-bit variable in the distinguisher state depends on eight consecutive key bits (that is key bytes), and with the existence of variable sets having cardinality four, one must consider $8 \times 4 = 32$ key bits simultaneously, for guessing (i.e., $|MKS_h| = 32$). Further, the key bytes associated with each variable set are independent, and hence each VG_h will contain only a single variable set. Overall, $Comp(\mathcal{T}) = 2^{32}$.

Example 2.7 (PRESENT with 28th Round Fault Injection) The distinguisher here is formed at the input of the 30th round S-Box. As it can be seen from Fig. 2.5, each distinguisher bit (actually each nibble) here depends on 20 key bits. However, due to the overlappings present in different nibble-wise key sets (KSs), the distinguisher evaluation process can eventually be partitioned into four independent (MKS, VG) pairs, each having 32 key bits involved—16 from the last round and rest from the penultimate round. The size of corresponding VG_hs become 4 state differential variables each. As a result, $Comp(\mathcal{T})$ becomes 2^{32}.

2.5.3 Complexity Evaluation of the Remaining Keyspace $\mathcal{R}$

The final step in finding a successful DFA is the evaluation of the remaining keyspace size ($|\mathcal{R}|$) after the fault injection. Often, the complexity remains beyond the practical exhaustive search complexity with a single fault injection and as a result, one might require multiple faulty ciphertexts. Nevertheless, the required number of faults for the successful attack can be estimated from the remaining space complexity of a single injection, and hence we specifically focus on the remaining search space with a single fault. A distinguisher $\mathcal{D}^i_j$ and the corresponding key parts obtained in the last two steps can be utilized to figure out the remaining keyspace complexity efficiently. Another important component of this computation is the *differential characteristic* of the S-Boxes. Differential characteristic (DC) of an S-Box S basically reports the average number of solutions an S-Box differential equation may have. They can be calculated from the difference distribution tables (DDT) of S-Boxes. The DC values for different ciphers can be found in [15].

The algorithm for remaining search space evaluation is presented in Algorithm 3. The main idea in this step is to figure out the probability, with which the distinguishing property occurs during distinguisher enumeration with random key guesses. This probability is then multiplied with the total keyspace in the corresponding MKS, giving the remaining search space complexity. Referring to the algorithm, the input consists of the corresponding distinguisher $\mathcal{D}^i_j$ and a set of tuples with cardinality M, which contains the MKSs and corresponding VGs. As an additional component, the DC characteristic of the S-Boxes $\mathcal{H}^h_S$ corresponding to each MKS_h, VG_h pair is also supplied. The $\mathcal{H}^h_S$ is the DC value corresponding to each (MKS_h, VG_h) pair. In some cases, a distinguisher may involve multiple S-Box layers and as a result $\mathcal{H}^h_S$ should be multiplied many times for each distinguisher variable (or variable set) evaluation. To keep things simple we directly provide the algorithm with properly tailored values within $\mathcal{H}^h_S$. Values of the $\mathcal{H}^h_S$ with above-mentioned tailoring can be trivially obtained from the CDGs described in the last subsection, just by keeping track of the S-Boxes encountered with the distinguisher.

Example 2.8 (IDFA Attack on AES) In this case, it turns out that $|\mathcal{R}|_{VG_1} = 2^{32} \times \left(\frac{255}{2^8}\right)^4$ (roughly equal to $2^{32} - 2^{26}$). This is because each state differential variable in VG_1 assumes the distinguishing property with probability $\frac{255}{2^8}$ and there are four such variables in VG_1. The size of the remaining keyspaces is the same for other three $\langle MKS_h, VG_h \rangle$ pairs. The large size of the remaining keyspace indicates the need of multiple fault injection. Although the estimation of the required number of faults here is slightly nontrivial due to the impossible differential inequalities involved, it can be estimated using the construction from [9]. Overall, the attack complexity is $O(2^{32})$ and total 2^{11} faults will be required to extract the key uniquely [9].

Example 2.9 (AES with Eighth Round Fault Injection) The MKS and VGs, which are the inputs to the Algorithm 3 are 4 in number in this case. Further, each VG contains a single variable set and 32 key bits corresponding to that. One needs to consider the number of itemsets corresponding to each variable set (or variable group, as in this case each group contains a single variable set) in this case. For each of the four variable sets, the probability of occurrence of the distinguishing criterion is $\mathbb{P}[VG_h] = \frac{255}{2^{32}}$. The DC characteristic of AES S-Box is found to be 1 and the total number of key possibilities is 2^{32}. The remaining keyspace corresponding to each variable set thus becomes $2^{32} \times 2^{-24} = 2^8$, leading to a complete remaining search space complexity of $(2^8)^4 = 2^{32}$. One should exhaustively search this remaining keyspace for the correct key. The total complexity of the attack, considering both $Comp(\mathcal{T})$ and $|\mathcal{R}|$ thus remains 2^{32}.

In [22], Tunstall et al. presented a two-step approach for the attack described here, which eventually reduces the remaining keyspace size to 2^8. The idea is to complete the attack just described, and then to exploit another distinguisher δ^1_9 which was previously costly to evaluate on the complete keyspace. However, one should notice that the attack complexity still remains 2^{32}. The distinguisher identification framework of ExpFault detects both the distinguishers. The two-step attack requires the existence of the inverted key schedule equations. The proposed tool in its current form can automatically identify the proper sequence of distinguishers for the attack. However, the algorithm for multi-step attack will not be outlined in this chapter. Instead, we shall describe it by means of examples on the GIFT cipher in subsequent sections.

Example 2.10 (Attack on PRESENT) The distinguisher evaluation process, in this case, can eventually be partitioned into 4 independent $\langle MKS_h, VG_h \rangle$ pairs, each having evaluation complexity of 2^{32}. For each of the 4 (MKS_h, VG_h) pair, the probability of occurrence of the distinguishing criterion is $\mathbb{P}[VG_h] = (\frac{2}{16})^4 = 2^{-12}$, and the remaining keyspace size is $|\mathcal{R}|_{VG_h} = 2^{20}$. With a single fault injection, thus the keyspace reduces to 2^{80} from 2^{128} [16] in this case, and the attack demands the

---

Algorithm 3: Procedure *EVAL_ SEARCH_ SPACE*

Input : $\mathcal{D}^i_j$, $\{\langle MKS_h,\ VG_h, \mathcal{H}^{h}_{S}\rangle\}_{h=1}^{M}$
Output : Complexity of the remaining search space $\mathcal{R}$, after one fault injection ($|\mathcal{R}|$)

$|\mathcal{R}| := 1$;
for ***each*** VG_h **do**
    $\mathbb{P}[VG_h] := 1$;
    for ***each*** $g_h \in VG_h$ **do**
        if $(VS_{\delta^i_j} == \phi)$; `// if` $\mathcal{D}^i_j$ `includes no variable sets`
        then
            $count := |Rng_{g_h}|$;
            $b_c := m$;
        else
            $count := |IS^{g_h}_{\delta^i_j}|$;
            $b_c :=$ `VarCount`$(g_h) \times m$;
        $\mathbb{P}[VG_h] := \mathbb{P}[VG_h] \times \frac{count}{2^{b_c}}$;
    $k_{size} :=$ `BitCount`(MKS_h) ▷ [17];
    $|\mathcal{R}|_{VG_h} := 2^{k_{size}} \times \mathbb{P}[VG_h] \times (\mathcal{H}^h_S)^{|VG_h|}$;
    $|\mathcal{R}| := |\mathcal{R}| \times |\mathcal{R}|_{VG_h}$;
return: $|\mathcal{R}|$.

---

[16] The distinguisher here simultaneously extracts round keys from the last two rounds of PRESENT. Total 128 key bits are extracted which can uniquely determine the 80 bit master key by using key scheduling equations.

[17] `BitCount` returns the number of bit variables in (MKS_h).

injection of at least another fault (complexity becomes $(2^{32} \times (2^{-12})^2)^4 = 2^{32}$, which is fairly reasonable). In summary, with two fault injections of 16 bit width, the 80 bit key can be figured out with $Comp(\mathcal{T}) = 2^{32}$ and $|\mathcal{R}| = 2^{32}$.

Discussion

Before going to the case studies, let us summarize the ExpFault framework in nutshell. For each fault instance, the tool analyzes all state differentials starting from the fault injection point and figures out a set of distinguishers from them. Each of these distinguishers is then analyzed with the graph-based framework and the evaluation and offline complexities are determined. The best performing distinguisher can be instantly chosen to realize the attack. However, in certain cases, multiple distinguishers can be combined to get better attacks. In the next section, we provide a typical example of such situations. Our tool was able to figure out the optimal attacks in these cases. Further details on the framework are provided in the appendix.

2.6 Case Studies

Until now we have provided proof-of-concept evaluations of the proposed framework on two well-known ciphers—AES-128 and PRESENT-80. Three known attacks have been elaborated step by step in the form of examples. However, both of the ciphers have been examined thoroughly for exploitable faults. It has been observed that exploitable faults are limited up to 7th round in AES and 28th round in PRESENT, which agrees with the existing literature. In this section, we evaluate the ExpFault framework on a recently proposed cipher called GIFT [1]. To the best of our knowledge, GIFT has never been considered explicitly in the context of DFA. With the help of the framework, we were able to figure out several interesting attack instances, which establishes the effectiveness of the proposed framework in the context of exploitable fault characterization.

2.6.1 Differential Fault Attack on GIFT Block Cipher

GIFT [1] is a lightweight block cipher proposed in CHES 2017. The basic construction of the algorithm bears resemblance to the PRESENT block cipher. However, specific changes were made to make it even more lightweight while ensuring improved resistance against certain attacks like linear hulls. More specifically, GIFT utilizes a different 4×4 S-Box and an improved bit permutation layer along with a new key addition layer which uses only 32 and 64 bit round keys for GIFT-64 and GIFT-128, respectively. The round keys are derived from a 128-bit key state utilizing a simple linear key schedule operation. It has been demonstrated in [1] that even with these simplifications, GIFT is able to provide comparable (and sometimes improved) security margins than that of PRESENT, SIMON [3], and SKINNY, for classical attacks.

In this work, we focus on GIFT-64 which iterates 28 times to generate the ciphertext. The round structure consists of three sub-operations `SubCells`, `PermBits`, and `AddRoundKey`, where `SubCells` is the nonlinear S-Box layer, `PermBits` is the bit permutation, and `AddRoundKey` is the key addition layer. In order to optimize hardware resources, which is, in fact, the main design goal of GIFT, the `AddRoundKey` layer XORs only 32 round key bits and 6 round constant bits in each round. More specifically, two round key bits are XORed with each nibble. The simple key schedule operation of GIFT deserves special mention in this context. It is observed that *for any four consecutive rounds, the round keys used are completely independent of each other*. This observation has a significant impact on the complexities of the fault attacks. In this subsection, we aim to examine the security of GIFT against differential fault attacks, which to the best of our knowledge has never been considered explicitly. In [18] authors proposed a side channel assisted DFA on PRESENT, which is claimed to be equally applicable for GIFT. In [7] Breier et al. proposed a side channel assisted differential plaintext attack on PRESENT which seems to be applicable to GIFT as well. None of these attacks, so far, have tried classical DFA attacks on GIFT, which makes GIFT a perfect candidate to be evaluated with our proposed framework.

After an extensive evaluation, we figured out several interesting attacks on GIFT-64, mostly while faults were injected at 25, 26, and 27th rounds. However, due to incomplete diffusion of the fault at the beginning of the propagation as well as independent round keys at four consecutive rounds, we found that none of the aforementioned fault locations can extract the complete 128 bit key alone. Another injection at a deeper round is thus necessary, and still only one of such fault pairs can get the complete 128-bit key. Evaluation of these attacks was done based on three parameters—the distinguisher evaluation complexity $Comp(\mathcal{T})$, size of the remaining key search space $|\mathcal{R}|$, and the number of injected faults, all of which are outputted by our tool for each fault location. A summary of these attacks is presented in Table 2.3. In this subsection, we shall elaborate one attack where a 4-bit fault is injected before the S-Box operation at the 25th round, followed by another nibble fault at the input of the S-Box operation at the 23rd round. This is the most efficient attack found so far with our framework. Further details of the attacks can be found in appendix section "More on the DFA of GIFT" of this chapter.

Injection of one 4-bit fault at the beginning of 25 round constructs several distinguishers. Examples of these distinguishers are provided in Table 2.4. Here the fault is injected at nibble location 0 from left. The attack will be described based on these distinguishers, as we have observed that injections at other nibble locations result in equivalent situations. Notably, none of these distinguishers contain any variable sets (i.e., variables are independent within the state differentials). However, the values assumed by the state differential variables differ which eventually

Table 2.3 Summary of DFA attacks on GIFT

Fault width	Round	Attack results Evaluation complexity	$\lvert\mathcal{R}\rvert$	No. faults per location	Keys extracted	Comments
4	24	–	–	–	–	No attack found
	25, 23	$2^{17.53}$	$2^{7.06}$	1	128	Best attack found
	26, 24	2^{6}	$2^{3.53}$	1	104	Cannot extract full key
	27, 25	2^{6}	$2^{3.53}$	1	72	Cannot extract full key
8	24	–	–	–	–	No attack found
	25, 23	$2^{17.53}$	$2^{7.06}$	1	128	Best attack found
	26, 24	2^{6}	$2^{3.53}$	1	104	Cannot extract full key
	27, 25	2^{6}	$2^{3.53}$	1	72	Cannot extract full key

We consider a fault injection a successful attack only if both the evaluation complexity and $|\mathcal{R}|$ is less than the size of the keyspace

Table 2.4 Distinguishers of the best attack (for the first fault injection)

Distinguisher	Location	Description
$\mathcal{D}^2_{27}$	Input of	$w_1^{227} \in \{0, 3, 5, 7, 9, 13\}$, $w_2^{227} \in \{0, 3, 5, 7, 9, 13\}$,
	`PermBits`	$w_3^{227} \in \{0, 3, 5, 7, 9, 13\}$, $w_4^{227} \in \{0, 3, 5, 7, 9, 13\}$,
	in round 27	$w_5^{227} \in \{0, 5, 6, 9, 10, 13, 14\}$,
		$w_6^{227} \in \{0, 5, 6, 9, 10, 13, 14\}$,
		$w_7^{227} \in \{0, 5, 6, 9, 10, 13, 14\}$,
		$w_8^{227} \in \{0, 5, 6, 9, 10, 13, 14\}$,
		$w_9^{227} \in \{0, 5, 6, 8, 9, 10, 11, 12, 15\}$,
		$w_{10}^{227} \in \{0, 5, 6, 8, 9, 10, 11, 12, 15\}$,
		$w_{11}^{227} \in \{0, 5, 6, 8, 9, 10, 11, 12, 15\}$,
		$w_{12}^{227} \in \{0, 5, 6, 8, 9, 10, 11, 12, 15\}$,
		$w_{13}^{227} \in \{0, 3, 7, 11, 15\}$, $w_{14}^{227} \in \{0, 3, 7, 11, 15\}$,
		$w_{15}^{227} \in \{0, 3, 7, 11, 15\}$, $w_{16}^{227} \in \{0, 3, 7, 11, 15\}$
$\mathcal{D}^1_{27}$	Input of	$w_1^{127} \in \{0, 4\}$, $w_2^{127} \in \{0, 4\}$, $w_3^{127} \in \{0, 4\}$,
	`SubCells` in	$w_4^{127} \in \{0, 4\}$, $w_5^{127} \in \{0, 2\}$, $w_6^{127} \in \{0, 2\}$,
	round 27	$w_7^{127} \in \{0, 2\}$, $w_8^{127} \in \{0, 2\}$, $w_9^{127} \in \{0, 1\}$,
		$w_{10}^{127} \in \{0, 1\}$, $w_{11}^{127} \in \{0, 1\}$,$w_{12}^{127} \in \{0, 1\}$,
		$w_{13}^{127} \in \{0, 8\}$, $w_{14}^{127} \in \{0, 8\}$, $w_{15}^{127} \in \{0, 8\}$,
		$w_{16}^{127} \in \{0, 8\}$
$\mathcal{D}^1_{26}$	Input of	$w_1^{126} \in \{0, 4\}$, $w_2^{126} \in \{0\}$, $w_3^{126} \in \{0\}$, $w_4^{126} \in \{0\}$,
	`SubCells` in	$w_5^{126} \in \{0, 2\}$, $w_6^{126} \in \{0\}$, $w_7^{126} \in \{0\}$, $w_8^{126} \in \{0\}$,
	round 26	$w_9^{126} \in \{0, 1\}$, $w_{10}^{126} \in \{0\}$, $w_{11}^{126} \in \{0\}$, $w_{12}^{126} \in \{0\}$,
		$w_{13}^{126} \in \{0, 8\}$, $w_{14}^{126} \in \{0\}$, $w_{15}^{126} \in \{0\}$, $w_{16}^{126} \in \{0\}$

create the distinguishing properties. The very first distinguisher utilized is being constructed in round 27 at the output of the S-Boxes ($\mathcal{D}_{27}^2$). This distinguisher (Row 1 of Table 2.4) results in the extraction of the round keys of the 28th round. The evaluation complexity of this distinguisher is $Comp(\mathcal{T}) = 2^8$ and the $|\mathcal{R}| = 2^{11.53} \approx 2^{12}$, with a single fault injection. In other words, one needs to guess the 32 bit round key in 8 bit chunks to reduce the key search space from 2^{32} to 2^{12}.[18] The cause of obtaining such $Comp(\mathcal{T})$ is elaborated in appendix section "More on the DFA of GIFT" with graphs outputted by the tool.

In order to obtain the master-key of GIFT, one must extract all the 128-bits. Unfortunately, the equations from the key schedule cannot be used in extracting rest of the key-bits with the knowledge of 32 bit, due to the very special nature of GIFT key schedule mentioned above. One must extract all four consecutive round keys in order to get a complete attack. In this scenario, our framework efficiently exploits all other distinguishers obtained. The second distinguisher that we utilize is constructed at the input of the S-Boxes at round 27 ($\mathcal{D}_{27}^1$). This distinguisher, which can extract the round keys of round 27, has an evaluation complexity of 2^2 provided the keys of round 28 are known. For each of the 2^{12} choices of the 28th round key, the size of the remaining keyspace (i.e., the keyspace of the round keys from 27th round) is 1, which leaves us with total 2^{12} choices for the keys of last two rounds.

In the third phase of the attack, we utilize the distinguisher formed at the input of the S-Box layer of round 26 ($\mathcal{D}_{26}^1$). With the knowledge of the keys from round 27 and 28, keys of round 26 can be determined uniquely with an evaluation complexity of 2^2 only. However, it is worth noting that only 8-bit keys of round 26 can be extracted in this case, as the fault only affects the input of 4 S-Boxes here (incomplete diffusion). As a result, the evaluation this distinguisher leaves us with total 2^{12} ($2^{11.53}$) choices for 72 bits of the last 3 round keys (32 key bits from round 28, 32 bits from round 27, and 8 bits from round 26). Also, no other distinguisher is left which necessitates the injection of another fault. In the final step of the attack, another nibble fault is injected at the input of the S-Box layer of round 23. One should note that with the knowledge of 27th and 28th round keys, the last two rounds of the cipher can be decrypted. As a result, the injection at round 23 results in an exactly same pattern of distinguishers as for round 25, now up to round 26. In other words, we now exploit the state differentials $\mathcal{D}_{25}^2$, $\mathcal{D}_{25}^1$, and $\mathcal{D}_{24}^1$ as distinguishers. With the knowledge of 8 key bits from round 26, the remaining key complexity (round key 26) after the evaluation of $\mathcal{D}_{25}^2$ becomes $2^{3.53}$ (distinguisher evaluation complexity is 2^6). Next, $\mathcal{D}_{25}^1$ is utilized to uniquely extract the keys of round 25 (with evaluation complexity 2^2 for each choice of the round 26 keys). By means of these two distinguishers, the remaining keyspace complexity $|\mathcal{R}|$ becomes $2^{11.53} \times 2^{3.53} = 2^{15.06}$ for the whole 128 bit secret key. However, there is still scope of further improvements. It is to be noted that the key material

[18]Following the terminology used in this chapter, here we have $|MKS_h| = 8$ and $|VG_h| = 4$.

at some round r_i of GIFT is repeated once again at round $r_i + 4$. As a result, the key material at round 24 and 28 becomes the same (actually a permutation of each other). The attack described until now has already reduced the entropy of the 28th round key to $2^{11.53}$, which makes the effective entropy of the 24th round key to be $2^{11.53}$ (instead of 2^{32}). Now using the distinguisher $\mathcal{D}^1_{24}$ (which is structurally same as $\mathcal{D}^1_{26}$), 8 key bits from round 24 can be determined uniquely. The complexity of this key material, which happened to be $2^{11.53}$ without this reduction now becomes $2^{11.53-8} = 2^{3.53}$. To summarize, two consecutive nibble fault injections at round 25 and 23 recover the complete 128-bit key. The remaining keyspace complexity $|\mathcal{R}|$ is $2^{11.53-8} \times 2^{3.53} = 2^{7.06}$.[19] The overall value of $Comp(\mathcal{T})$ of this attack is given by $max(2^8, 2^2 \times 2^{11.53}, 2^2 \times 2^{11.53}, 2^6 \times 2^{11.53}, 2^{11.53} \times 2^{3.53} \times 2^2, 2^{11.53} \times 2^{3.53} \times 2^2) = 2^{17.53}$. Each of the components of $Comp(\mathcal{T})$ corresponds to an independent computation chunk which can be executed on a single thread, in parallel to other similar chunks running in other threads. Such independent computation chucks result from the divide-and-conquer strategy automatically identified by our tool.

2.7 Chapter Summary

In this chapter, we have proposed an automated framework for exploitable fault identification in modern block ciphers. The main idea is to estimate the attack complexity without doing the attack in the original sense. Moreover, the proposed framework is fairly generic to cover most of the existing block ciphers, and provides high fault coverage and degree of automation. Three step-by-step case studies on different ciphers and fault attack instances were presented to establish the claims. Further, the tool has been utilized to figure out DFA attacks on a recently proposed block cipher GIFT. Future works will target further automation and generalization of the proposed framework as well as comprehensive analysis of different existing ciphers using it. Some obvious future extensions include the attack automation on key schedule and round counters. Another extremely important goal could be the detection of integral attacks, DFIA attacks, and MitM attacks [16, 17] which also seems to be feasible in this data-analysis based framework. Design of countermeasures with the assistance from this tool could be another research direction.

Acknowledgements We would like to acknowledge Synopsys Inc, USA for partially supporting the research through the grant entitled "Formal Methods for Physical Security Verification of Cryptographic Designs Against Fault Attacks."

[19]For each choice of 28th and 27th round keys we have $2^{3.53}$ choices for 26th and 25 round keys combined.

Appendix 1: Implementation Details of ExpFault

This section elaborates the implementation details of an initial prototype of the ExpFault framework. The tool is mostly written in Python-3, with an exception for the data mining algorithm (Apriori) for which we use the WEKA [13] toolbox implemented in Java. In the following subsections, several attributes of the tool will be elaborated. We shall also point out some limitations of our current implementation in nutshell.

Assumptions

The main reason behind the development of ExpFault is a fast and cipher-oblivious characterization of exploitable faults. As we shall show in later subsections, even in its prototype implementation, ExpFault is able to characterize individual fault instances within a very reasonable time. In this work, we have characterized each fault location of the ciphers under consideration in an exhaustive manner (only up to a reasonable number of rounds.). However, it is worth mentioning that the characterization can be made significantly faster considering the structural symmetries present in standard block ciphers. For example, it is well-known that all 16 byte locations of AES in a specific stage of computation are equivalent as fault injection points. The presence of such symmetries should extensively reduce the number of fault locations to be checked. However, a systematic analysis of such equivalences is out of scope for this chapter.

Exploitable fault analysis is expected to be performed by an evaluator. During the construction of the framework, we made assumptions which are only consistent in the context of an evaluator. For example, the fault locations are assumed to be known, which is indeed a reasonable assumption in our case. However, it is worth mentioning that the attacks discovered by our tool can be extended in an unknown fault location context with a reasonable penalty incurred on the attack complexities.

Inputs and Outputs

In its current form, ExpFault takes an executable of the cipher algorithm as input, which is mainly used to generate fault simulation traces. Simulation traces are dumped in `.arff` format which is the default data format for the WEKA toolset. The framework also expects an input file describing the cipher, which is used to generate the CDG and other internal data structures. We call this input as the *Cipher Description file*. A cipher description file mainly contains abstract descriptions of the sub-operations as specified in Fig. 2.4. As a concrete example, we provide the description for two 4×4 S-Boxes in Fig. 2.8. It can be observed that the S-Box

```
BEGINBLOCK SLAYER
OPTYPE NONLINEAR
OPINPUT 64
OPOUTPUT 64
% 0 = 0,1,2,3
% 1 = 0,1,2,3
% 2 = 0,1,2,3
% 3 = 0,1,2,3

% 4 = 4,5,6,7
% 5 = 4,5,6,7
% 6 = 4,5,6,7
% 7 = 4,5,6,7
:
:
:
:
ENDBLOCK SLAYER
```

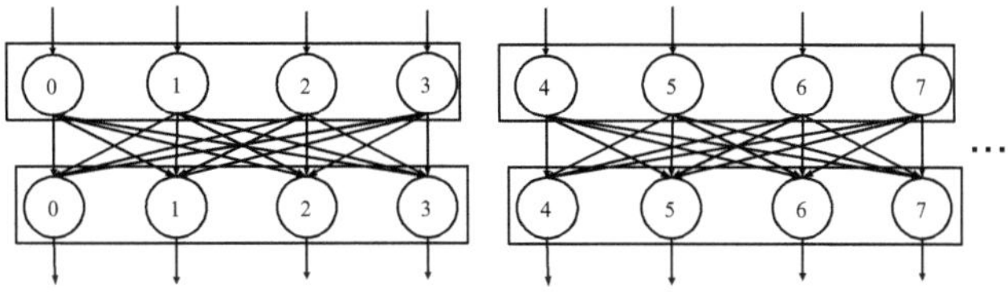

Fig. 2.8 Code snippet from cipher description file and its corresponding graphical representation [20]

is specified in a bitwise manner, where the number at the left side of the "=" sign specifies the sink node of a directed edge, and the number at the right side of the "=" specifies the source nodes. The complete S-Box layer is required to be specified within a single <BEGINBLOCK>· · ·<ENDBLOCK> construct. Likewise one can define other sub-operations.

The outputs corresponding to each fault injection is printed in a text file. The standard output contains the description of all the distinguishers found along with their entropy. The evaluation complexity of each distinguisher and the size of the remaining keyspace after distinguisher evaluation is also outputted along with that. Attack evaluation with multiple distinguishers and key schedule equations is not fully automated yet. However, the outputs provided are sufficient to figure out attacks using multiple distinguishers. Example of a single distinguisher is provided in Fig. 2.9, just to illustrate the output format and the information provided. It is also possible to extract the independent and parallely executable computation chunks (resulted from the divide-and-conquer based distinguisher evaluation) in the form of subgraphs from ExpFault. Such subgraphs are extremely useful for interpreting and implementing the attacks. Examples of such subgraphs will be provided in the appendix section "More on the DFA of GIFT".

Setup for Distinguisher Identification

Frequent itemset mining is crucially dependent on the *support* parameter of the mining algorithm. The implementation of the Apriori algorithm we used (from WEKA package [13]) iteratively decrements the support from a value of 1.0 to

```
-----------------------------------------------------------
Distinguisher Evaluation Complexity (in log scale) 8
Remaining Key Space Complexity (in log scale)
11.53668207643374

Distinguisher Level 79
Round_no 27
Subop_no 2
Has_associations False
Entropy  43.536682076433735

V2 [0, 3, 5, 7, 9, 13, ]
V0 [0, 3, 5, 7, 9, 13, ]
V12 [0, 3, 7, 11, 15, ]
V6 [0, 5, 6, 9, 10, 13, 14, ]
V8 [0, 5, 6, 8, 9, 10, 11, 12, 15, ]
V14 [0, 3, 7, 11, 15, ]
V3 [0, 3, 5, 7, 9, 13, ]
V1 [0, 3, 5, 7, 9, 13, ]
V5 [0, 5, 6, 9, 10, 13, 14, ]
V10 [0, 5, 6, 8, 9, 10, 11, 12, 15, ]
V9 [0, 5, 6, 8, 9, 10, 11, 12, 15, ]
V13 [0, 3, 7, 11, 15, ]
V7 [0, 5, 6, 9, 10, 13, 14, ]
V4 [0, 5, 6, 9, 10, 13, 14, ]
V11 [0, 5, 6, 8, 9, 10, 11, 12, 15, ]
V15 [0, 3, 7, 11, 15, ]

No Variable sets exist..

----------------------------------
```

Fig. 2.9 Description of a distinguisher from the output file [20]

a predefined lower bound. To generate all desired maximal frequent itemsets, the support lower bound of Apriori was experimentally decided to be $\frac{1}{2^m}$ (m: bit length of each variable). The maximality of the itemsets was ensured experimentally by varying the support threshold as well as the data set size, which also nullifies the risk of generating an insufficient number of itemsets. For ciphers having MDS or similar operations we found that the dataset size of 12, 750 (that is, 10 plaintexts, 5 different keys, and all 255 possible fault values) for 128-bit ciphers, and 750 for 64-bit ciphers (10 plaintexts, 5 different keys, and all 15 possible fault values) are sufficient to discover all possible itemsets. The relatively small dataset size is attributed to their deterministic fault propagation patterns (i.e., number of active S-Boxes are same for all fault values at a specific location). However, for ciphers with bit permutation operations, the dataset size should be larger as the fault propagation pattern becomes probabilistic. We found that a dataset of 25, 500 works well for both PRESENT and GIFT. Varying the keys, plaintexts, and the fault values ensure that the discovered rules/itemsets are independent of all these factors, which is essential for a DFA

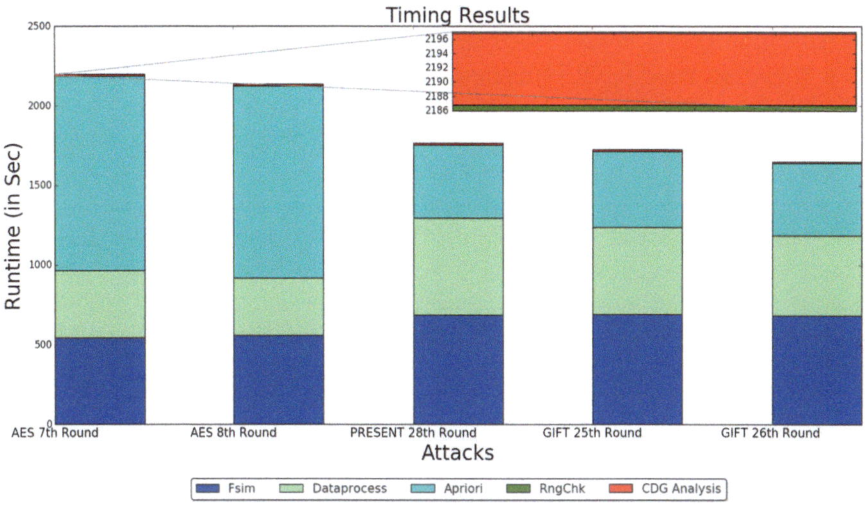

Fig. 2.10 Runtime analysis: different components of the runtime are shown. Runtimes of `RngChk` and CDG based complexity analysis are found to be negligibly small (shown in zoom) [20]

distinguisher. An interesting feature of the itemset generation algorithm is that it returns null when all the variables are independent. The `RngChk` function does not require any parameter setting and works fine with the dataset sizes provided above.

There always exists a risk of generating spurious itemsets in frequent itemset mining, especially with very low support values. However, detecting such spurious tuples is not very difficult as the variables from consecutive state differentials have well-defined mathematical relations. Any itemset not obeying these relations can be easily removed as spurious. In our experiments, we observed some spurious tuples for GIFT with low support values which were successfully eliminated using the structural knowledge from the cipher.[20] Such structural knowledge is already available in the form of the CDG and thus can be exploited without incurring any significant computational cost.

Analysis of Runtime

The runtime of the framework is a crucial factor for exploitable fault identification. In order to get a clear idea about different subparts of the framework, extensive runtime analysis was performed. The results are summarized in Fig. 2.10 for different attack examples considered in this chapter. All the experiments were

[20]In fact, all the itemsets found for GIFT were spurious and the cipher does not have variable associations. In such cases, the support can be increased to stop spurious itemset generation.

performed on a laptop with Intel Core i5 processor, 8 Gb RAM running Ubuntu 16.04 as the OS.

It is evident from Fig. 2.10 that the Apriori algorithm dominates the runtime which is about 1220 s. for the AES 7th round attack example and moderately less in other examples. Fortunately, this step can be extensively parallelized as the analysis of state differential datasets is completely independent of each other. Although the current prototype does not implement any such parallelization, improvement of runtime can be anticipated. Similar arguments can be made for the fault simulation which is another dominating factor in the tool runtime. As an alternative strategy for runtime improvement, knowledge about the cipher structure can be exploited. For example, it is quite well understood that ciphers with bit permutation layers cannot have variable associations. One may opt to skip the itemset mining step while analyzing such ciphers, in order to improve the runtime. The third dominating component of the runtime is the data processing operation which is an implementation specific overhead. This timing overhead is attributed to the python wrapper for reading `.arff` files. In future versions of the tool, we shall try to get rid of such unnecessary overheads. To summarize, the framework takes less than an hour to analyze each fault instance which can be improved further. The usability of ExpFault for exploitable fault space characterization is thus evident.

Current Limitations

The limitations of the initial prototype have already been elaborated in the previous subsection. The tool at its current state cannot directly handle key schedule relations and attacks on the key schedules. Handling key schedule relations during complexity calculation is not very difficult and extension of the tool can handle it. However, handling key schedule attacks will require algorithmic improvements in the complexity analysis step. It also worth to mention that the current implementation does not directly handle attacks using multiple distinguishers and minor human intervention is still required. However, the algorithm for using multiple distinguishers is straightforward and is planned to be incorporated in the next version of the tool.[21]

Appendix 2: More on the DFA of GIFT

In this section, we provide further details about the DFA attacks discovered by the ExpFault framework. Specifically, we shall elaborate the distinguisher evaluation complexity of the attack described in our case study with the help of graphs

[21]One should note that the tool is still in its initial phase and we shall try to address all the above-mentioned issues before making it open source.

generated by our tool. This will be followed by a brief discussion on some other attack instances on the cipher.

Calculation of the Attack Complexity

Let us refer to the attack on GIFT described in Sect. 2.6.1 of the chapter. For each of the distinguishers found (for injections at round 25 and 23) we perform a series of BFS searches on the CDG of the cipher, which eventually provides independently and parallely computable sub-parts (as described in Sect. 2.5.2) along with the attack complexity (described in Sect. 2.5.3).[22] For each of the distinguishers, we provide the associated sub-graphs from the CDG, generated by our tool. As an example, Fig. 2.11a corresponds to the sub-graph associated with the distinguisher $\mathcal{D}_{27}^2$—the first distinguisher utilized for the attack. For convenience, we refer to a magnified part of this sub-graph in Fig. 2.11b.[23] The graph in Fig. 2.11b corresponds to one independently computable chunk for the $\mathcal{D}_{27}^2$ with 8 associated key bits, which results in an evaluation complexity of 2^8. Each numbered node corresponds to a bit variable of the cipher. The first layer of this graph (shown with its nodes colored) corresponds to state differential variables from distinguisher $\mathcal{D}_{27}^2$. It can be observed from Fig. 2.11b that total four state differential variables are associated with this independently computable chunk (their corresponding bits are shown with four different colors). These four variables construct a *variable group* VG_h according to the terminology used in the chapter. The 16 key nodes in the graph are highlighted with black circles and squares. Interestingly, the square key nodes are not required to be guessed to evaluate the distinguisher, as there exists no nonlinear layer between these keys and the distinguisher. This leaves us with 8 key bits for this computing part which are the members of the MKS_h corresponding to the VG_h. Evidently, there are 4 such independent computing chunks in total, each with the same evaluation complexity of 2^8. As a result, $Comp(\mathcal{T})$ becomes 2^8.

The offline complexity $|\mathcal{R}|$ of the attack can now be computed individually for each computing subpart/chunk. Referring to the first computing subpart/chunk, there are four associated state differential variables $w_{13}^{227}, w_{14}^{227}, w_{15}^{227}, w_{16}^{227}$ in VG_4, each of which can take at most five possible values from the set $\{0, 3, 7, 11, 15\}$. As a result, the size of the remaining keyspace becomes $|\mathcal{R}|_{VG_1} = \left(\frac{5}{16}\right)^4 \times 2^8 = 2.4414$. In a similar manner, remaining keyspace corresponding to three other computing chucks can be estimated as $|\mathcal{R}|_{VG_1} = \left(\frac{6}{16}\right)^4 \times 2^8 = 5.0625$, $|\mathcal{R}|_{VG_2} = \left(\frac{7}{16}\right)^4 \times 2^8 = 9.3789$, and $|\mathcal{R}|_{VG_3} = \left(\frac{9}{16}\right)^4 \times 2^8 = 25.6289$. The offline complexity $|\mathcal{R}|$ thus becomes $2.4414 \times 5.0625 \times 9.3789 \times 25.6289 = 2^{11.53}$.

[22] Refer to Table 2.4 for the description of the distinguishers.

[23] The round constant bits of GIFT cipher are not shown in the graphs as they are found to have no effect on the DFA complexity calculation.

Fig. 2.11 Graph corresponding to the distinguisher $\mathcal{D}_{27}^2$ for the attack with a nibble fault injection at the 25th round on GIFT; (**a**) The complete graph from the tool (used to extract the keys of round 28), (**b**) One independent computation chunk/subpart [20]

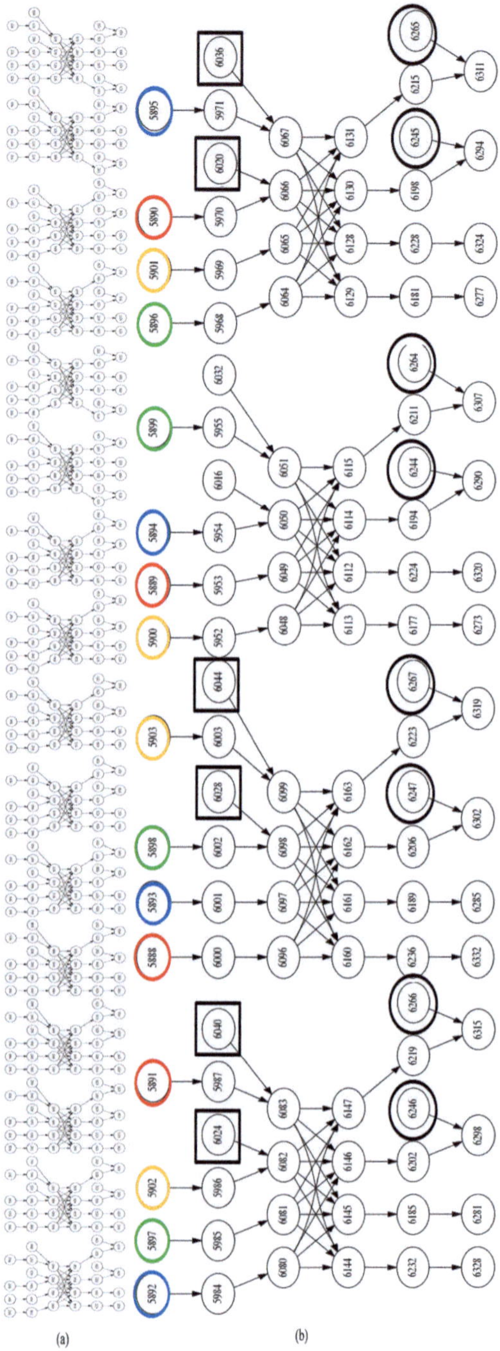

Similar computations can be performed for the two other distinguishers from 25th round fault injection. The complexity calculations are even simpler in these cases as the sub-graph from the CDG is already segregated into 16 independent components which make the complexity calculations trivial. Figure 2.12 presents the graph corresponding to $\mathcal{D}_{27}^1$ outputted by ExpFault framework. Here we assume that the 28th round keys are already set, and as a result, they are not shown in the graph. It can be noticed from Fig. 2.12b that the computing chuck only contains one state differential variable and 2 key bits (that is $|VG_h| = 1$ and $|MKS_h| = 2$). Each state differential variable in $\mathcal{D}_{27}^1$ may take two different values. As a result $|\mathcal{R}|_{VG_h} = \left(\frac{2}{16}\right) \times 2^2 = 0.5$, which essentially means that the key can be uniquely determined. Thus 27th round keys can be uniquely extracted, provided some values are set for the 28th round keys. Extraction of the round keys corresponding to round 26, however, becomes tricky. Due to incomplete diffusion of the fault at this distinguisher ($\mathcal{D}_{26}^1$) input of 4 S-Boxes get corrupted and as a result, only 8 associated key bits can be extracted uniquely. With the injection of another nibble fault at round 23, rest of the key bits from round 26 as well as the keys from round 25 can be extracted. The distinguishers obtained are exactly the same in this case, but occur two round earlier. Assuming the key bits from last two rounds and 8 bits of 26th round key to be known, rest of the keys from round 26 can be extracted with a complexity $\left(\left(\frac{6}{16}\right)^4 \times 2^6\right) \times \left(\left(\frac{7}{16}\right)^4 \times 2^6\right) \times \left(\left(\frac{9}{16}\right)^4 \times 2^6\right) \times \left(\left(\frac{5}{16}\right)^4 \times 2^6\right) = 2^{3.53}$. The evaluation complexity at this step is 2^6 (it is not 2^8 as two key bits from each of the independent computation chunk is already known). For each choice of 26th round keys, the round keys of 25th round can be extracted uniquely. The evaluation complexity here is 2^2. Finally, we utilize the fact that in GIFT round keys r_i and $r_i + 4$ are permutations of each other with the same entropy. Given this property, we can further reduce the keyspace using distinguisher $\mathcal{D}_{24}^1$, which has incomplete diffusion and thus can extract only 8 key bits corresponding to round 24 and 28 uniquely. Combining all these complexities, the whole attack extracts the 128 bit keys of GIFT with $|\mathcal{R}| = 2^{3.53} \times 2^{11.53-8} = 2^{7.06}$, with two faults at two different locations. If the multiplicities of the faults are increased, the keys can be extracted uniquely. In other words, with two injections at round 25 and two injections at round 23 (same locations), the full key can be extracted uniquely.

Other Attacks on GIFT

Nibble fault injection at the 26th (and 24th) round of GIFT also results in a similar attack. However, in this case, 104 key bits can be extracted. The attack results targeting different rounds of the cipher with varying fault widths are presented in Table 2.3. It is interesting to observe that even with increased fault width of 8 bits the attacks remain the same. We found that the distinguishers corresponding to 8 bit faults and 4 bit faults are similar (they are same complexity-wise). This is due

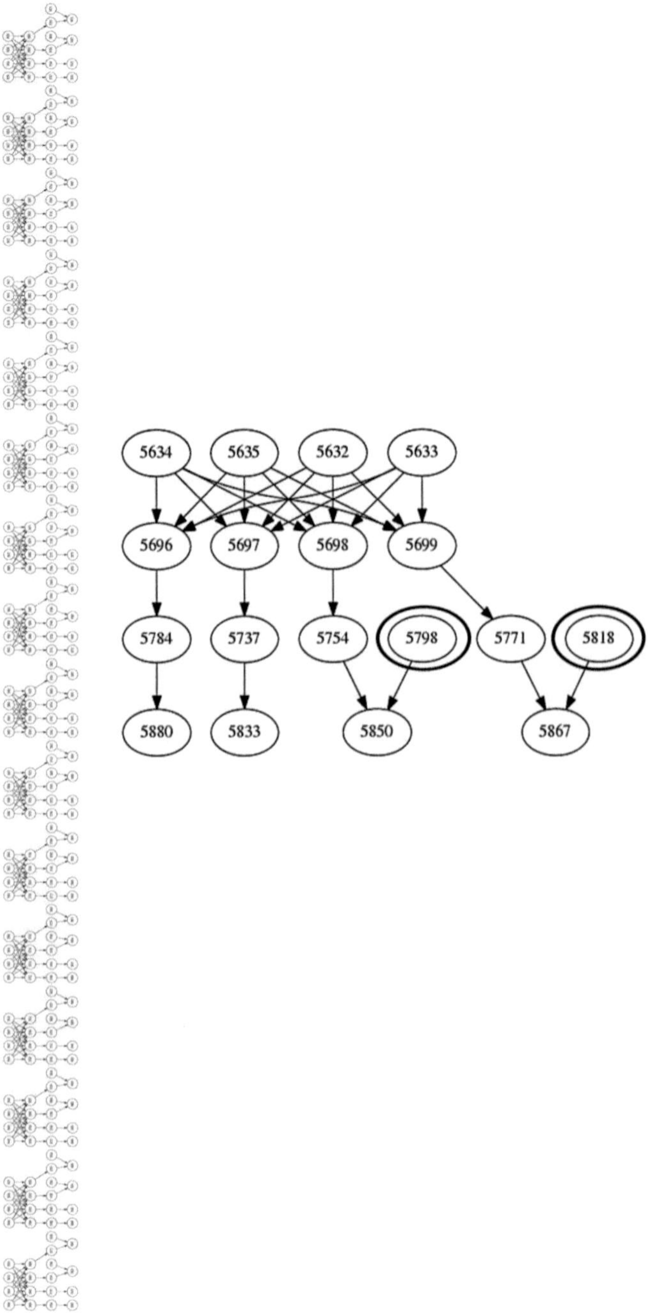

(a) (b)

Fig. 2.12 Graph corresponding to the distinguisher $\mathcal{D}^1_{27}$ for the attack with a nibble fault injection at the 25th round on GIFT; (**a**) The complete graph from the tool (used to extract the keys of round 27), (**b**) One independent computation chunk/subpart [20] to the fact that *the same set of S-Boxes become active by a nibble fault and a byte fault in GIFT*. This fact can be easily verified from the permutation table of the cipher. It is also worth mentioning that the fault propagation in GIFT (and in any bit permutation based cipher in general) is probabilistic as the number of the active S-Boxes (and thus the complexity of the resulting attack) depends on the value of the plaintext and the injected fault. The complexity figures returned by our tool are actually the average-case results and the worst-case attack complexity might vary up to some extent. From the perspective of a cipher evaluator, such average-case complexity figures are of utmost importance as they typically represent the average advantage the attacker can have. However, from the perspective of an attacker, the average success rate is also very interesting. The future versions of this tool will include measures for calculating such average-case complexities. In the context of attacks on GIFT cipher, we observed that the increasing width of the fault up to 8 bits actually makes the attacks more likely to happen. This is quite obvious as a larger fault width always makes the fault propagation more rapid. However, a very wide fault window may not work as the fault paths will become overlapped resulting in the destruction of some distinguishing properties. Although we have not done any such analysis in this chapter, it is worth mentioning that doing such analysis is quite straightforward using the ExpFault framework.

Appendix 3: Comparison with the AFA and ML-AFA

Automation of fault attacks has gained significant attention from the research community in the recent past. In this section, we provide a detailed comparison of the ExpFault framework with the AFA. The AFA is a powerful method for automated fault analysis [24]. The main idea of AFA is to construct an algebraic equation system representing the cipher and injected faults. This system is then solved by means of state-of-the-art SAT solvers, which are sufficiently robust and powerful to handle such large problem instances. Although the AFA approach is fairly easy to implement and quite generic in nature, it is not very suitable for exploitable fault analysis. This is attributed to the fact that AFA has to explicitly perform an attack to evaluate the exploitability status of a fault instance, which can be extremely time-consuming. Evaluation of a fault instance in AFA requires solving a SAT problem. The time required for solving SAT problems often depends on the size of the search space. The key fact behind the success of an AFA (or any DFA attack) is that the size of keyspace of a cipher reduces significantly with the injection of faults. Solving an AFA instance sometimes suffers from serious scalability issues if the size of the keyspace after fault injection is still large. Moreover, computing the exact attack complexity in AFA requires enumerating all solutions of the corresponding SAT instance. This strategy may incur huge computational overhead. Although in [24] Zhang et al. handled such scalability issues of AFA by assuming some of the key bits to be known, the complexity evaluation process still takes a significant amount of time. Perhaps, the most critical problem with AFA lies in its lack of interpretability. From the perspective of an evaluator, a clear understanding of the attack cause is essential, because it may help him to improve the cipher design or design good countermeasures. The CNF based abstraction used in AFA hides all structural information of the attack. In contrast, it is evident that from the outputs provided by the ExpFault one can have a precise understanding of each possible attack instance.

Recently, Saha et al. [21] have proposed an alternative approach for exploitable fault characterization which combines AFA with machine learning to achieve the speedup desired for exploitable fault characterization. The scheme proposed by them utilizes a ML model to classify exploitable faults from unexploitable ones. The ML model is constructed by extracting features from CNF representations of certain exploitable and unexploitable fault instances already known for a given cipher. A small set of exploitable and unexploitable fault instances can always be constructed in an initial profiling phase of the cipher by means of AFA. This specific framework is somewhat complementary to the ExpFault proposed in this chapter. In particular, the AFA-ML combination can explicitly characterize the exploitability status of different fault values corresponding to a specific fault location. This property is interesting for ciphers without MDS layers, as the exploitability of a fault instance for these ciphers critically depends upon the value of the fault. As a result, the success rate of a fault attack corresponding to a specific fault location can be estimated by the AFA-ML framework exploiting this property, which is still not possible in the ExpFault framework. However, the attacks identified by the AFA-ML based framework lack interpretability, which is a strong point of ExpFault. Moreover, the calculation of attack complexity is not possible with the AFA-ML tool, which is the main goal of this chapter. In some sense, these two frameworks are complementary. The AFA-ML tool will be described in the next chapter of this book.

References

1. S. Banik, S.K. Pandey, T. Peyrin, Y. Sasaki, S.M. Sim, Y. Todo, GIFT: a small PRESENT, in *International Conference on Cryptographic Hardware and Embedded Systems* (Springer, New York, 2017), pp. 321–345
2. G. Barthe, F. Dupressoir, P.-A. Fouque, B. Grégoire, J.-C. Zapalowicz, Synthesis of fault attacks on cryptographic implementations, in *Proceedings of the 2014 ACM SIGSAC Conference on Computer and Communications Security* (ACM, New York, 2014), pp. 1016–1027
3. R. Beaulieu, S. Treatman-Clark, D. Shors, B. Weeks, J. Smith, L. Wingers, The simon and speck lightweight block ciphers, in *2015 52nd ACM/EDAC/IEEE Design Automation Conference (DAC)*, pp. 1–6, June 2015
4. C. Beierle, J. Jean, S. Kölbl, G. Leander, A. Moradi, T. Peyrin, Y. Sasaki, P. Sasdrich, S.M. Sim, The SKINNY family of block ciphers and its low-latency variant MANTIS, in *Annual Cryptology Conference* (Springer, Berlin, 2016), pp. 123–153
5. E. Biham, A. Shamir, Differential fault analysis of secret key cryptosystems, in *Advances in Cryptology - CRYPTO '97*, ed. by B.S. Kaliski Jr. Lecture Notes in Computer Science, vol. 1294 (Springer, Berlin, 1997), pp. 513–525

6. A. Bogdanov, L.R. Knudsen, G. Leander, C. Paar, A. Poschmann, M.J. Robshaw, Y. Seurin, C. Vikkelsoe, PRESENT: an ultra-lightweight block cipher, in *Proceedings of the 9th International Workshop on Cryptographic Hardware and Embedded Systems, CHES '07* (Springer, Berlin, 2007), pp. 450–466
7. J. Breier, D. Jap, S. Bhasin, SCADPA: side-channel assisted differential-plaintext attack on bit permutation based ciphers, in *Design, Automation & Test in Europe Conference & Exhibition (DATE), 2018* (IEEE, Piscataway, 2018), pp. 1129–1134
8. J. Daemen, V. Rijmen, *The Design of Rijndael* (Springer, New York, 2002)
9. P. Derbez, P.-A. Fouque, D. Leresteux, Meet-in-the-middle and impossible differential fault analysis on AES, in *International Workshop on Cryptographic Hardware and Embedded Systems* (Springer, Berlin, 2011), pp. 274–291
10. P. Dusart, G. Letourneux, O. Vivolo, Differential fault analysis on AES, in *International Conference on Applied Cryptography and Network Security* (Springer, Berlin, 2003), pp. 293–306
11. N.F. Ghalaty, B. Yuce, M. Taha, P. Schaumont, Differential fault intensity analysis, in *Workshop on Fault Diagnosis and Tolerance in Cryptography (FDTC), 2014* (IEEE, Piscataway, 2014), pp. 49–58
12. J. Guo, T. Peyrin, A. Poschmann, M. Robshaw, The LED block cipher, in *Cryptographic Hardware and Embedded Systems – CHES 2011* (Springer, Berlin, 2011), pp. 326–341
13. M. Hall, E. Frank, G. Holmes, B. Pfahringer, P. Reutemann, I.H. Witten, The WEKA data mining software: an update. SIGKDD Explor. **11**(1), 10–18 (2009)
14. K. Jeong, Y. Lee, J. Sung, S. Hong, Improved differential fault analysis on present-80/128. Int. J. Comput. Math. **90**(12), 2553–2563 (2013)
15. P. Khanna, C. Rebeiro, A. Hazra, XFC: a framework for eXploitable fault characterization in block ciphers, in *Proceedings of the 54th Annual Design Automation Conference 2017, DAC '17* (ACM, New York, 2017), pp. 8:1–8:6
16. C.H. Kim, Efficient methods for exploiting faults induced at AES middle rounds. IACR Cryptol. ePrint Arch. **2011**, 349 (2011)
17. Z. Liu, Y. Liu, Q. Wang, D. Gu, W. Li, Meet-in-the-middle fault analysis on word-oriented substitution-permutation network block ciphers. Secur. Commun. Netw. **8**(4), 672–681 (2015)
18. S. Patranabis, J. Breier, D. Mukhopadhyay, S. Bhasin, One plus one is more than two: a practical combination of power and fault analysis attacks on PRESENT and PRESENT-like block ciphers, in *Workshop on Fault Diagnosis and Tolerance in Cryptography (FDTC), 2017* (IEEE, Piscataway, 2017), pp. 25–32
19. D. Saha, D. Mukhopadhyay, D.R. Chowdhury, A diagonal fault attack on the advanced encryption standard. IACR Cryptol. ePrint Arch. **2009**, 581 (2009)
20. S. Saha, D. Mukhopadhyay, P. Dasgupta, ExpFault: an automated framework for exploitable fault characterization in block ciphers. IACR Trans. Cryptogr. Hardw. Embed. Syst. **2018**(2), 242–276 (2018)
21. S. Saha, D. Jap, S. Patranabis, D. Mukhopadhyay, S. Bhasin, P. Dasgupta, Automatic characterization of exploitable faults: a machine learning approach. IEEE Trans. Inf. Forensics Secur. **14**(4), 954–968 (2019)
22. M. Tunstall, D. Mukhopadhyay, S. Ali, Differential fault analysis of the advanced encryption standard using a single fault, in *IFIP International Workshop on Information Security Theory and Practices* (Springer, Berlin, 2011), pp. 224–233
23. G. Wang, S. Wang, Differential fault analysis on PRESENT key schedule, in *International Conference on Computational Intelligence and Security (CIS), 2010* (IEEE, Piscataway, 2010), pp. 362–366
24. F. Zhang, S. Guo, X. Zhao, T. Wang, J. Yang, F.-X. Standaert, D. Gu, A framework for the analysis and evaluation of algebraic fault attacks on lightweight block ciphers. IEEE Trans. Inf. Forensics Secur. **11**(5), 1039–1054 (2016)
25. X. Zhao, S. Guo, T. Wang, F. Zhang, Z. Shi, Fault-propagate pattern based DFA on PRESENT and PRINT cipher. Wuhan Univ. J. Nat. Sci. **17**(6), 485–493 (2012)

Chapter 3
Exploitable Fault Space Characterization: A Complementary Approach

Sayandeep Saha, Dirmanto Jap, Sikhar Patranabis, Debdeep Mukhopadhyay, Shivam Bhasin, and Pallab Dasgupta

3.1 Introduction

Chapter 2 presented the ExpFault framework for constructing fault attacks automatically in a rather cipher oblivious manner. ExpFault has been constructed by systematizing and generalizing the traditional fault analysis techniques where the core part is to figure out a set of wrong key distinguishers. The ultimate goal of ExpFault is to formulate the attack algorithm automatically for a given fault. The average complexity of the attack is also estimated in this process. However, in order to achieve these goals, several structural abstractions are made in ExpFault. For example, the cipher dependency graph (CDG) data structure assumes the S-Boxes as complete graphs, which abstracts out certain mathematical details of the S-Boxes. Likewise, the fault model in ExpFault does not explicitly take the fault value and plaintext value into account. Although the simulation-based distinguisher identifier captures most of the necessary mathematical properties of sub-operations, it does not capture them explicitly in the form of equations. Such abstractions are consistent with traditional fault analysis approaches and do not affect the average case complexity of an attack in general. Most importantly abstractions enable scalable solutions to the exploitable fault characterization problem.

An alternative approach to such abstractions is to consider the exact mathematical structure of the cipher as a whole. Consideration of the exact structure guarantees

S. Saha (✉) · S. Patranabis · D. Mukhopadhyay · P. Dasgupta
Department of Computer Science and Engineering, Indian Institute of Technology, Kharagpur, India
e-mail: sahasayandeep@iitkgp.ac.in; sikhar.patranabis@iitkgp.ac.in; debdeep@iitkgp.ac.in; pallab@iitkgp.ac.in

D. Jap · S. Bhasin
Temasek Laboratories, Nanyang Technological University, Singapore, Singapore
e-mail: djap@ntu.edu.sg; sbhasin@ntu.edu.sg

J. Breier et al. (eds.), *Automated Methods in Cryptographic Fault Analysis*,
https://doi.org/10.1007/978-3-030-11333-9_3

accurate analysis of the attacks. In fact, an exact approach is supposed to capture intricate bit-level constraints over the keyspace, which may not get identified explicitly by ExpFault through its word-level analysis. The first challenge in realizing the exact approach lies in the encoding of the cipher and faults. Fortunately, such an encoding is fairly straightforward given the fact that any given block cipher can be represented as a system of multivariate polynomial equations over the finite field $GF(2)$.[1] Faults can also be encoded as extra equations within this system. Generation of such a system of equations from a given cipher specification is trivial and can be done in an automated manner. Recently, there has been significant progress in designing automated fault analysis tools using such algebraic representations [6, 14, 16, 29, 30, 32]. The most prominent among these automated frameworks is the so-called algebraic fault attack (AFA), which encodes a given cipher and an injected fault as an equation system in algebraic normal form (ANF) [6, 14, 29, 30, 32]. The ANF system is then converted to an equivalent system in conjunctive normal form (CNF) and fed to a Boolean satisfiability (SAT) solver with the aim of extracting the key by solving the system. The SAT solver is used as a black box and handles the mathematical constraints internally making the AFA framework fairly simple to implement.

Although the algebraic approach of fault attack seems simple and tempting, it has certain critical drawbacks. In particular, AFA involves solving a SAT problem for each individual fault instance. Although SAT solvers are remarkably good at finding solutions to a large class of NP-Complete problem instances, the time taken for solving is often prohibitively high. In fact, in the context of AFA attacks, the solver may not stop within a reasonable time for many fault instances. Although setting a proper timeout seems to be a reasonable fix for such cases, the variation of solving times is often very high. As a result, the timeout threshold for SAT must be reasonably high as well, to guarantee the capture of every possible attack classes. It was already pointed out in the last chapter that fault spaces in block ciphers are of incredibly large size.[2] It is thus quite evident that an informative characterization of the entire fault space by means of AFA is impractical within reasonable limits of time. Even the enumeration of the fault space becomes an issue for the algebraic approaches of fault attack. Unfortunately, even a statistical characterization of this fault space is difficult, and one must obtain a sufficiently large number of samples, as the distribution of the fault space may be unknown, even for well-studied ciphers. Given the time complexity of SAT solving in this case, characterization of a sufficiently large sample becomes challenging. Also, the accurate estimation of attack complexity is almost infeasible as it involves a $\#P$-class problem.

---

[1]A block cipher is nothing but a Boolean function, and for every Boolean function, we can have such an algebraic representation. In fact, this representation is a normal form known as algebraic normal form (ANF).

[2]In this chapter, we shall present a quantification of the fault space size. It is worth mentioning that ExpFault handles the fault space by means of abstraction, which makes the fault space exploration problem rather scalable.

This chapter will address the issues of AFA, and eventually, provide a reasonable solution for (algebraic) exploitable fault space characterization. We shall mainly focus on the machine learning (ML) based approach proposed in [24]. However, a reader may question the necessity of such a framework given the existence of ExpFault. As we shall explain in this chapter, the proposed framework provides new insights which are not feasible to obtain with ExpFault at its current state. For example, the algebraic constructions in this chapter expose the fact that certain attacks are critically dependent on fault and plaintext values. Most importantly, the success rate of an attack can be estimated which is currently not straightforward to calculate with ExpFault. One should note that the success rate is different from the average case complexity estimation as it also encompasses the attack instances for which the complexity is lower than the average and also cases for which the complexity is higher than average but still practical for an attacker. In several occasions, the attack complexities show significant variation from the average complexity. The success rate can be interpreted as an accumulation of all attack complexities corresponding to a specific attack location, which are within the practical limits of exhaustive search and can extract the secret key. This quantity can be utilized as a metric for estimating the influence of each sub-operation of a cipher on fault attacks. All these features make this tool indispensable for the cipher designers for estimating the overall robustness of a construction. In summary, the new framework is not a replacement for ExpFault, but rather a complementary one as both of them have different powers.

Before going into the details of the fault characterization framework, we provide an overview of the approach to motivate the readers. The main crux of the framework is to make the exploitable fault space characterization feasible by means of AFA. No abstraction is made for encoding the cipher or faults. In other words, the cipher encoding encompasses every detail up to the level of bits and the fault model takes the fault value and the plaintext value into account. The result of these is an extremely complex equation system and a fault space of formidable size, which naturally leads towards a statistical solution. To accumulate a large sample quickly for a sound statistical characterization, we seek the assistance of ML. In AFA, a fault instance is manifested as mathematical constraints. An exploitable fault reduces the size of key-space by a significant extent so that exhaustive key search becomes trivial. Based on the intuition that constrained search spaces for different exploitable fault instances on a cipher may have certain structural similarities, a machine learning (ML) classifier is adapted, which, if trained with some already known exploitable fault instances on a cipher, can predict new attacks on the same. In essence, one can identify a fault instance to be exploitable with high confidence without solving a SAT problem. This strategy allows one to characterize an arbitrarily large number of samples of faults for a cipher within a reasonable time, making the statistical characterization practically feasible. One should note that characterization of such arbitrarily large number of samples solely with SAT solvers is not possible within a reasonable time.

In the main proposal of ML-based fault space characterization in [24], authors presented experimental evaluations of the framework over two state-of-the-art lightweight ciphers—PRESENT [4] and LED [10]. In this chapter, we repeat these two examples in a more informative manner. One important observation in this context is that the exploitability of the faults is critically dependent on fault and plaintext values in the PRESENT block cipher, which further strengthens the motivation behind estimating the success rates. In the final part of this chapter, we provide success rate estimates using the framework and show how these estimates can be utilized to assess the robustness of different S-Boxes with respect to fault attacks.

The rest of the chapter is organized as follows. Some necessary preliminaries are presented in the next section. We elaborate on the proposed framework in Sect. 3.3, along with supporting case studies and a potential application scenario in Sect. 3.4. Concluding remarks are presented in Sect. 3.5.

3.2 Preliminaries

3.2.1 General Model for Block Cipher and Faults

Let $\mathcal{E}_k$ be a block cipher defined as a tuple $\mathcal{E}_k = \langle Enc, Dec \rangle$, where Enc and Dec denote the encryption and decryption functions, respectively. Further, the Enc function (and similarly the Dec function) is defined as $Enc(p) = \mathcal{A}_R \circ \mathcal{A}_{R-1} \circ \ldots \circ \mathcal{A}_1(p) = c$, for a plaintext $p \in \mathcal{P}$, ciphertext $c \in C$, and a key $k \in \mathcal{K}$. Each $\mathcal{A}_j$ denotes a round function in a R round cipher. Further, each $\mathcal{A}_j = o^l_j \circ o^{l-1}_j \circ \ldots \circ o^1_j$ is a composition of certain functions of the form o^i_j, generally denoted as *sub-operations* in this work. Each $\mathcal{A}_j$ is thus assumed to have l sub-operations. A sub-operation may belong to the key schedule or the datapath of the cipher. In other words, o^i_j denotes either a key schedule sub-operation or a datapath sub-operation at any round j. We shall use the term o^i_j throughout this work to denote a sub-operation, without mentioning whether it belongs to key schedule or datapath. Table 3.1 lists the notations used throughout this chapter.

An injected fault in DFA usually corrupts the input of some specific sub-operation during the encryption or decryption operation of the cipher. Given the cipher model, we denote the set of faults as $\mathcal{F} = \{F_1, F_2, \ldots, F_H\}$, where each individual fault $F_h \in \mathcal{F}$ is specified as follows:

$$F_h = \langle o^i_r, \lambda, w, T, N, \{f\}^N_{n=1}, \{p\}^N_{n=1} \rangle \tag{3.1}$$

Here $r < R$ is the round of injection, and o^i_r denotes the sub-operation, input of which is altered with the fault. The parameter λ denotes the data-width of the sub-operation (more specifically, the bit-length of the input of the sub-operation). The parameter w is the width of the fault which quantifies the maximum number of bits affected by a fault. In general, *bit-based*, *nibble-based*, and *byte-based* fault models are considered which corresponds to w =1, 4, and 8, respectively. The *position*

Table 3.1 List of notations

Symbol	Definition
"+"	Bitwise XOR
$\mathcal{E}_k$	A block cipher
o^i_j	ith *sub-operation* in the jth round
F_h	A fault instance
R	Number of cipher rounds
l	Number of sub-operation in each round
r	Round of fault injection
N	Fault multiplicity
w	Fault width
T	Fault position
λ	Bit-width of a sub-operation
$\{f\}_{n=1}^{N}$	Set of fault values for a fault injection
$\{p\}_{n=1}^{N}$	Set of plaintext values for a fault injection
$M_{\mathcal{F}}$	Set of exploitable faults
τ	Timeout for SAT solvers
$\mathcal{S}$	Sensitivity threshold

of the fault at the input of a sub-operation is denoted by T with $t \in \{0, 1, \ldots \frac{\lambda}{w}\}$. For practical reasons, w and T are usually defined in a way so that the injected faults always remain localized within some pre-specified block-operations of the corresponding sub-operation o^i_r. The parameter N represents the number of times a fault is injected at a specific location to obtain a successful attack within a reasonable time. N is called the *fault multiplicity*.

The sets $\{f\}_{n=1}^{N}$, and $\{p\}_{n=1}^{N}$ denote the values of the injected faults and the plaintexts processed during each fault injection, respectively. In the most general case, the diffusion characteristics (and thus the exploitability) of an injected fault critically depend upon the value of the fault and the corresponding plaintext on which the fault is injected. A typical example is PRESENT cipher, where the number of active S-Boxes due to the fault diffusion depends on the plaintext and the fault value and as a result, many faults injected at a specific position with the same multiplicity may become exploitable, whereas some of them at the same position may become unexploitable. According to the fault model in Eq. (3.1), the total number of possible faults for a specific position T in sub-operation o^i_r is $2^{N(w+\lambda)}$, for a given fault width w. The total number of possible faults for a sub-operation o^i_r is $(2^{N(w+\lambda)} \times \frac{\lambda}{w})$, and that for the whole cipher o^i_r is $(2^{N(w+\lambda)} \times \frac{\lambda}{w} \times R \times l)$.[3]

In certain cases the fault space can be pruned significantly utilizing the fact that a large number of faults may be actually equivalent. A prominent example is the

[3]The fault values and the plaintext values are not explicitly considered (i.e., abstracted) in ExpFault. The fault space size becomes relatively reasonable to be exhausted without these two parameters. The flip side of this abstraction is that ExpFault returns the best case attack complexity (from attacker's perspective) for certain ciphers like PRESENT.

AES where every byte fault at some specific position is equivalent irrespective of its value. However, there exists no automatic procedure to figure out such equivalences, till date, and the only way is to manually analyze the cipher. As a result, it is reasonable to adapt the above calculation of the size of the fault space while analyzing a general construction.

3.2.2 Algebraic Representation of Ciphers

Multivariate polynomial representation, which is quite well known in the context of AFA [30], is considered one of the most generic and informative representations for block ciphers. In this work, we utilize the polynomial representations to encode both the ciphers and the faults. The usual way of representing block ciphers algebraically is to assign a set of symbolic variables for each iterative round, where each variable represents a bit from some intermediate state of the cipher. Each cipher sub-operation is then represented as a set of multivariate polynomial equations over the polynomial ring constructed on these variables, with $GF(2)$ being the base ring. The equation system should be sparse and low-degree in addition, to make the cipher representation easy to solve.

In order to elaborate the process of polynomial encoding, we consider the example of the PRESENT block cipher. PRESENT is a lightweight block cipher proposed by Bogdanov et al. in CHES 2007 [4]. It has a substitution-permutation network (SPN) based round function which is iterated 31 times to generate the ciphertext. The basic version PRESENT-80 has a block size of 64-bits and a master key of size 80 bits, which is utilized to generate 64-bit round keys for each round function by means of an iterated key schedule. Each round of PRESENT consists of three sub-operations, namely *addRoundKey*, *sBoxlayer*, and *pLayer*. The *addRoundKey* sub-operation, computing bitwise XOR between the state bits and round key bits, is represented as:

$$y_i = x_i + k_i, \text{ for } 1 \leq i \leq 64 \tag{3.2}$$

where x_i, k_i represent the input state bits and round key bits, respectively, and y_i represents the output bits of the *addRoundKey* sub-operation. Similarly, the *pLayer* operation, which is a 64-bit permutation, can be expressed as:

$$y_{\pi(i)} = x_i, \text{ for } 1 \leq i \leq 64 \tag{3.3}$$

where $\pi(i)$ is the permutation table. The non-linear substitution operation *sBoxlayer* of PRESENT consists of 16 identical 4×4 bijective S-Boxes, each of which can be represented by a system of non-linear polynomials. The solvability of a typical cipher polynomial system critically depends on the S-Box representation, which is expected to be sufficiently sparse and consisting of low-degree polynomials. One way of representing the PRESENT S-Boxes is the following:

$$
\begin{aligned}
y_1 &= x_1x_2x_4 + x_1x_3x_4 \\
&\quad + x_1 + x_2x_3x_4 + x_2x_3 + x_3 + x_4 + 1 \\
y_2 &= x_1x_2x_4 + x_1x_3x_4 + x_1x_3 + x_1x_4 + \\
&\quad x_1 + x_2 + x_3x_4 + 1 \\
y_3 &= x_1x_2x_4 + x_1x_2 + x_1x_3x_4 + x_1x_3 + \\
&\quad x_1 + x_2x_3x_4 + x_3 \\
y_4 &= x_1 + x_2x_3 + x_2 + x_4
\end{aligned}
\tag{3.4}
$$

Here x_is $(1 \leq i \leq 4)$ and y_is $(1 \leq i \leq 4)$ represent the input and output bits of a 4×4 S-Box, respectively.

Each injected fault instance can be added in the cipher equation system in terms of new equations. Let us assume that the fault is injected at the input state of the ith sub-operation o_r^i at the rth round of the cipher. For convenience, we denote the input of o_r^i as $X^i = x_1||x_2||\dots||x_\lambda$, where λ is the bit-length of X^i. In the case of PRESENT $\lambda = 64$. Let, after the injection of the fault, the input state changes to $Y^i = y_1||y_2||\dots||y_\lambda$. Then the state differential can be represented as $D^i = d_1||d_2\dots||d_\lambda$, where $d_z = x_z + y_z$ with $1 \leq z < \lambda$. Further, depending on the width of the fault w, there can be $m = \frac{\lambda}{w}$ possible locations in X^i, which might have got altered. Let us partition the state differential D^i in m, w-bit chunks as $D^i = D_1^i||D_2^i||\dots||D_m^i$, where $D_t^i = d_{w\times(t-1)+1}||d_{w\times(t-1)+2}||\dots||d_{w\times t}$ for $1 \leq t \leq m$. Assuming T be the location of the fault, the fault effect can be modeled with the following equations:

$$D_t^i = 0, \text{ for } 1 \leq t \leq m, t \neq T \tag{3.5}$$

$$(1 + d_{w\times(t-1)+1})(1 + d_{w\times(t-1)+2})\dots(1+d_{w\times t}) = 0, \quad \text{for } t = T \tag{3.6}$$

It is notable that the location T of a fault can be unknown in certain cases, and this can also be modeled with equations of slightly complex form [30]. However, for exploitable fault characterization, it is reasonable to assume that the locations are known as we are working in the evaluator mode.

3.3 ML-Based Fault Space Characterization: Methodology

3.3.1 Motivation

The goal of the present work is to efficiently filter out the exploitable faults for a given cryptosystem. It is apparent that the ANF polynomials provide a reasonable way for modeling the ciphers and the faults [30]. Although, the ANF description and its corresponding CNF is easy to construct, solving them is non-trivial as the decision problem associated with the solvability of an ANF system is NP-Complete. In practice, SAT solvers are used for solving the associated CNF systems, and it is observed that the solving times vary significantly depending on the instance.

One key observation regarding the cipher equation systems is that they are never unsatisfiable, which is due to the fact that for a given plaintext–ciphertext pair there always exists a key. However, it is not practically feasible to figure out the key without fault injections, as the size of the key search space is prohibitively large. The search space complexity reduces with the injection of faults. The size of the search space is expected to reach below some certain limit which is possible to search exhaustively with modern SAT solvers within reasonable time, if a sufficient number of faults are injected at proper locations.

The above-mentioned observation clearly specifies the condition for distinguishing the exploitable faults from the non-exploitable ones. To be precise, if a SAT solver terminates with the solution within a pre-specified time limit, the fault instance is considered to be exploitable. Otherwise, the fault is considered non-malicious. Setting a proper time-limit for the SAT solver is, however, a critical task. A relatively low-time limit is unreliable as it may fail to capture some potential attack instances. As an example, for the PRESENT cipher we observed that most of the 1-bit fault instances with fault multiplicity 2, injected at the inputs of 28-th round S-Box operation, are solvable within 3 min. This observation is similar to that mentioned in [30]. However, we observed that when nibble faults are considered at the 28th round, the variation of solving time is significantly high; in fact, there are cases with solving times around 16–24 h. These performance figures are obtained with Intel Core i5 machines running CryptominiSAT-5 [25] as the SAT solver in a single threaded manner. Moreover, such cases comprise nearly 12% of the total number of samples considered. This is not insignificant in a statistical sense, where failure in detecting some attack instances cannot be tolerated. Such instances do not follow any specific pattern through which one can visually characterize them without solving them. This observation necessarily implies that one has to be more careful while setting solver timeouts and a high value of timeout is preferable. However, setting high timeout limits the number of instances one can acquire through exhaustive SAT solving within a practically feasible time span.

According to the fault model described in Sect. 3.2.1, the size of the fault space in a cipher is prohibitively large. As a concrete example, there are total $2^{(64+4)} = 2^{68}$ possible nibble fault instances, with *fault multiplicity* $N = 1$, for any specific *position* T, on any sub-operation o_r^i in the PRESENT cipher. The number is even larger if one considers other positions, sub-operations, fault multiplicity, and fault models. Moreover, the ratio of exploitable faults to the total number of faults is unknown a priori. The whole situation suggests that in order to obtain a reliable understanding of the exploitable fault space even in a statistical sense, one must test a significantly large sample from fault space. Also, to obtain a sufficiently large set of exploitable faults for testing purpose, a large number of fault instances must be examined. With a high timeout required for SAT solvers, exhaustive SAT solving is clearly impractical for fault space characterization and a fast mechanism is required. **Our aim in this chapter is to prepare an efficient alternative to the exhaustive enumeration of the fault space via SAT solving.**

3.3.2 Empirical Hardness Prediction of Satisfiability Problems

NP-Complete problems are ubiquitous in computer science, especially in AI. While they are hard-to-solve on worst case inputs, there exist numerous "easy" instances which are of great practical value. In general, the algorithms for solving NP-Complete problems exhibit extreme runtime variations even across the solvable instances, and there is no describable relationship between the instance size and the algorithm runtime as such. Over the past decade, a considerable body of work has shown how to use supervised ML models to answer questions regarding solvability or runtime using features of the problem instances and algorithm performance data [12, 13, 22, 23, 28]. Such ML models are popularly known as empirical hardness models (EHM). Some applications of EHMs include proper algorithm portfolio selection for a problem instance [28], algorithm parameter tuning [12], hard benchmark construction [17], and analysis of algorithm performance and instance hardness [17].

In the context of the present work, we are interested in EHMs which predict the hardness of SAT instances. The most prominent result in the context of empirical runtime estimation of SAT problems is due to Xu et al., who constructed a portfolio-based SAT solver SATzilla [28] based on EHMs. The aim of SATzilla was to select the best solver for a given SAT instance, depending upon the runtime predictions of different EHMs constructed for a set of representative SAT solvers. The SATzilla project also provided a large set of 138 features for the model construction depending on various structural properties of the CNF descriptions of the problem instance as well as some typical features obtained from runtime probing of some basic SAT solvers. In this work, we utilize some of these features for constructing EHMs which will predict the exploitability of a given fault instance without solving it explicitly. Brief description of our feature set will be provided later in this section.

3.3.3 ML Model for Exploitable Fault Identifier

In this subsection, we shall describe the ML-based framework in detail. In nutshell, our aim is to construct a binary classifier, which, if trained with certain number of exploitable and unexploitable fault instances, can predict the exploitability of any fault instance queried to it. Before going to further details, we formally describe the exploitable fault space for a given block cipher and an exploitable fault in the context of SAT solvability.

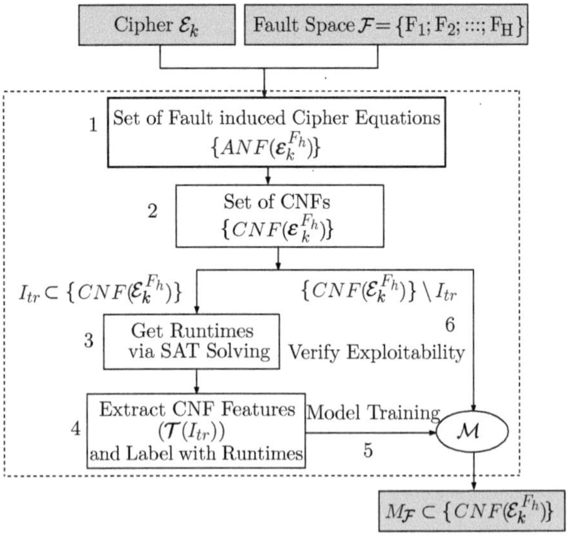

Fig. 3.1 The exploitable fault characterization framework: basic idea [24]

Definition (Exploitable Fault Space) *Given a cipher $\mathcal{E}_k$ and a corresponding fault space $\mathcal{F}$, the exploitable fault space $M_{\mathcal{F}} \subset \mathcal{F}$ for $\mathcal{E}_k$ is defined as a set of faults such that $\forall F_h \in M_{\mathcal{F}}$, it is possible to extract n_e bits of the secret key k, where $0 < n_e \leq |k|$.* □

In other words, exploitable fault space denotes the set of faults for which the combination of the injected fault and a plaintext results in the extraction of n_e bits of the secret key. From the perspective of a cipher evaluator, two distinct scenarios can be considered at this point. In the first one, it is assumed that none of the key bits are known a priori and faults are inserted to extract the complete master key of the cipher. Indeed, one may increase the number of injections to reduce the complexity of the search space in this scenario. However, it is practically reasonable to assume some upper bound on the number of injections. In other words, the fault multiplicity N in the fault model is always $\leq$ some pre-specified threshold. The second scenario in this context occurs when some specific key bits are assumed to be known. This model is extremely useful when only a subset of the key can be extracted by the fault injection due to incomplete diffusion of the faults. In a typical AFA framework, it is not possible to obtain a unique solution for the incompletely defused faults unless some of the key bits are known. However, in this work, we mainly elaborate the first scenario. It is worth mentioning that the second scenario can be dealt with the framework we are going to propose, without any significant changes.

The framework for exploitable fault space characterization is depicted in Fig. 3.1. Referring to the figure, let $\mathcal{E}_k^{F_h}$ indicate the cipher $\mathcal{E}_k$, with a fault F_h from its fault space $\mathcal{F}$ injected in it. This can be easily modeled as an ANF equation system denoted as $ANF(\mathcal{E}_k^{F_h})$. The very next step is to convert $ANF(\mathcal{E}_k^{F_h})$ to the corresponding CNF model denoted by $CNF(\mathcal{E}_k^{F_h})$. At this point, we specify the exploitable faults in terms of solvability of SAT problems, with the following definition:

Definition (Exploitable Fault) *A fault $F_h \in \mathcal{F}$ for the cipher $\mathcal{E}_k$ is called exploitable if the $CNF(\mathcal{E}_k^{F_h})$ is solvable by a SAT solver within a pre-specified time bound τ.* □

Given fault instances from the fault space of a cipher, we construct CNF encoding for each of them. A small fraction I_{tr} of these CNFs are solved exhaustively with SAT solver and labeled accordingly depending on whether they are solvable or not within the threshold τ. Next, a binary classifier $\mathcal{M}$ is trained with these labeled instances, which is the EHM in this case. The ML model is defined as:

$$\mathcal{M} : \mathcal{T}\left(CNF\left(\mathcal{E}_k^{F_h}\right)\right) \mapsto \{0, 1\} \tag{3.7}$$

Here, $\mathcal{T}$ is an abstract function which represents the features extracted from the CNFs. In the present context, $\mathcal{T}$ outputs the feature vectors from the SATzilla feature set [28]. For convenience, we use the following nomenclature:

Class 0 : Denotes the class of exploitable faults.
Class 1 : Denotes the class of benign/unexploitable faults.

One important difference of our EHM model with the conventional EHM models is that we do not predict the runtime of an instance but use the labels 0 and 1 to classify the faults into two classes. In other words, we solve a classification rather than a regression problem solved in conventional EHMs [13]. The reason is that we just do not exploit the runtime information in our framework. The main motive of ours is to distinguish instances whose search space size is within the practical search capability of a solver, from those instances which are beyond the practical limit. It is apparent that our classifier-based construction is sufficient for this purpose. In the next section, we describe the feature set utilized for the classification.

3.3.4 Feature Set Description

In this work, we use the features suggested by the SATzilla—a portfolio-based SAT solving tool [28]. The SATzilla project proposed a rich set of 138 features to be extracted from the CNF description of a SAT instances for the construction of runtime predicting EHMs. The feature set of SATzilla is a compilation of several algorithm-independent properties of SAT instances made by the artificial intelligence (AI) community on various occasions [17]. A widely known example of such algorithm-independent properties is the so-called *phase-transition* of random 3-SAT instances. In short, SAT instances, generated randomly on a fixed number of variables, and containing only 3-variable clauses, tend to become unsatisfiable as the clause-to-variable ratio crosses a specific value of 4.26 [22]. Intuitively, the reason for such a behavior is that instances with fewer clauses are underconstrained and thus almost always satisfiable, while those with many clauses are overconstrained and unsatisfiable for most of the cases. The SATzilla feature set is divided into 12 groups. Some of the feature groups consist of structural features like the one described in the example, whereas the others include features extracted from runtime behaviors of the SAT instances on solvers from different genre—like Davis-Putnam-Logemann-Loveland (DPLL) solvers, or local search solvers [13, 28].

The structural features of SATzilla are divided into several feature groups, which include simple features like various variable-clause statistics as well as features based on complex clause–variable interactions in the CNF formula obtained through different graph-based abstractions. The first group includes properties related to the problem size, measured in terms of the number of clauses, variables, and the ratio of the two. The next three feature groups consist of features extracted from various graph representations of SAT instances, namely variable-clause graphs (VCG), variable graphs (VG), and clause graphs (CG). Graphical abstraction of complex clause–variable interactions quite efficiently represents the difficulty in solving an instance. Statistics (mean, standard deviation, min, max, entropy) on the degrees of the nodes, the graph diameters, and special clustering-coefficients are extracted as features from these graphs. Intuitively, these statistical measurements quantify the difficulty of an instance. The fourth feature group of SATzilla is called balanced features which include some simple statistical measurements on the variables and clauses of an instance. The next feature group, which is the last one among structural features, measures the proximity of an instance to a Horn formula, which is a class of SAT instances relatively easier to deal with.

The so-called runtime features in SATzilla, also known as "probing" features, are computed with short-time runs of the instances on different genres of candidate solvers. The seventh feature group consists of DPLL probing features, which include the unit propagation statistics on the corresponding DPLL search tree, as well as an unbiased estimation of the size of the search space estimated with the average depth of contradictions in DPLL search trees [20]. The eighth group of features is obtained by solving a linear programming relaxation of an integer program representing the SAT instance under consideration, whereas the ninth group consists of probing features from two stochastic local search algorithms, GSAT and SAPS. The next feature group contains the statistics of learned clauses for an instance obtained with a 2 s run on ZChaff SAT solver. Finally, there are the survey propagation features which are based on estimates of variable bias in a SAT formula obtained using probabilistic inference [11]. The computation times for these 12 groups of features are not uniform and there exist both structural and runtime features which are computationally expensive. The computation time of the features also provides significant information regarding the instance hardness and as a result, they are included as the final feature group in SATzilla. Further details on SATzilla feature set can be found in [13, 23].

3.3.5 Handling the False Negatives

The ML model proposed in this work provides quick answers regarding the exploitability of the fault instances queried to it. Such a quick answering system has an enormous impact on the exploitable characterization problem as it makes the problem tractable from a practical sense. However, the efficiency comes at the cost of accuracy. Being a ML-based approach, there will always be some *false positives* (a benign fault instance classified as exploitable) and *false negatives* (an exploitable instance classified as benign). While a small number of false positives can still be tolerated, false negatives can be crucial for some applications, for example, generating a test set of exploitable faults for testing countermeasures. If some typical exploitable faults are missed, they may lead to successful attacks on the countermeasure.

In this work, we provide a potential solution for the misclassification issue. More precisely, we try to statistically eliminate the chances of false negative cases—that is the chances of an attack getting misclassified. The main idea is to first determine the cases for which the classification confidence of the classifier is not very high. We denote such cases as *sensitive instances*. Note that, sensitive instances are determined on the validation dataset once the classifier is trained and deployed for use. Intuitively, such sensitive instances are prone to misclassification (we have also validated this claim experimentally.). Each sensitive instance is exhaustively tested with SAT solver. Figure 3.2 presents a conceptual schematic of what we mean by sensitive instances. Typically, we assume that the two classes defined in terms of the feature vectors can be overlapping, and the region of overlap constitutes the set of sensitive instances.

Determination of the sensitive instances or this region of overlap is, however, not straightforward and could be dealt in many ways. In this chapter, we take a very simple albeit effective strategy. We use *random forest* (RF) of decision trees as our classification algorithm [5]. Random forest is constructed with several decision trees, each of which is a weak learner. Usually, such *ensemble methods* of learning perform majority voting among the decisions of the constituent weak learners (decision trees in the present context) to determine the class of the instance. Here, we propose a simple methodology for eliminating the false negatives using

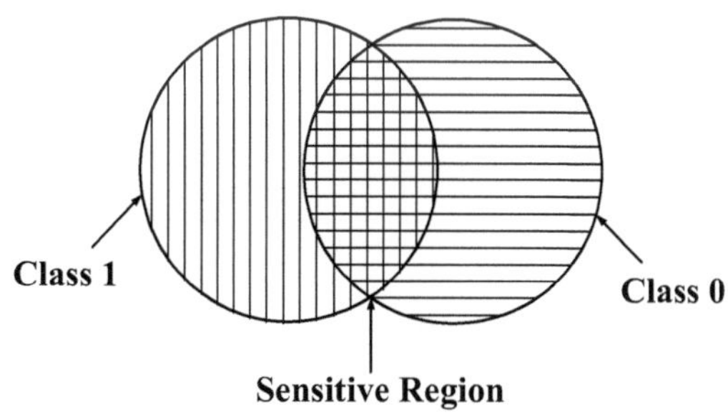

Fig. 3.2 Sensitive region: conceptual illustration [24]

the properties of the RF algorithm. The proposed approach is reminiscent of classification with reject—a well-studied area in ML, where a classifier can reject some instances if the classification confidence is low for them [26]. Let Cl be the random variable denoting the predicted class of a given instance x in the two-class classification problem we are dealing with. For any instance x, we try to figure out the quantities $Pr[Cl = 0 \mid x]$ and $Pr[Cl = 1 \mid x]$, which are basically the probabilities of x lying in any of the two classes. Evidently, the sum of these two quantities is 1. Note that the probabilities are calculated purely based on the decisions made by the classifier. In other words, it is calculated exploiting the properties of the classification algorithm. Next, we calculate the following quantity:

$$\delta = (Pr[Cl = 0 \mid x] - Pr[Cl = 1 \mid x]) \tag{3.8}$$

It is easy to observe that having a large value for δ implies the classifier is reasonably confident about the class of the instance x. In that case, we consider the decision of the classifier as the correct decision. For the other case, where δ is less than some predefined threshold $\mathcal{S}$, we invoke the SAT solver to determine the actual class of the instance. The overall flow for false negative removal is summarized in Algorithm 1.

Algorithm 1: Procedure *CLASSIFY_FAULTS*

Input : A random fault instance F_h
Output : Exploitability status of F_h
Construct $ANF(\mathcal{E}_k^{F_h})$ and then $CNF(\mathcal{E}_k^{F_h})$;
Compute $x = \mathcal{T}(CNF(\mathcal{E}_k^{F_h}))$;
Compute $\langle Pr[Cl = 0 \mid x], Pr[Cl = 1 \mid x] \rangle = \mathcal{M}(x)$;
Compute δ using Equation (3.8);
if $(|\delta| < \mathcal{S})$; // $\mathcal{S}$ is a predefined threshold
then
    Query the SAT engine with $CNF(\mathcal{E}_k^{F_h})$;
    if $(CNF(\mathcal{E}_k^{F_h})$ *is solvable within* $\tau)$ **then**
        return: $F_h \in M_{\mathcal{F}}$;
    else
        return: $F_h \notin M_{\mathcal{F}}$;
else
    if $(\delta > 0)$ **then**
        return: 0 ; // $F_h \in M_{\mathcal{F}}$
        ;
    else
        return: 1 ; // $F_h \notin M_{\mathcal{F}}$
        ;

Each tree in an RF returns the class probabilities for a given instance. The class probability of a single tree is the fraction of samples of the same class in a leaf node of the tree. Let the total number of trees in the forest be t_r. The class probability of a random instance x is defined as:

$$Pr[Cl = c \mid x] = \frac{1}{t_r} \sum_{h=1}^{t_r} Pr_h[Cl = c \mid x] \tag{3.9}$$

where $Pr_h[Cl = c \mid x]$ denotes the probability of x being a member of a class c according to the tree h in the forest.

The success of this mechanism, however, critically depends on the threshold $\mathcal{S}$, which is somewhat specific to the cipher under consideration, and is determined experimentally utilizing the validation data. Ideally, one would expect to nullify the false negatives without doing too many exhaustive validations. Although no theoretical guarantee can be provided by our mechanism for this, experimentally we found that for typical block ciphers, such as PRESENT and LED, one can reasonably fulfill this criterion. Detailed results supporting this claim will be provided in Sect. 3.4.

3.4 Case Studies

This section presents the experimental validation of the proposed framework by means of case studies. Two state-of-the-art block ciphers, PRESENT and LED, are selected for this purpose. The motivation behind selecting these two specific ciphers is that they utilize the same non-linear, but significantly distinct linear layers. One main application of the proposed framework is to quantitatively examine the effect of different cipher sub-operations in the context of fault attacks, and in this chapter, we mainly elaborate this application. The structural features of PRESENT and LED allow us to make a fair comparison between their diffusion layers. In order to evaluate the effect of the non-linear S-Box layer, we further perform a series of experiments on the PRESENT block cipher by replacing its S-Box with three alternative S-Boxes of similar mathematical properties. In the following two subsections, we present the detailed study of the PRESENT and LED ciphers with the proposed framework. The study involving the S-Box replacement will be presented after that.

3.4.1 Learning Exploitable Faults for PRESENT

The basics of PRESENT block cipher has already been described in Sect. 3.2.2. Several fault attack examples have been proposed on PRESENT, mostly targeting the 29-th and 28-th round of the cipher as well as the key schedule of PRESENT [1, 7, 9, 27, 30, 31]. Zhang et al. [30] presented an AFA on PRESENT, requiring 2

Table 3.2 Setup for the ML on PRESENT and LED

Cipher	PRESENT	LED
Target rounds	27–30	29–32
Maximum number of times a fault is injected (N)	2	2
Timeout for the SAT solver (τ)	24 h	48 h

bit-fault instances on average, at the 28-th round of the cipher in the best case. The solving times of the corresponding CNFs are mostly around 3 min.

3.4.1.1 Experimental Setup

In order to validate the proposed framework, we create random AFA instances following different fault models. In order to make the ML classifier generic, we decided to train it on instances from different fault models. Two competitive fault models for PRESENT are the bit and nibble fault models, both of which can generate plenty of exploitable fault instances. In any case, we end up getting a CNF, the solvability of which determines the exploitability of an instance. So the ML classifier is supposed to learn to estimate the search complexity of an instance in some way. Hence, there is no harm in combining instances from two fault models as such. Table 3.2 presents the basic setup we used for the experiments on PRESENT and LED. Experiments on any given cipher begin with an initial *profiling phase*, where the parameters mentioned in Table 3.2 are determined and attack samples for training are gathered. For PRESENT, we mainly targeted the rounds 27–30 in our experiments as one can hardly find any exploitable fault beyond these rounds. Further, the fault multiplicity (N) was restricted to 2 (that is, N can assume values 1 and 2) considering low-fault complexities of DFAs. Interestingly, it was observed that the nibble fault instances (injected 2 times in succession) at 28-th round do not result in successful attacks, even after 2 days. Further, many of these instances (almost 12%) take 16–24 h of solving time. No successful attack instances were found taking time beyond 24 h in our experiments, which were conducted on a machine with Intel Core i5 running CryptominiSAT-5 [25] as the SAT solver in a single threaded manner. We thus set the SAT timeout $\tau = 24$ h for PRESENT. For the sake of experimentation, we exhaustively characterized a set of 1000 samples from the fault space of PRESENT and LED, individually. However, one should note that such exhaustive characterization was only required to prove the applicability of the proposed methodology, and in practice, a much smaller number of instances are required for training the ML classifier, as we will shall show later. For every new cipher, such *profiling* should be performed only once just to build the ML classifier.

3.4.1.2 Feature Selection

The first step in our experiments is to evaluate the feature set. Although we started with a well-accepted feature set, it is always interesting to know how these features impact the learning process and which are the most important features in the present context. Identification of the main contributing features for a given problem may also lead to significant reduction of the feature space dimensionality by the selection of actually useful features, which also reduces the chances of overfitting. We therefore perform a quantitative assessment of the importance of various features using the RF algorithm. Before proceeding further, it is worth mentioning that some of the SATzilla features might be computationally expensive depending on problem instances. It was found that the unit propagation features (which belong to the group of DPLL probing features) and the linear programming features in our case take even more than 15 min of computing time for certain instances. As a result, we did not consider them in our experiments which left us with 123 features in total.

In this work, we evaluate the feature importance based on the *mean decrease of gini-impurity* of each feature during the construction of the decision trees [5]. Every node ζ, in a given decision tree γ, of an RF Γ imposes a partition on the dataset by putting some threshold condition on a single feature, so that similar samples end up in the same partition. The optimal split at a node is calculated based on a statistical measure which quantifies how well a potential split is separating the samples of different classes at this particular node. The *gini-impurity* is one of the most popular measure for such purposes and is actually used in random forests [5]. Let us assume that a node ζ in some decision tree $\gamma \in \Gamma$ has total $|\zeta|$ samples among which the subset ζ^q consists of samples from class $q \in \{0, 1\}$. Then the *gini-impurity* of ζ is calculated as:

$$\mathcal{G}(\zeta) = 1 - (p_\zeta^0)^2 - (p_\zeta^1)^2 \tag{3.10}$$

where $p_\zeta^q = \frac{|\zeta^q|}{|\zeta|}$ for $q \in \{0, 1\}$. Let the node ζ partitions the dataset into two nodes (subsets) ζ_{left} and ζ_{right}, using some threshold condition t_θ on some feature θ, and the gini-impurity of these two nodes are $\mathcal{G}(\zeta_{left})$ and $\mathcal{G}(\zeta_{right})$, respectively. Then the decrease in impurity at the node ζ, due to this specific split, is calculated as:

$$\Delta\mathcal{G}(\zeta) = \mathcal{G}(\zeta) - p_{left}\mathcal{G}(\zeta_{left}) - p_{right}\mathcal{G}(\zeta_{right}) \tag{3.11}$$

where $p_{left} = \frac{|\zeta_{left}|}{|\zeta|}$, and $p_{right} = \frac{|\zeta_{right}|}{|\zeta|}$. In an exhaustive search over all variables θ available at ζ and the space of corresponding t_θs, the optimal split at ζ, for a particular tree γ, can be determined which is quantified as $\Delta^\theta\mathcal{G}(\zeta, \gamma)$. In the calculation of feature importance, decrease in gini-impurities is accumulated in a per-variable basis and the importance value of the feature θ is calculated as follows:

$$\mathcal{I}_\mathcal{G}(\theta) = \sum_{\gamma \in \Gamma} \sum_{\zeta \in \gamma} \Delta^\theta\mathcal{G}(\zeta, \gamma) \tag{3.12}$$

The result of the feature importance assessment experiment is presented in Fig. 3.3a, where the X-axis represents the index of a feature and the Y-axis represents its importance scaled within an interval of [0, 1]. It is interesting to

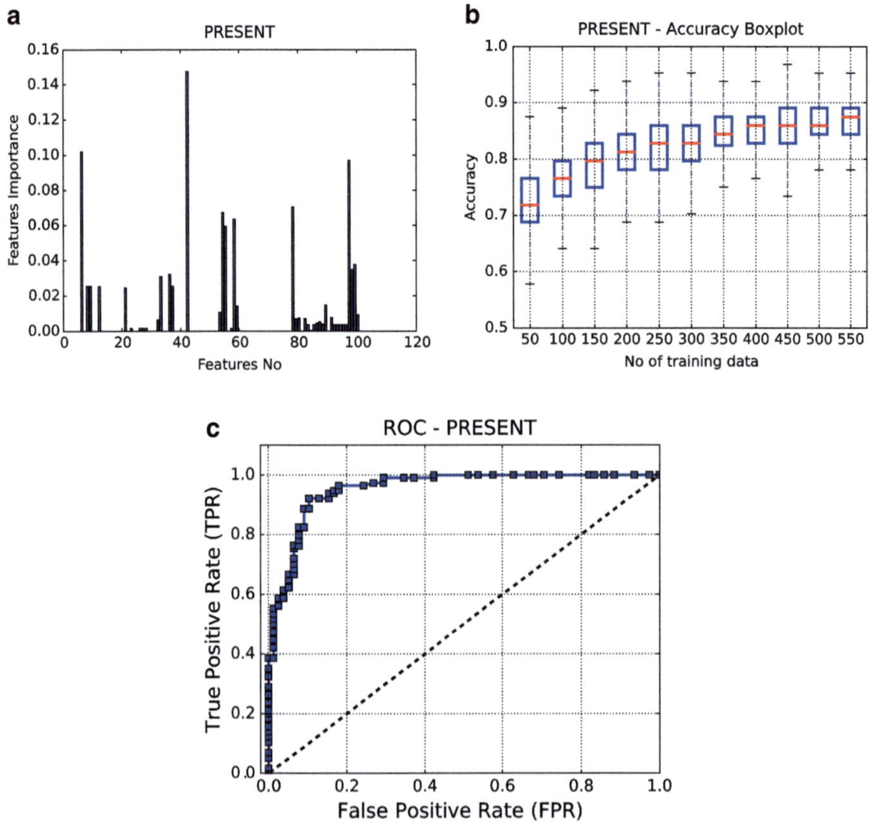

Fig. 3.3 Machine learning results for PRESENT. **(a)** Feature importance, **(b)** ROC curve, and **(c)** variation of accuracy with the size of training set [24]

observe that there are almost 66 features, for which the importance value is 0. Further investigation reveals that these features obtain constant values for all the instances. As a result, they can be safely ignored for the further experiments.

It can be observed from Fig. 3.3a that the feature no. 42 is the most important one for our experiments. This feature corresponds to the aggregated computation time for the variable-clause graph (VCG) and variable-graph (VG) graph-based features. A VCG is a bipartite graph, with nodes corresponding to each variable and clause. The edges in this graph represent the occurrence of a variable in a clause. The VG has a node for each variable and an edge between variables that occur together in at least one clause. Intuitively, the computation time is a crude representative for the dense-nature of these graphs, which is usually high if the search space is very large and complex. However, it is difficult to directly relate this feature with quick solvability of an instance as other selected features also play a significant role. In fact, it was observed that every structural feature group have some contribution in the classification, which is somewhat expected (feature no. 0–59 in Fig. 3.3a).

In contrast, the contributions from the runtime features were not so regular. In particular, only the survey propagation features (based on estimates of variable bias in a SAT formula obtained using probabilistic inference [11]) were found to play some role in the classification (feature no. 79–96 in Fig. 3.3a). Interestingly, the features 98–100, which correspond to the approximate search-space size (estimated with the average depth of contradictions in DPLL search trees [20]), were found to play some role in the classification. This is indeed expected, as the classification margin in this work is defined based on the search space size.

3.4.1.3 Classification

We next measured the classification accuracy of the RF classifier with the reduced set of features. In order to check the robustness of the learning, we ran each of our experiments several times. For each repetition, new training and validation sets were chosen from a set of 640 labeled samples (the remaining 360 samples collected at the profiling phase were utilized for further validation and false negative removal experiments), where the sizes of them are in the ratio 7:3. The sample set consists of 320 exploitable and 320 unexploitable fault instances in order to achieve an unbiased training. The average accuracy obtained in our experiment was 85%. We also provide the receiver operating characteristics (ROC) curve for the RF classifier, which is considered to be a good representative for the quality of a classifier. The area under curve (AUC) represents the goodness of a classifier, which ranges between 0 and 1 with higher values representing a better classifier. The ROC curve for the PRESENT example is provided in Fig. 3.3c, which shows that the classifier performs reasonably well in this case. Figure 3.3b presents the variation of accuracy with the size of training dataset as a box plot. It can be observed that reasonable accuracy can be reached within 450 training instances (which is around 70% of our dataset size) and accuracy does not improve much after that.

3.4.1.4 Handling False Negatives

Although our classifier reaches a reasonable good accuracy of 85%, there are almost 15% instances which get misclassified in this process, which contains both false positives and false negatives. As pointed out in Sect. 3.3.5, false negatives are not acceptable in certain scenarios. The approach presented in Sect. 3.3.5 critically depends on the threshold parameter $\mathcal{S}$, which must be set in a way so that the percentage of false negatives becomes 0 or at least negligibly small. If the percentage of instances below $\mathcal{S}$ is too high, it would be costly to estimate all of them via exhaustive SAT solving. However, the reasonably good accuracy of our classifier suggests that the percentage of such sensitive instances may not be very high. We tested our proposed fix from Sect. 3.3.5 on a new set of 250 test instances with different $\mathcal{S}$ values. Table 3.3 presents the outcome of the experiment. The percentage of instances to be justified via SAT solving and the percentage of

Table 3.3 Misclassification handling for PRESENT

S Value	0.10	0.12	0.14	0.16	0.18	0.20	0.22	0.24	0.26	0.28	0.30
% sensitive instances	6.0	6.0	10.0	13.2	17.2	20.4	20.4	22.0	22.8	24.8	27.6
% false negatives beyond S	4.1	4.1	3.0	1.8	0.6	0.2	0.0	0.0	0.0	0.0	0.0

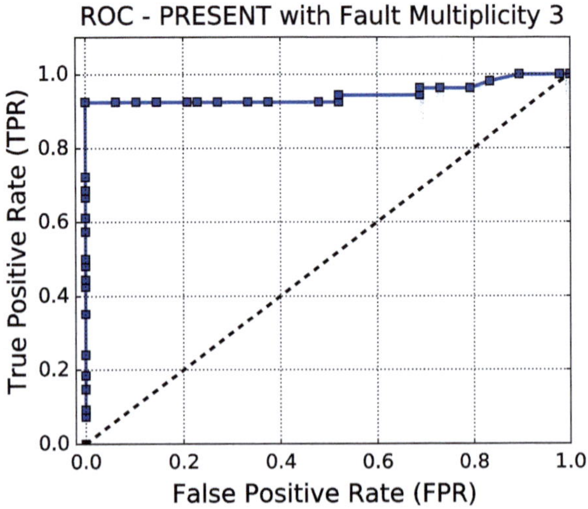

Fig. 3.4 ROC curve for PRESENT cipher with fault multiplicity of three [24]

false negatives beyond S are presented in the table for each choice of S. It can be observed from Table 3.3 that a threshold of 0.22 nullifies the number of false negatives and keeps the percentage of sensitive instances (see Sect. 3.3.5) to 20%, which is indeed reasonable.

Classically in DFA, increase in fault multiplicity is considered costly and the target is to extract the complete key with the minimum number of injections possible. Due to this fact, we intentionally set the operating point of our experiments to a fault multiplicity 2, which is the minimum number of faults required to extract the master key of PRESENT. The multiplicity of a fault, however, plays a crucial role in the success of fault attacks. In order to evaluate the effect of fault multiplicity on the classification accuracy and false negatives of the tool, we validated it with fault multiplicity 3 and 4. It has been observed that 94% classification accuracy can be achieved with fault multiplicity 3 and the value is even higher for higher fault multiplicities. Figure 3.4 presents the ROC curve corresponding to fault multiplicity 3. Further, roughly 12% of the instances have to be checked with SAT solvers in order to nullify the false negatives.

The trend observed in classification accuracy and the false negative rate is consistent with the theoretical understanding of DFA. The increase in fault multiplicity increases the number of equations in the cipher equation system which eventually results in a constrained search space of reasonable size to be exhausted by a SAT solver. As a result, most of the instances from round 28 become solvable and unsolvable instances mostly belong to other rounds. This phenomenon leads to a stronger separation between two classes making the classification accuracy significantly high.

3.4.1.5 Gain over Exhaustive SAT Solving

It would be interesting to estimate the overall gain of our ML assisted methodology compared to exhaustive characterization via SAT solving. For the sake of elaboration, let us consider a scenario where only nibble faults are injected at the 28th round of PRESENT. Further, each fault is assumed to be of multiplicity 2. The size of the resulting fault space is $2^{2\times(64+4)} = 2^{136}$, which is impossible to enumerate. Even if one considers a reasonable-sized sample of 10,000 fault instances, the exhaustive characterization with SAT solving only would be impractical. Considering a timeout threshold of 24 h ($\tau = 24$ h), characterization of these many instances even with a parallel machine with a reasonable number of cores would take an impractical amount of time. For example, if one considers a 24 core system, the characterization would require 416 days, in the worst case. Even with an optimistic consideration of roughly 50% of the instances hitting the timeout threshold, the time requirement is still high. In contrast, the proposed framework can provide a fairly reasonable solution. Firstly, the size of the training set is extremely small, and also saturates after reaching a reasonable accuracy. One can rapidly characterize any number of fault instances after training with a reasonable error probability and the time requirement for that is insignificant.

For a statistical understanding of the exploitable fault space, small error bounds can be reasonably tolerated. Let us consider an application scenario where two S-Boxes are compared in terms of their sensitivity to fault attacks. Such a comparison can be made by estimating what fraction of faults corresponding to a given fault model is exploitable for each of the S-Boxes. A concrete example of such a scenario will be presented later in this chapter. Even in the presence of a small number of false negatives, such comparative analyses remain reasonably accurate. On the other hand, a small fraction of the false negatives can easily be ensured, while the classification accuracy is sufficiently high.

For certain security-critical applications, like evaluating a countermeasure or quantification of security bounds, the misclassified attack instances can be crucial. A typical example in this context is the test generation for the evaluation of low-cost countermeasures. The evaluator should store the plaintext and fault values for a **reasonable-sized albeit representative** set of attack instances, in this case, corresponding to each fault location. Misclassification of attack instances, in this case, may leave some critical corner cases unexplored. Given the practical feasibility of repeating such corner cases by an attacker, this may eventually lead to a successful attack with reasonably high probability. Quite evidently removal of false negatives is essential for such a scenario. It is, however, evident from the experimental results that the proposed method works fine in such critical scenarios with a reasonable overhead of characterizing 22% of the instances exhaustively.

This is indeed better than exhaustive SAT-based characterization, which is the only alternative, otherwise. For a set of 10,000 fault instances, the proposed approach, including the false negative removal step, would require 83 days, even in the worst case which is much better than the figures (416 days) obtained with exhaustive characterization.

3.4.1.6 Discussion

One of the goals of the ML framework is to discover new attacks while trained on a set of known attack instances. It was found that the proposed framework is able to do that with reasonably high accuracy. More specifically, we found that if the training set contains only of fault instances injected at even-numbered nibbles at the 28th round, it can successfully predict all attacks from odd-numbered nibbles. This clearly indicates the capability of discovering new attacks. The proposed framework also successfully validated the claim that with the bit permutation-based linear layer of PRESENT, the fault diffusion (and thus the attack) strongly depends on the plaintext and the value of the injected fault. Although this might not be a totally new observation, our framework figures it out, automatically, and can quantify this claim statistically within reasonable amount of time.

3.4.2 Exploitable Fault Space Characterization for LED

LED is a 64-bit block cipher proposed in CHES 2011 [10]. LED utilizes a round function which is similar to that of AES; more specifically it has the following sub-operations in sequence—*SubByte ShiftRow*, *MixColumn*, and *addRoundKey*. In contrast to AES, the 64-bit key is added once in each 4 rounds. All the diffusion layer operations have identifiable nibble-wise structures. The 4×4 S-Box of PRESENT is used as the confusion layer. Interestingly LED has no key schedule and the same key is used in all rounds. Like PRESENT, LED has also been subjected to DFA and DFIA [9, 15, 18]. Most of the DFA attempts on LED targeted the last 3 rounds of LED [14, 15, 18, 32]. Recently, Li et al. [19] have proposed an impossible differential fault analysis attack on the 29-th round of the cipher which requires 43 nibble faults to be injected at a particular nibble. Jovanovic et al. [14] and Zhao et al. [32] independently presented AFA attacks on LED, where they show that it is possible to attack the cipher at 30th round with a single fault instance.

3.4.2.1 ML Experiments

In this work, we mainly focus on the last 5 rounds of the LED cipher. However, unlike the previous experiment on PRESENT, a slightly different strategy was adopted. In order to examine the proper potential of the ML model in discovering

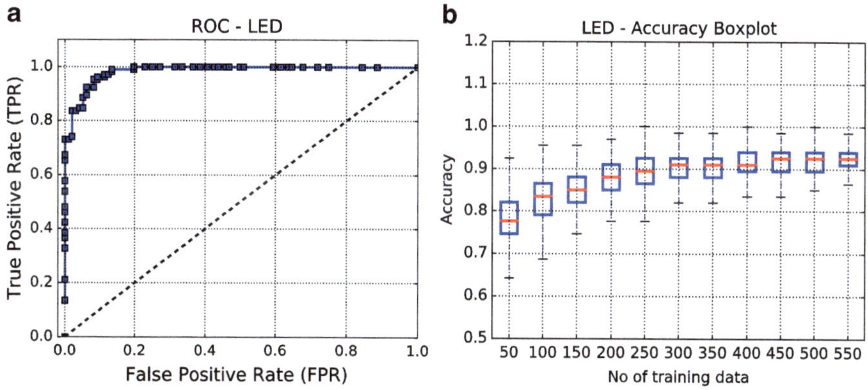

Fig. 3.5 Machine learning results for LED. **(a)** ROC curve and **(b)** variation of accuracy with the size of training set [24]

Table 3.4 Misclassification handling for LED

S Value	0.10	0.12	0.14	0.16	0.18	0.20	0.22	0.24	0.26	0.28	0.30
% sensitive instances	4.2	6.0	6.0	11.6	13.2	15.2	17.6	17.6	21.2	23.8	23.8
% false negatives beyond S	2.4	1.6	0.9	0.3	0.18	0.0	0.0	0.0	0.0	0.0	0.0

newer attack instances across different rounds, we intentionally trained it with samples from the 30 and the 31st rounds and tested it on instances from rounds 29 and 32. The RF model is trained with a total of 450 instances from the 30 and the 31st rounds and tested on 190 instances from rounds 29 and 32. The setup for the data acquisition is given in Table 3.2. The accuracy box plot and ROC curve for the classifier are provided in Fig. 3.5a, b, respectively. It can be observed that the accuracy is almost 93%. The features used were similar to the PRESENT experiments. Handling of misclassification was also performed and the result is presented in Table 3.4.

3.4.2.2 Discovery of New Attacks

We observed a quite interesting phenomenon in this experiment which clearly establishes the capability of the ML tool in discovering newer attack instances. More specifically, we found that the ML tool can identify attacks on 29th round of the cipher, even if it not trained with any instances from the 29th round. The attack instances observed at the 29th round of LED are mainly bit-fault instances with 2 fault injections. **Attacks on the 29th round of LED are usually difficult than the 30th round attacks [19, 33], and it is quite remarkable that the ML model can figure out difficult attacks just by learning easier attack instances.**

3.4.2.3 Discussion

So far we have discussed two ciphers with same S-Box and different diffusion layers. A comparative study of these two experiments establishes that compared to PRESENT, the fault space of LED is quite regular in nature. For example, almost all of the 30 round nibble faults in LED resulted in a successful attack, whereas for PRESENT there was a significant number of unexploitable instances at 28th round. From the perspective of an adversary, targeting a cipher having bit-permutation-based diffusion layers thus becomes a little more challenging as he/she must attack it with more number of fault injections in order to obtain a successful attack.

3.4.3 Utilizing the Success Rate: Analyzing the Effect of S-Boxes on Fault Attacks

So far in this chapter, we have discussed how to realize the ML-based fault space characterization efficiently. However, one of the main targets of this characterization is to estimate the success rate of an attack, which has not been demonstrated yet. In this subsection, we shed light on this issue. The estimation of the success rate is fairly straightforward once we have the ML model. One needs to characterize a large sample of faults with the model and calculate the percentage of exploitable faults among them. As we have already pointed out in the introduction, this quantity indicates what percentage of faults corresponding to a given location in the cipher practically results in an attack. Quite evidently if this percentage is fairly high, the location is extremely vulnerable.

The success rate metric, in fact, can shed light on something more crucial. It can provide relative grading among different fault locations and cipher structures. As a consequence, we can decide which structure is more vulnerable towards attacks automatically. Let us establish this claim by means of examples. The S-Boxes are one of the most important resources in a block cipher construction. However, till date, no quantitative analysis was performed to evaluate the effect of S-Boxes on the fault attacks as such. In classical DFA, the attack complexity is related with the average number of solutions of the S-Box difference equations having the form $S(x) \oplus S(x \oplus \alpha) = \beta$. However, S-Boxes were never characterized in the context of fault attacks considering the cipher as a whole. The characterization of the exploitable fault space (i.e., the success rate) in this work gives us the opportunity to perform such analysis.

In this experiment, we study the effect of three different S-Boxes on the PRESENT cipher, with respect to fault attack. More specifically, we replace the original S-Box of PRESENT with the S-Box of SKINNY [2], and the S_0 S-Box of the SERPENT [3], and study their effect on the exploitable fault space. The algebraic characteristics of these 3 S-Boxes are almost identical and presented in Table 3.5. The exploitable fault space in each case was characterized with our ML-based framework. For the sake of simplicity, we only tested with nibble faults

Table 3.5 Mathematical properties of PRESENT, SERPENT, and SKINNY S-Boxes

Property	PRESENT	SERPENT	SKINNY
Size	4×4	4×4	4×4
Differential branch number	3	3	2
Differential uniformity	4	4	4
Max. degree of component functions	3	3	3
Min. degree of component functions	2	2	2
Linearity	8	8	8
Nonlinearity	4	4	4
Max. differential probability	0.25	0.25	0.25
Max. degree of polynomial representation (with lexicographic variable ordering)	3	3	3

injected with $N = 2$ (that is two times for each fault instances) at the 28th round. The obtained test accuracies were similar to that of the PRESENT experiment and so we do not repeat them here. Further, for each of the S-Box case, we consider 1000 fault instances for each nibble location (there are total 16 nibble locations). The characterized fault spaces of the three S-Box test cases are depicted in Fig. 3.6a–c, respectively. It is interesting to observe that although the PRESENT and SERPENT S-Box result in almost similar behavior, the SKINNY S-Box results in a significantly different fault distribution. More specifically, whereas most of the fault instances for the PRESENT and SERPENT are exploitable (60% exploitable faults on average), the situation is reverse in the case of SKINNY (23% exploitable faults on average). In other words, the attack success rate is less in the case of SKINNY S-Box.

3.4.3.1 Analysis of the Observations

In order to explain the observations made in this experiment, we had an in-depth look in the 3 S-Boxes as well as the diffusion layer of PRESENT. The fault diffusion in PRESENT linear layer depends on the number of active S-Boxes (S-Boxes whose inputs are affected by the faults). *For a multi-round fault propagation, the number of active S-Boxes in the* ith *round depends on the* Hamming weights *(HW) of the output S-Box differential in the* $(i-1)$th *round.* Figures 3.7 and 3.8 emphasize this claim with a very simple example. The lines colored red indicate non-zero differential value and the red S-Boxes are the active S-Boxes.

Now let us consider the fault diffusion tree for the 28th round nibble fault injection in the PRESENT structure, shown in Fig. 3.8 up to 30th round, for convenience. It can be observed that most of the S-Boxes obtain an input difference of 1 bit. In other words, the inputs of the S-Boxes will have a single bit flipped. With this observation, the investigation boils down to the following question—*If the HW of the input difference of an S-Box is 1, what is the HW of the output difference?* For all three S-Boxes considered, the average HW of the output difference should

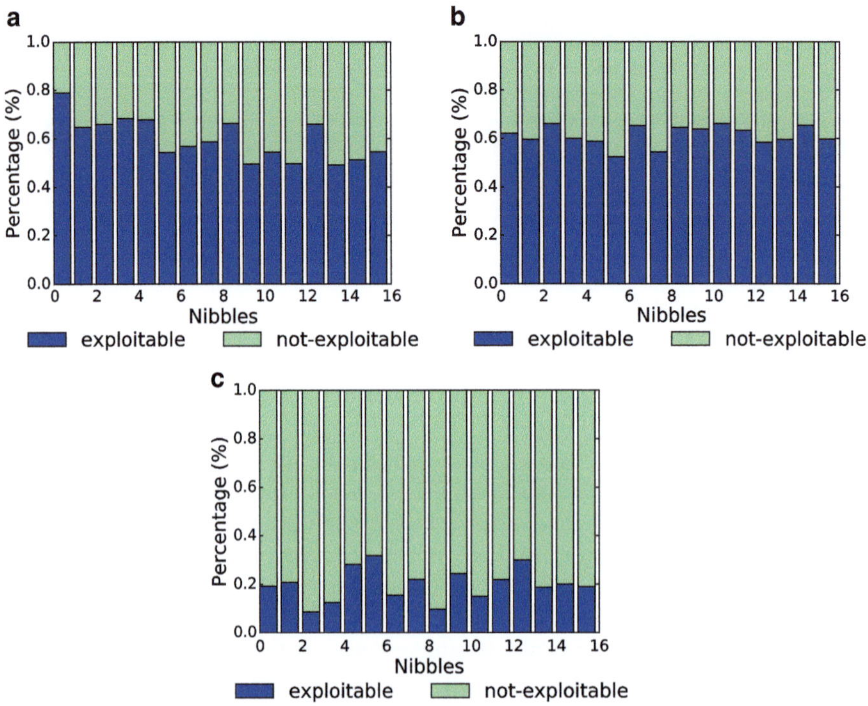

Fig. 3.6 Exploitable fault spaces with. **(a)** PRESENT S-Box, **(b)** SERPENT S-Box, and **(c)** SKINNY S-Box [24]

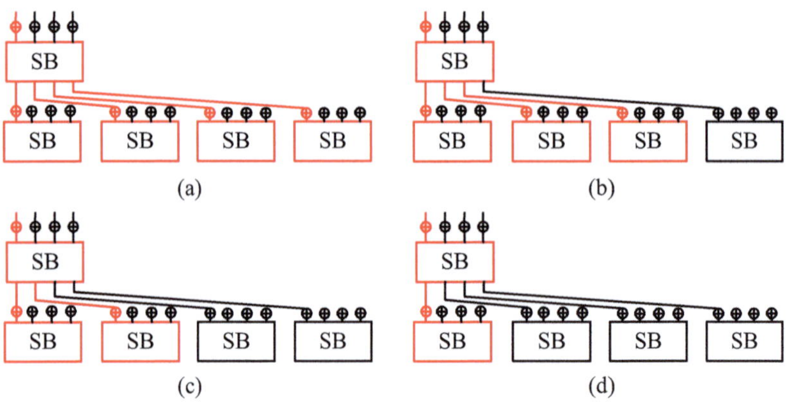

Fig. 3.7 Relation between the HW of S-Box output differential and fault diffusion in PRESENT. **(a)** 4 S-Boxes activated, **(b)** 3 S-Boxes activated, **(c)** 2 S-Boxes activated, and **(d)** 1 S-Box activated at the ith level [24]

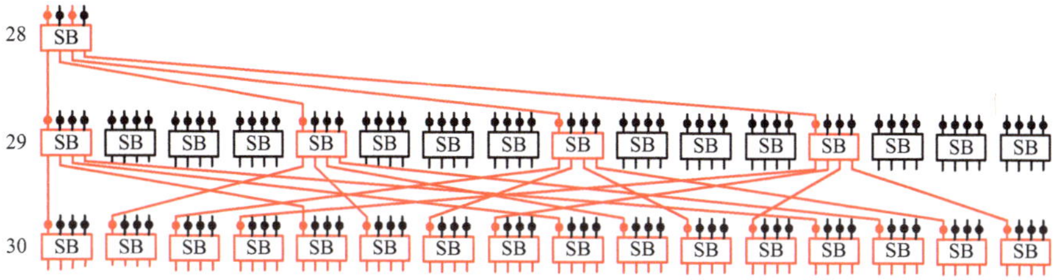

Fig. 3.8 Fault propagation up to 30th round in PRESENT structure [24]

be 2 when the average is considered over all possible input differences (this is due to the strict avalanche criteria (SAC)). However, for the typical case, where the HW of the input difference is restricted to 1, the average HW of the output differences vary significantly. More specifically, the average is quite low for the SKINNY S-Box where it attains a value of 2.2. For the PRESENT S-Box, the value is 2.45 and for SERPENT it is 2.5. This stems from the fact that, *for the SKINNY S-Box, there exist input difference values, for which the HW of the output differences become 1. Whereas for PRESENT and SERPENT S-Boxes, the minimum HW of the output differences is 2 for any given input difference. In essence*, the fault diffusion with PRESENT and SERPENT S-Box is more rapid on average than the SKINNY S-Box, which got reflected in the profile observed for exploitable fault spaces (and over the success rate).

The result presented in this subsection is unique from several aspects. Firstly, it shows empirically that even if the S-Boxes are mathematically equivalent, they may have different effects in the context of fault attacks. Secondly, the proposed framework of ours can identify such interesting phenomenon for different cipher sub-blocks, which are otherwise not exposed from standard characterization. This clearly establishes the efficacy of the proposed approach and success rate estimation.

3.4.3.2 Discussion

One of the main applications of exploitable fault characterization is a quantitative evaluation of block ciphers. One may consider the scenario presented in this section as a concrete example of such evaluation. The fraction of exploitable faults for a specific fault location, model, and multiplicity can be utilized as a metric for cipher evaluation. For example, Fig. 3.6a clearly indicates the high vulnerability of PRESENT against DFA attacks while a random nibble fault with multiplicity 2 is injected at the 28th round of the cipher. The attacker will succeed with probability 0.6. The complete extraction of the master key is possible within a reasonable time for all of these exploitable faults.[4] Similar quantitative evaluations can be performed

[4]However, it is worth mentioning that, by assigning some of the correct key bits in the equation system, a cipher evaluator can also handle the cases where key extraction by means of a fault is partial.

at other locations and fault models. In particular, if a cipher contains a significant number of such vulnerable locations for any of the practically achievable random fault models, it cannot be considered as a good design in the context of DFA attacks.

Relative grading of ciphers or various sub-operations is another potential application in the context of exploitable fault characterization. For example, among the three variants of PRESENT with three different S-Boxes, the one with SKINNY S-Box seems to be the most resilient alternative so far. Such experiments also have significant practical value for designing proprietary ciphers, where a common trend is to pick a reasonably good design and then replacing the sub-operations with similar ones. Also among PRESENT and LED, the former one seems relatively better as for the LED almost all nibble faults at round 30 are found to be exploitable; that too with a fault multiplicity of 1.

One critical aspect in statistical exploitable characterization is the choice of the fault multiplicity. Increasing fault multiplicity often makes the fault propagation more deterministic which in turn improves the accuracy of the ML model from all aspects and makes most of the faults at certain locations exploitable. However, one of the main practical requirements of DFA attacks is low fault multiplicity. Thus, for a fair comparison, one should choose a proper operating point where the attacks are possible with the lowest fault multiplicity. This fact justifies our choice of fault multiplicity 2 throughout our experiments with PRESENT. This fault multiplicity can be easily determined while acquiring the training data for the ML model.

3.5 Chapter Summary

Exploitable fault space characterization is an extremely relevant but relatively less explored topic in the fault attack research. We address this problem in the context of block ciphers, in this chapter, and eventually, come up with a reasonable solution. The proposed solution is able to efficiently handle the prohibitively large fault space of a cipher with reasonable computational overhead. The ML-based framework proposed here is not limited to block ciphers only as is quite well-known that even stream ciphers, public key algorithms [8], and hash functions can be mapped to algebraic systems [21].

Fault attacks are somewhat obvious for block ciphers, and there exists no construction so far, which is inherently resistant against such attacks. However, the degree of vulnerability may vary which we estimate from different perspectives in ExpFault and the ML-based framework. Countermeasures are inevitable against such attacks but they do incur huge overheads. One of the major application scenarios for such frameworks can be systematic and optimal countermeasure design. One simple optimization could be to deploy strong countermeasures to locations which are extremely vulnerable, and less robust but lightweight countermeasures to moderately critical locations. However, countermeasures do leak information if not designed properly. One of the relevant future works in this direction is to automate the vulnerability assessment of countermeasures.

Both the ML-based framework and ExpFault work at the algorithm level and do not consider certain implementation level issues which may leak information. Although the fault models considered in both of these chapters can incorporate the faults resulting from implementation aspects like instruction skips, the incorporation is not obvious. One of the major future works will be to bridge the gap between implementation and high-level faults in a fully automated manner so that implementation faults can be directly considered within the rich frameworks developed in these two chapters. The next couple of chapters in this book, however, will focus on handling implementation faults directly at the implementation level. These approaches do not exploit the cipher structures for estimating attacks and can figure out rather simpler implementation vulnerabilities as a result.

References

1. N. Bagheri, R. Ebrahimpour, N. Ghaedi, New differential fault analysis on present. EURASIP J. Adv. Signal Process. **2013**(1), 145 (2013)
2. C. Beierle, J. Jean, S. Kölbl, G. Leander, A. Moradi, T. Peyrin, Y. Sasaki, P. Sasdrich, S.M. Sim, The SKINNY family of block ciphers and its low-latency variant MANTIS, in *Annual Cryptology Conference* (Springer, Berlin, 2016), pp. 123–153
3. E. Biham, R. Anderson, L. Knudsen, Serpent: a new block cipher proposal, in *Proc. 5th Int. Workshop Fast Software Encryption (FSE)* (Springer, Paris, 1998), pp. 222–238
4. A. Bogdanov, L.R. Knudsen, G. Leander, C. Paar, A. Poschmann, M.J. Robshaw, Y. Seurin, C. Vikkelsoe, PRESENT: an ultra-lightweight block cipher, in *Proceedings of the 9th International Workshop on Cryptographic Hardware and Embedded Systems*, CHES '07 (Springer, Berlin, 2007) pp. 450–466
5. L. Breiman, Random forests. Mach. Learn. **45**(1), 5–32 (2001)
6. N.T. Courtois, K. Jackson, D. War, Fault-algebraic attacks on inner rounds of des, in *e-Smart'10 Proceedings: The Future of Digital Security Technologies* (Strategies Telecom and Multimedia, Montreuil, 2010)
7. F. De Santis, O.M. Guillen, E. Sakic, G. Sigl, Ciphertext-only fault attacks on present, in *International Workshop on Lightweight Cryptography for Security and Privacy* (Springer, Cham, 2014), pp. 85–108
8. J.-C. Faugere, A. Joux, Algebraic cryptanalysis of hidden field equation (HFE) cryptosystems using Gröbner bases, in *Proc. 23rd Annual Cryptology Conference (CRYPTO)*, vol. 2729 (Springer, Santa Barbara, 2003), pp. 44–60
9. N.F. Ghalaty, B. Yuce, P. Schaumont, Differential fault intensity analysis on present and led block ciphers, in *International Workshop on Constructive Side-Channel Analysis and Secure Design* (Springer, Cham, 2015), pp. 174–188
10. J. Guo, T. Peyrin, A. Poschmann, M. Robshaw, The LED block cipher, in *Cryptographic Hardware and Embedded Systems—CHES 2011* (Springer, Berlin, 2011), pp. 326–341
11. E. Hsu, C. Muise, J. Beck, S. McIlraith, Probabilistically estimating backbones and variable bias: experimental overview, in *Proc. 14th Int. Conf. Principles Practice Constraint Programming (CP)* (Springer, Sydney, 2008), pp. 613–617
12. F. Hutter, Y. Hamadi, H.H. Hoos, K. Leyton-Brown, Performance prediction and automated tuning of randomized and parametric algorithms, in *Proc. 12th Int. Conf. Principles Practice Constraint Programming (CP)*, (Springer, Nantes, 2006), pp. 213–228
13. F. Hutter, L. Xu, H.H. Hoos, K. Leyton-Brown, Algorithm runtime prediction: methods & evaluation. Artif. Intell. **206**, 79–111 (2014)

14. P. Jovanovic, M. Kreuzer, I. Polian, An algebraic fault attack on the led block cipher. IACR Cryptology ePrint Archive **2012**, 400 (2012)
15. P. Jovanovic, M. Kreuzer, I. Polian, A fault attack on the LED block cipher, in *Proc. 3rd Int. Workshop Constructive Side-Channel Analysis Secure Design (COSADE)*, (Springer, Darmstadt, 2012), pp. 120–134
16. P. Khanna, C. Rebeiro, A. Hazra, XFC: a framework for eXploitable fault characterization in block ciphers, in *Proceedings of the 54th Annual Design Automation Conference 2017*, DAC '17 (ACM, New York, 2017), pp. 8:1–8:6
17. K. Leyton-Brown, E. Nudelman, Y. Shoham, Empirical hardness models: methodology and a case study on combinatorial auctions. J. ACM **56**(4), 22 (2009)
18. W. Li, D. Gu, X. Xia, C. Zhao, Z. Liu, Y. Liu, Q. Wang, Single byte differential fault analysis on the LED lightweight cipher in the wireless sensor network. Int. J. Comput. Intell. Syst. **5**(5), 896–904 (2012)
19. W. Li, W. Zhang, D. Gu, Y. Cao, Z. Tao, Z. Zhou, Y. Liu, Z. Liu, Impossible differential fault analysis on the LED lightweight cryptosystem in the vehicular ad-hoc networks. IEEE Trans. Dependable Sec. Comput. **13**(1), 84–92 (2016)
20. L. Lobjois, M. Lemaître, Branch and bound algorithm selection by performance prediction, in *Proc. 15th National Conf. AI, 10th Innovative Applications AI Conf. (AAAI/IAAI)* (AAAI, Wisconsin, 1998), pp. 353–358
21. P. Luo, K. Athanasiou, Y. Fei, T. Wahl, Algebraic fault analysis of sha-3, in *Proc. 20th Design, Automation & Test in Europe Conference & Exhibition (DATE)* (IEEE, Lausanne, Mar 2017), pp. 151–156
22. D. Mitchell, B. Selman, H. Levesque, Hard and easy distributions of SAT problems, in *Proc. 10th National Conf. AI (AAAI)*, vol. 92 (AAAI, San Jose, 1992), pp. 459–465
23. E. Nudelman, K. Leyton-Brown, H.H. Hoos, A. Devkar, Y. Shoham, Understanding random SAT: beyond the clauses-to-variables ratio, in *Proc. 10th Int. Conf. Principles Practice Constraint Programming (CP)* (Springer, Toronto, 2004), pp. 438–452
24. S. Saha, D. Jap, S. Patranabis, D. Mukhopadhyay, S. Bhasin, P. Dasgupta, Automatic characterization of exploitable faults: a machine learning approach. IEEE Trans. Inf. Forensics Secur. **14**(4), 954–968 (2019)
25. M. Soos, K. Nohl, C. Castelluccia, Extending SAT solvers to cryptographic problems, in *Proc. 12th Int. Conf. Theory Applications Satisfiability Testing (SAT)* (Springer, Wales, 2009), pp. 244–257
26. K.R. Varshney, A risk bound for ensemble classification with a reject option, in *Proc. 14th IEEE Workshop Statistical Signal Process. (SSP)* (IEEE, Nice, June 2011), pp. 769–772
27. G. Wang, S. Wang, Differential fault analysis on PRESENT key schedule, in *2010 International Conference on Computational Intelligence and Security (CIS)* (IEEE, Nanning, 2010), pp. 362–366
28. L. Xu, F. Hutter, H.H. Hoos, K. Leyton-Brown, SATzilla: portfolio-based algorithm selection for SAT. J. Artif. Intell. Res. **32**, 565–606 (2008)
29. F. Zhang, X. Zhao, S. Guo, T. Wang, Z. Shi. Improved algebraic fault analysis: a case study on piccolo and applications to other lightweight block ciphers, in *Proc. 4th Int. Workshop Constructive Side-Channel Analysis Secure Design (COSADE)* (Springer, Paris, 2013), pp. 62–79
30. F. Zhang, S. Guo, X. Zhao, T. Wang, J. Yang, F.-X. Standaert, D. Gu. A framework for the analysis and evaluation of algebraic fault attacks on lightweight block ciphers. IEEE Trans. Inf. Forensics Secur. **11**(5), 1039–1054 (2016)
31. X. Zhao, S. Guo, T. Wang, F. Zhang, Z. Shi. Fault-propagate pattern based DFA on PRESENT and PRINT cipher. Wuhan Univ. J. Nat. Sci. **17**(6), 485–493 (2012)
32. X. Zhao, S. Guo, F. Zhang, Z. Shi, C. Ma, T. Wang, Improving and evaluating differential fault analysis on LED with algebraic techniques, in *Proc. 10th IEEE Workshop Fault Diagnosis Tolerance Cryptogr. (FDTC)* (IEEE, Santa Barbara, 2013), pp. 41–51
33. G. Zhao, R. Li, L. Cheng, C. Li, B. Sun, Differential fault analysis on LED using Super-Sbox. IET Inform. Secur. **9**(4), 209–218 (2014)

Chapter 4
Differential Fault Analysis Automation on Assembly Code

Jakub Breier, Xiaolu Hou, and Yang Liu

4.1 Introduction

As mentioned in Chap. 1, when it comes to attacking cryptographic algorithms, fault injection attacks are among the most serious threats, being capable of revealing the secret information by just one single disturbance in the execution [22, 35, 44]. Differential fault analysis (DFA) [11] has become the most commonly used fault analysis method for attacking symmetric block ciphers. It is the first method of choice when it comes to testing fault resilience of new cryptographic algorithms because of its simplicity and power to recover the secret key by a low number of faulty encryptions.

In practice, the attack always has to be mounted on a real-world device, in an implementation that is either hardware- or software-based. When we focus on software, there are many different ways to attack such implementations—one can

---

This research was conducted when the author "Jakub Breier" was with Temasek Laboratories, NTU.
This research was conducted when the author "Xiaolu Hou" was with Nanyang Technological University.

J. Breier
Underwriters Laboratories, Singapore, Singapore
e-mail: jbreier@jbreier.com

X. Hou (✉)
Acronis, Singapore, Singapore
e-mail: ho0001lu@e.ntu.edu.sg

Y. Liu
School of Computer Science and Engineering, Nanyang Technological University, Singapore, Singapore
e-mail: yangliu@ntu.edu.sg

J. Breier et al. (eds.), *Automated Methods in Cryptographic Fault Analysis*,
https://doi.org/10.1007/978-3-030-11333-9_4

corrupt the instruction opcodes resulting to instruction change, skip the instructions completely, flip the bits in processed constant values or register addresses, or change the values in the registers and memories directly [3, 6]. These attacks can be achieved by various fault injection methods, such as clock/voltage glitch, laser fault injection, or electromagnetic fault injection [3, 5].

As a consequence, different implementations of the same encryption algorithm do not necessarily share the same vulnerabilities. Some attacks that work in theory might either not be possible or be hard to execute in practice (e.g., precisely setting particular bits of the cipher state to some value [26]). On the other hand, there might be an exploitable spot in the implementation that is not visible from the specification of the encryption algorithm and can only be found by analyzing the assembly code. Up to now, the known DFA on PRESENT all aim before the execution of Sbox layer of the last round. In Sect. 4.5 we will show a DFA attack on one PRESENT implementation which aims at the end of player of the last round. The vulnerability comes from the implementation technique and is not intrinsic to the cipher.

Chapters 2 and 3 presented a thorough automated analysis of DFA on cipher design level. Since the real attack is always done on a concrete implementation [15, 18], it is also important to have automated tools to find the vulnerabilities of specific implementations. Our work aims to contribute to the automation of DFA on assembly code.

Since DFA makes use of the data dependency between the intermediate values of the algorithm and the secret key, we represent an assembly code as a customized data flow graph in static single assignment (SSA) form to record the operations (edges) and memory structures (nodes) holding the data, as well as their relations. Hence, assuming a fault is injected in one node (equivalently, the fault is injected in the instruction where this node is an output operand), we can construct the corresponding subgraphs that represent propagation of faults from this node to the ciphertext, while also showing its relation to the round key(s). Ultimately, this allows us to automatically generate differential fault analysis equations, by solving which we can mount a successful fault injection attack.

Different DFA attacks make use of different properties of the cryptosystems. The simplest attacks aim at the last round, requiring higher number of faults. More sophisticated attacks utilize properties of the cipher permutation layer, enabling them to aim at second or third last round, while lowering the number of faults. Hence, we allow the users to specify what kind of vulnerable spots to look for, which is realized as a user input variable called *output criteria*.

Our methodology was implemented in a tool named *DATAC—DFA Automation Tool for Assembly Code* [16] that takes an assembly code as input, and outputs subgraphs and equations for each vulnerable node according to user-specified output criteria.

Main Focus of the Chapter In this work we design and implement an automated tool DATAC which automatically finds the vulnerable instructions with respect to DFA in an assembly implementation of cryptosystem and outputs the related DFA equations for further analysis.

DATAC does not require any specific inputs of the cryptosystem and works independently of plaintext and secret key, providing a general evaluation of the underlying implementation. To the best of our knowledge, there is no tool working on assembly level that could automate the whole process to such extent. We emphasize that DFA vulnerabilities specific to software implementations are not visible from cipher design level.

We provide a case study on PRESENT-80 cipher implementation for 8-bit AVR microcontroller that shows capabilities of DATAC by finding a new DFA attack which is implementation specific, being able to recover the last round key by 16 fault injections.

To show the usage of output criteria, we provide an analysis of SPECK 64/128 and SIMON 64/128 lightweight ciphers, as well as current industry standard, AES-128. Time required for the analysis of AES is less than a second, by using a standard laptop computer. The results show that the whole process is computationally feasible.

We would like to point out that our tool is modular and enables an easy extension to the instruction set, making it easy to evaluate implementations for different devices.

Organization of the Chapter The rest of the chapter is structured as follows. Section 4.2 provides preliminaries for this chapter. Section 4.3 specifies our approach, by detailing each step of the evaluation used in DATAC. Section 4.4 provides implementation details of DATAC. Section 4.5 explains an implementation-specific DFA attack on PRESENT-80 found with DATAC. Section 4.6 provides a discussion and finally, Sect. 4.7 concludes this chapter.

4.2 Preliminaries

In this section, we first detail several related works in Sect. 4.2.1, focusing on automated fault analysis on design level and on evaluating implementations. In Sect. 4.2.2, we explain necessary basics of intermediate representations that are related to our work. We continue with stating assumptions and scope for DATAC usage in Sect. 4.2.3. Finally Sect. 4.2.4 closes this part with formal definitions and notations that are used later in the chapter.

4.2.1 Related Work

Agosta et al. [2] utilized an intermediate representation in order to check for single bit-flip vulnerabilities in the code to point out the exploitable parts. However, the approach aims at detecting cipher design vulnerabilities instead of low-level implementation-specific vulnerabilities. That is also why instruction skip model is not considered, although it is being one of the most powerful and easy to implement attacks.

Khanna et al. [37] proposed XFC—a method that checks exploitable fault characteristics of block ciphers considering the DFA attack method. Their approach takes a cipher specification as an input and then indicates the fault propagation through the cipher. The main drawback of this work is its focus on a high-level cipher representation, and therefore, being unable to check the security of a particular cipher implementation.

Goubet et al. [32] developed a framework that generates a set of equations for an SMT solver from assembly code. Then it uses this representation for evaluating robustness of countermeasures against fault injection attacks. The evaluation is based on comparison of two code snippets: one that represents a code without any protection, and a hardened code. These snippets are then represented as finite automata, unfolded, and analyzed. The main drawback of this work is its focus on code snippets instead of real implementations of cryptosystems and the fact that analyzing 10 lines of code requires 10.7 s. Therefore it is not feasible to analyze the full cipher in a reasonable time.

Dureuil et al. [25] proposed an approach using fault model inference—they first determine fault models that can be achieved on a target hardware, together with probability of occurrence of these models. Based on this information, they compute a "vulnerability rate" that gives an estimate of the software robustness. The focus of this chapter is to estimate time required in order to successfully inject the required fault model.

Gay et al. [30] took several hardware implementations of AES and provided their algebraic representations and translation into formulas in conjunctive normal form (CNF). These can be directly used by a SAT solver to mount an algebraic fault analysis.

A different approach to fault analysis automation was presented by Endo et al. [28]. Their work does not focus on finding a DFA vulnerability. However, it automates the way to find a pre-defined vulnerable spot in a black-box implementation with a countermeasure by checking whether the cipher output matches the requirement for the attack.

Our approach analyzes the assembly code directly, by building a customized data flow graph, allowing users to tailor the requirements for vulnerabilities according to desired fault models. Thanks to this, we can identify the points of interest efficiently and design DFA equations automatically, so that only the solving part is left to the user. DATAC is also scalable and can be used for analyzing a whole cipher implementation efficiently.

4.2.2 Intermediate Representation

Compiler construction normally depends on program analysis, where statements from an abstract high-level language are translated into a binary form that is understandable by the underlying processor. This process is not done in a single step but there are several subprocesses involved, in order to optimize the resulting program. One of these steps involves creation of an intermediate form that can be represented as a directed graph. There are various intermediate forms that can represent a program. Here we detail three intermediate representations which are most related to our work.

Data Dependency Graph Data dependency graph [1, 19] represents dependencies between (arithmetic assignment) statements that exist within program loops. For assembly code, the data dependency graph is also called instruction dependency graph, where each node corresponds to one instruction and each edge represents a dependency [24].

Data Flow Graph A data flow graph is a representation of data flow in a program execution. The nodes correspond to operations. Input data of an operation are represented as inward edges of the corresponding node and output data of an operation are represented as outward edges of the node [41].

Static Single Assignment Static single assignment (SSA) form requires that each variable is assigned exactly once and every variable is defined before it is used. As an example, we can take an assignment of the form $x = exp$. Then, the left-hand side variable x is replaced by using a new variable, e.g., x_1. After this point, any reference to x, before x is assigned again, is replaced with x_1 [7].

DATAC In our work, we are dealing with the dependency between different memory units that can be affected by performing operations on these units. Thus, DATAC constructs a graph where each node corresponds to a variable and each edge corresponds to an instruction. There is an edge f from a variable a to a variable b if and only if a is one input operand of instruction f and b is one output operand of f.

Furthermore, we consider an unrolled implementation. Hence, we can adopt the SSA form to facilitate DFA. Therefore, a graph constructed by DATAC is a customized data flow graph in SSA form.

4.2.3 Assumptions and Scope

Obviously, there are many aspects that have to be taken into account when analyzing a cipher implementation. It would be a daunting task to make a general-purpose analyzer that could work on unrestricted space of programs and fault analysis methods. Therefore, in the following we specify the scope for the tool usage and the rationale behind the design and implementation.

- As most of the DFA proposals, our method assumes known-ciphertext model and a single fault adversary (i.e., injecting one fault per encryption).

- We assume the implementation is available to the user and he can add annotations to the assembly code for the purpose of distinguishing different rounds, round keys, ciphertext words, etc.
- Majority of DFA proposals assume either bit-flip or random byte fault models. Additionally, some works utilize a powerful instruction skip model that can change the program flow in a way that some operations are avoided completely, resulting into a trivial cryptanalysis [29]. Therefore, we assume these three fault models in our analysis.
- The analysis automates the process of DFA that focuses on finding the vulnerable spot in the program and creating the difference equations.
- The set of vulnerable nodes that are outputs from the analysis is dependent on parameters that are set by the user. These are referred to as the *output criteria*. While we give some suggestions on tuning these so that the user can get readily exploitable outputs, in the future there might be more efficient DFA methods available that would require different parameters. By defining these as an input variable, it makes it easy in the future to adjust the tool to these new attacks without rebuilding the analysis subsystem.
- While the program outputs the difference equations, the final step of the analysis has to be done manually. One of the reasons for this is that our tool does not operate on concrete values, only on dependencies between the variables.
- For the analysis in this chapter, we have chosen Atmel AVR instruction set.[1] However, for analyzing different instruction sets, only the parsing subsystem of the analyzer has to be redefined (`AsmFileReader` class stated in the class diagram in Fig. 4.8).

We would like to point out that the analysis is done on unrolled implementations. The main reason for this is that fault attack can always exploit a jump/branch operation in a way to render the computation vulnerable to DFA [46]. For example, if a round counter is tampered with, attacker can easily change the number of rounds to 1, making it trivial to recover the key. Or, if cipher operations are implemented as macros, one can skip the jump to such macro [23].

4.2.4 Formalization of Fault Attack

Definition 4.1 We define a *program* to be an ordered sequence of assembly instructions $\mathcal{F} = (f_0, f_1, \ldots, f_{N_\mathcal{F}-1})$. $N_\mathcal{F}$ is called the *number of instructions* for the program. For each instruction $f \in \mathcal{F}$, we associate f with a 4-tuple $(f^{seq}, f^{mn}, f^{io}, f^{do})$, where f^{seq} is the sequence number and f^{mn} is the mnemonic of f. f^{io} is the set of input operands of f, which can be registers,

[1] http://www.atmel.com/images/Atmel-0856-AVR-Instruction-Set-Manual.pdf.

Table 4.1 Assembly code $\mathcal{F}_{ex}$ for a sample cipher

#	Instruction	#	Instruction	#	Instruction
	`//round_1`	5	`EOR r1 r3`	10	`LD r3 key2+`
0	`LD r0 X+`	6	`ANDI r0 0x0F`	11	`EOR r0 r2`
1	`LD r1 X+`	7	`ANDI r1 0xF0`	12	`EOR r1 r3`
2	`LD r2 key1+`	8	`OR r0 r1`		`//store_ciphertext`
3	`LD r3 key1+`		`//round_2`	13	`ST x+ r0`
4	`EOR r0 r2`	9	`LD r2 key2+`	14	`ST x+ r1`

constant values, or pointers to memory addresses. f^{do} is the set of destination (output) operands of f, which can be registers or pointers to memory addresses.

Example The assembly implementation $\mathcal{F}_{ex}$ of a simple sample cipher in Table 4.1 has $N_{\mathcal{F}_{ex}} = 15$ instructions. Instruction $f_6 =$ `ANDI r0 0x0F` has input operands `r0` and `0x0F`, destination operand `r0`. Thus f_6 is associated with the 4-tuple $(6, \texttt{ANDI}, \{\texttt{r0}, \texttt{0x0F}\}, \{\texttt{r0}\})$.

We note that for an instruction $f =$ `ADD r0 r1`, the output operands of f are actually `r0` and `carry`, where `carry` is a flag, usually represented by a bit in the status register of a microcontroller. The `carry` itself does not appear in the assembly code directly; however, we consider it in our analysis as a standalone operand.

Fault attack is an intentional change of the original data value into a different value. This change can either happen in a register/memory, on the data path, or directly in ALU. In general, there are two main fault models to be considered—program flow disturbances and data flow disturbances. The first one is achieved by disturbing the instruction execution process that can result in changing or skipping the instruction currently being executed. The second one is achieved either by directly changing the data values in storage units or by changing the data on the data paths or inside ALU.

Formally, we define a fault injection in a program $\mathcal{F} = \{f_0, f_1, \ldots, f_{N_{\mathcal{F}}-1}\}$ to be a function $\vartheta_i : \mathcal{F} \mapsto \mathcal{F}'$, where $0 \le i < N_{\mathcal{F}}$ and $\mathcal{F}'$ is a program obtained from $\mathcal{F}$ with the instruction f_i being tampered. Thus ϑ_i represents a fault injection on the instruction with sequence number i in $\mathcal{F}$. There are different possible fault models, we focus on the following:

- **Instruction skip**: $\vartheta_i(\mathcal{F}) = \mathcal{F} \backslash f_i$, i.e., instruction i is skipped.
- **Bit flip**: $\vartheta_i(\mathcal{F}) = \{f_0, f_1, \ldots, f_i, f'_{i+1}, f'_{i+2}, \ldots, f'_{N_{\mathcal{F}}-1}, f'_{N_{\mathcal{F}}}\}$ such that $f'_{j+1} = f_j$ for $i < j < N_{\mathcal{F}}$ and $f'_{i+1} = \texttt{r xor } \Delta$, where $r \in f_i^{do} \cup f_i^{io}$ is either a destination operand or an input operand of instruction f_i and Δ is a pre-defined value which is called a *fault mask*. In the case $f_i^{do} = \emptyset$, $f'_{i+1} = \texttt{NOP}$.
- **Random byte fault**: $\vartheta_i(\mathcal{F}) = \{f_0, f_1, \ldots, f_i, f'_{i+1}, f'_{i+2}, \ldots, f'_{N_{\mathcal{F}}-1}, f'_{N_{\mathcal{F}}}\}$ such that $f'_{j+1} = f_j$ for $i < j < N_{\mathcal{F}}$ and $f'_{i+1} = \texttt{r xor } \Delta$, where $r \in f_i^{do} \cup f_i^{io}$ and Δ is a random value. In the case $f_i^{do} = \emptyset$, $f'_{i+1} = $NOP.

In the rest of this chapter we assume that the attacker has the knowledge of the fault model for the differential fault analysis.

4.3 Automated Assembly Code Analysis

In Sect. 4.3.1, we first provide an overview of the internal working of DATAC. Then, we explain the methodology used in DATAC in detail, following the same logical flow as the actual analysis. Section 4.3.2 defines the specifics of our customized data flow graph. Output criteria parameters are elaborated in Sect. 4.3.3. Subgraph and equation constructions are illustrated in Sects. 4.3.4 and 4.3.5, respectively.

4.3.1 Overview

The main goal of DATAC is to analyze the assembly code and find vulnerable spots w.r.t. DFA attack. This can be a challenging task, since the same instruction can be vulnerable in one part of the code, but secure in the other part, depending on the context.

For our purposes, we have to capture the following details when transforming the assembly code:

- Memory units holding the data (registers, RAM, flash, etc.) as well as direct operands (constants)—these will be represented as nodes.
- Transitions between the nodes.
- Operations (instructions) in the program—represented as edges.
- Properties of operations (linear/non-linear).
- Ability to distinguish important nodes, such as round keys and ciphertext.

Our evaluation method is depicted in Fig. 4.1. First, an assembly code is fetched as the input. Based on its structure, a data flow graph is created. Depending on the output criteria, vulnerable nodes are identified. Then, a subgraph is created for each vulnerable node—it specifies a propagation pattern of the fault from the vulnerable node to the ciphertext. Together with it, fault analysis equations are generated—based on them, a fault attack can be executed. In the rest of this section, we provide details on how our approach works.

4.3.2 From Assembly to Data Flow Graph

Given a program $\mathcal{F} = (f_0, f_1, \ldots, f_{N_{\mathcal{F}}-1})$, a data flow graph is a directed graph $G_{\mathcal{F},full} = (V, E)$, where the set of nodes $V = A \cup B$ is the union of two sets of labeled nodes. A consists of labeled nodes with labels "x (i)" such that x is

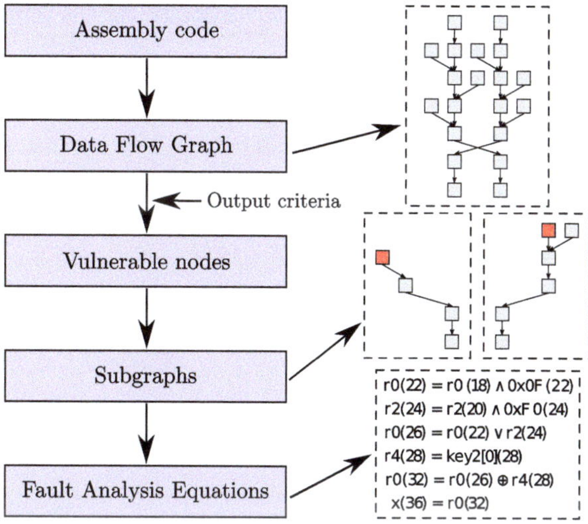

Fig. 4.1 Our evaluation method for analyzing assembly code w.r.t. fault injection vulnerabilities

a destination operand (also called output operand) of instruction i. B consists of labeled nodes with labels "y (i)" such that y is an input operand of instruction i and y is not a destination operand of any instruction.

- $A = \{$"x (i)" $: x \in f_i^{do}$ for some $0 \leq i < N_{\mathcal{F}}\}$;
- $B = \{$"y (i)" $: y \in f_i^{io}$ for some $0 \leq i < N_{\mathcal{F}}$ and $y \notin f_j^{do}$ for any $0 \leq j < N_{\mathcal{F}}\}$.

We draw an edge from node $a =$ "y (i)" to node $b =$ "x (j)" if and only if the following conditions are satisfied:

- $i \leq j$,
- x is a destination operand of instruction j,
- y is an input operand of instruction j,
- y is not an output operand for any instruction between instruction i and instruction j, which means the value in y is not changed between instructions i and j.

Formally, an edge $(a, b) \in E$ for $a =$ "y (i)", $b =$ "x (j)" $\in V$ iff $i \leq j$, $x \in f_j^{do}$, $y \in f_j^{io}$ and "y (k)" $\notin V$ $\forall i < k < j$. Furthermore, we label such an edge with "f_j^{mn} (j)" and we say that this edge is *associated with* instruction f_j. We also refer to a as an *input node* of f_j and b as an *output node* of f_j. Following the terminologies from graph theory, a is called the *tail* of the edge (a, b) and b is called the *head* of (a, b).

Example The data flow graph $G_{\mathcal{F}_{ex},full}$ corresponding to the assembly program $\mathcal{F}_{ex}$ in Table 4.1 is shown in Fig. 4.2. Instruction f_6 has input operands `r0` and `0x0F`, where `r0` is the output operand of f_4 and `0x0F` is not an output operand of any instruction. The output operand of f_6 is `r0`. Hence f_6 has two input nodes: "`r0` (4)," "`0x0F` (6)" and one output node "`r0` (6)". Furthermore, f_6 is related to two

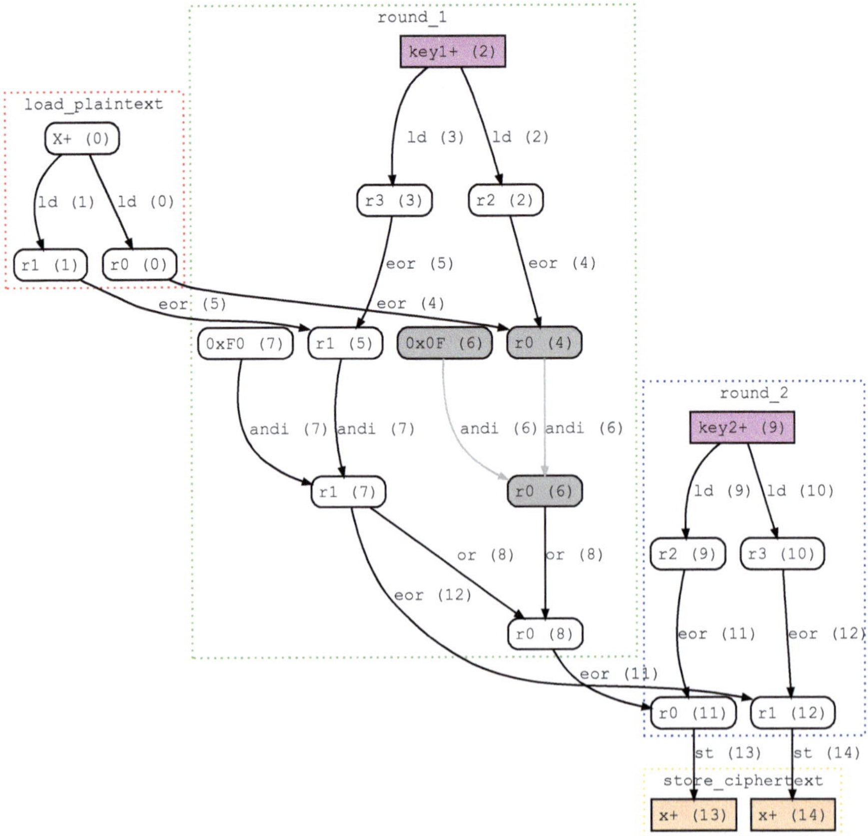

Fig. 4.2 Data flow graph $G_{\mathcal{F}_{ex},full}$ corresponding to the assembly program $\mathcal{F}_{ex}$ in Table 4.1 constructed by DATAC

edges in the graph, both labeled "`ANDI` (6)." Both edges have head "`r0` (6)," one with tail "`r0` (4)" and one with tail "`0x0F` (6)" (see the nodes and edges highlighted in gray).

Since we are dealing with implementations of ciphers, we highlight the round keys as well as the ciphertext in the graphs. As shown in Fig. 4.2, the node that corresponds to round key in round i will be denoted by "`keyi+` (j)," where j is the sequence number of the first instruction that loads key values to registers in this round. Furthermore, the output nodes of those key loading instructions are called *key word nodes*. Depending on which word is loaded first, they are more specifically called *the first key word node*, *the second key word node*, etc.[2]

[2]We note that this is only a naming convention to make the analysis consistent. In special cases, where the actual order of the key words is different from their loading sequence, user has to rearrange the key words after the analysis.

Table 4.2 Assembly code snippet 1

#	Instruction
0	`LD r0 key0+`
1	`EOR r1 r0`
2	`ST x+ r1`

Table 4.3 Assembly code snippet 2

#	Instruction
0	`LD r0 key0+`
1	`AND r1 r0`
2	`ST x+ r1`

Example In Fig. 4.2, "`r2` (2)" is the first key word node of round key for round one. "`r3` (10)" is the second key word node of round key for round two.

The nodes representing output operands that give us different words of the ciphertext are labeled "`x+` (j)," where "`x+` (j)" is an output node of instruction f_j, i.e., j is the sequence number of the instruction that stores this word. We refer to them as *the words of the ciphertext*. For example, in Fig. 4.2, "`x+` (13)" and "`x+` (14)" are the words of the ciphertext.

4.3.3 Output Criteria

For a directed graph G, a *directed path* from node v to node u is a sequence of edges $e_1, e_2, \ldots, e_k$ such that $e_1 = (v, x_1), e_2 = (x_1, x_2), e_3 = (x_2, x_3), \ldots, e_{k-1} = (x_{k-2}, x_{k-1}), e_k = (x_{k-1}, u)$. For any pair of nodes v and u, if there exists a directed path from v to u, we say u is a *Gchild* of v and v is a *Gparent* of u. For any edge e which appears in the sequence, we say e *belongs* to this directed path from v to u.

Now let us look at the following two simple scenarios in Tables 4.2 and 4.3.

1. In Table 4.2, let us assume a fault is injected at f_1 such that some bits in `r1` are flipped before the execution of `EOR`. Then the exact same bits will be changed in ciphertext `x+`. Knowing how `r1` is changed and values of ciphertext with and without fault injection will not give us any information about `r0`.
2. In Table 4.3, we assume a fault is injected in f_1 such that some bits in register `r1` are flipped before the execution of `AND`. For example, suppose the first bit of `r1` is changed. Then we look at the first bit of the ciphertext `x+`. If the first bit of the ciphertext is also changed, we know that the first bit of the key is `1`, otherwise, it is `0`.

In view of the above, we say an instruction f is *linear* if the following conditions are satisfied: Let n be the register size in bits. For any pair a, b ($a \in f^{io}, b \in f^{do}$), and for any $\Delta \in \mathbb{F}_2^n$, if a is changed to $a' = a \oplus \Delta$, then b will be changed to $b' = b \oplus \Delta$. Thus the linearity of an instruction f is determined by its mnemonics f^{mn}. For example, the following commonly used mnemonics correspond to linear instructions: `EOR, LD, MV, ST`. And we say an edge e is *non-linear* if the instruction associated with e is non-linear.

For a pair of nodes v and u such that u is a Gchild of v, the *Gdistance* between v and u, denoted by Gdistance(v,u), is defined to be the cardinality of the following set:

$$\{e : e \text{ belongs to a directed path from } v \text{ to } u \text{ and } e \text{ is non-linear}\}.$$

Example In Fig. 4.2, "`x+` (13)" is a Gchild of "`r0` (6)" with distance 1. "`r0` (8)" is a Gchild of "`key1+` (2)" with distance 4.

For each node a = "x (i)," we define *CTGchild* of a to be the set of ciphertext words which are Gchildren of a. Thus if a fault is injected in node a, the fault will be propagated to the ciphertext words that are in the set CTGchild of a.

To analyze the fault propagation that is useful, we need to focus on the nodes that are related to the key. We say a node a is *related to a key word node b* of a round key if b is not a Gparent of a and at least one of the Gchildren, say ch, of a is a Gchild of b with Gdistance$(b, ch) = 0$. The distance condition specifies that we are only looking at nodes that are linearly related to the key. The parent condition excludes nodes that would be related to several key words in a way that would combine these words linearly and therefore, making it impossible to recover the secret information. And we say a is *related to a round key* `key` if it is related to at least one key word node of `key`.

For a given node a which is to be examined, several possible parameters have to be specified. We refer to them as to *output criteria*, and they include the following:

- `minAffectedCT`: |CTGchild| ≥ `minAffectedCT`, i.e., the number of nodes in CTGchild is bigger or equal than `minAffectedCT`;
- `minDist`: the number of nodes in CTGchild with Gdistance at least `minDist` from a is at least `minAffectedCT`, $|\{ch : ch \in$ CTGchild, and Gdistance(a, ch) ≥ `minDist`$\}|$ ≥ `minAffectedCT`;
- `maxDist`: Gdistance(a, ch) ≤ `maxDist` $\forall ch$ ∈Gchild, i.e., the Gdistance between any Gchild and a should be at most `maxDist`;
- `maxKey`: the number of the round keys, counting from the last round key, that are related to node a is at most `maxKey`;
- `minKeyWords`: there exists at least one round key such that the number of its corresponding key word nodes related to a is at least `minKeyWords`.

DATAC takes a data flow graph G and an output criteria as input, then iterates through all the nodes in G and outputs the nodes of G that satisfy the output criteria. Recall, we assume that the information available to the attacker is the fault model, the correct and faulty ciphertext.

Below, we explain that the output criteria defined are necessary for DFA and are independent of the analyzed cipher. On the other hand, the choice of their values is dependent on the analyzed implementation.

4.3.3.1 Why Are the Output Criteria Mandatory

For DFA attack, `minAffectedCT` specifies how many words of the ciphertext are faulted after the fault injection. This value should be at least 1 so that the ciphertext values can be used. `minDist` reflects on how many non-linear operations are involved between the faulted node and the ciphertext. For DFA, `minDist` should be at least 1 so that there are non-linear operations involved and hence some information can be drawn. `maxDist` is an upper bound on how many non-linear operations are involved in the calculations. `maxDist` and `minDist` together restrict the number of non-linear operations to be solved for DFA. `maxKey` restricts which round key(s) are to be attacked. In most DFA attacks, the attacker focuses on the last round key (`maxKey`= 1) or the second last round key (`maxKey`= 2). `minKeyWords` is able to exclude nodes which are related to only small number of key words.

Remark 4.1 The choice of output criteria is essential to DFA, which is the scope of this chapter. In case a different fault analysis method is considered, the output criteria should be changed accordingly.

4.3.3.2 Choosing the Output Criteria Values

We note that the values of output criteria are closely related to each other and are highly dependent on the actual assembly program being analyzed. For example, if the program makes use of a high number of non-linear instructions right before storing the ciphertext, `maxDist` should be set higher so that there are actually key words related to the faulted node. On the other hand, `maxKey` should not be too big, otherwise some nodes in the output will be associated with too many round keys, making the analysis harder. Accordingly, `minKeyWords` should be set to a small value, but for obvious reasons, should be at least 1. Or, if the user would like to have all the ciphertext words being affected, i.e., setting `minAffectedCT`= the number of ciphertext words, the other conditions should be loosened. For example, for the data flow graph $G_{\mathcal{F}_{ex},full}$ in Fig. 4.2, with an output criteria (`minAffectedCT, minDist, maxDist, maxKey, minKeyWords`) = (2, 1, 1, 1, 1) we cannot get any output from DATAC.

For the data flow graph in Fig. 4.2, with an output criteria (`minAffectedCT, minDist, maxDist, maxKey, minKeyWords`)= (1, 1, 1, 1, 1), we get two nodes "`r0` (6)" and "`r1` (7)". For illustration purpose, we will focus on node "`r0` (6)" in the following.

4.3.4 Subgraph Construction

For a full cipher assembly implementation, the corresponding data flow graph involves plenty of nodes and edges. It is not easy to see the fault propagation

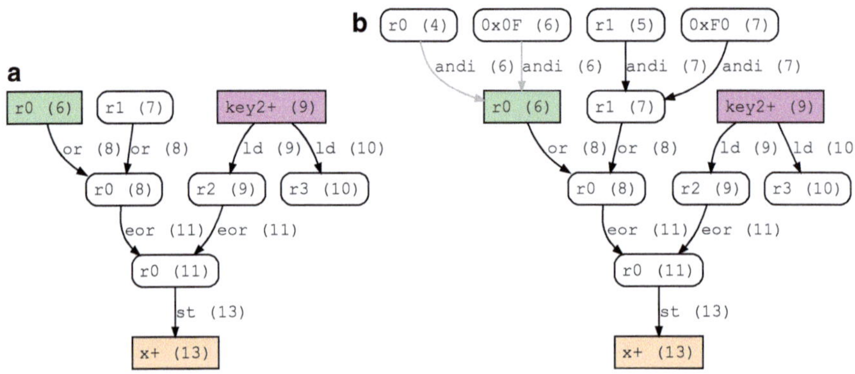

Fig. 4.3 Subgraph constructed from node "r0 (6)" of data flow graph in Fig. 4.2 with depth (**a**) 0 and (**b**) 1

properties from the full data flow graph. Thus we would like to construct a subgraph which shows the fault propagation clearly.

Given a data flow graph $G_{\mathcal{F},full}$ for an assembly program $\mathcal{F}$ and node a in $G_{\mathcal{F},full}$, we construct a graph G_a which is a subgraph of $G_{\mathcal{F},full} = (V, E)$, i.e., $G_a = (V_a, E_a)$ is a pair such that $V_a \subseteq V$ and $E_a \subseteq E$.

Sometimes, knowing how the faulted node relates to previous instructions will also help with the fault analysis. Keeping this in mind, we define a parameter called `depth` for the construction of the graph G_a.

We define *KNGchild* to be the set of key word nodes that are related to a. Then $V_a = \left(\bigcup_{i=0}^{\texttt{depth}} U_i\right) \bigcup \left(\bigcup_{j=1}^{4} V_j\right)$, where

- $U_0 = \{b : b$ is an input node of an instruction f for which a is an input node$\}$
- For $1 \leq i \leq \texttt{depth}$, $U_i = \{b : b$ is an input node for an instruction f such that v is an output node of f for some $v \in U_{i-1}\}$
- $V_1 = \{b : b$ is a child of $a\}$
- $V_2 = \{k : k$ is a round key that is related to $a\}$
- $V_3 = \{b : b$ is a key word node for a key $k \in V_2\}$
- $V_4 = \{b : b$ is a child of a node $v \in \texttt{KNGchild}$ and b is a parent of a child of $a\}$.

Let $V' = (V_a \backslash (V_2 \cup V_3)) \cup \texttt{KNGchild}$. Then $E_a = E_1 \cup E_2$, where $E_1 = \{e :$ both the head and the tail of e are in $V'\}$ and $E_2 = \{(k, b) : k \in V_2, b \in V_3\}$.

Example In Fig. 4.3a, b we present the subgraphs constructed from node "r0 (6)" of the data flow graph $G_{\mathcal{F}_{ex},full}$ (Fig. 4.2) with depths equal to 0 and 1, respectively. For this case, we have

- $U_0 = \{$"r0 (6)", "r1 (7)"$\}$
- $U_1 = \{$"r1 (5)", "0xF0 (7)", "r0 (4)", "0x0F (6)"$\}$
- $V_1 = \{$"r0 (8)", "r0 (11)", "x+ (13)"$\}$

- $V_2 = \{$"`key2+` (9)"$\}$
- $V_3 = \{$"`r2` (9)", "`r3` (10)"$\}$
- $V_4 = \{$"`r0` (11)"$\}$

From Fig. 4.3, we can see that with `depth` $=$ 1, we do get extra useful information: the two edges with label "`andi` (6)" show that the first four bits of "`r0` (6)" are 0.

4.3.5 Equation Construction

Having the subgraph, constructed from a potentially vulnerable node, we would like to construct equations out of the subgraph to connect different input/output nodes, which can be easily analyzed by algebraic methods.

Given any subgraph $G_a = (V_a, E_a) \subseteq G_{\mathcal{F},full}$, where $G_{\mathcal{F},full}$ is the data flow graph of an assembly program $\mathcal{F}$, we take all the instructions in $\mathcal{F}$ that are related to at least one edge $e \in E_a$. Next, we order these instructions according to their sequence numbers. The equations are then constructed according to the input/output nodes and the edges associated with the corresponding instructions.

In Table 4.4 we show some representations of equations for different mnemonics. Here, the symbol "|" represents concatenation. For example, take $f =$ `MUL r2 r3`, it calculates the product of values in registers `r2` and `r3`, then the high byte of the product is stored in `r1` and the low byte of the product is stored in `r0`. Hence the product in the equation is represented as a concatenation of `r1` and `r0`: `r1 | r0`.

In case the equation is related to an instruction that loads a round key, DATAC is designed to indicate which key word node is involved in the equation (see Remark 4.2).

Now let us look at the assembly program $\mathcal{F}_{ex}$ for our sample cipher from Table 4.1. $\mathcal{F}_{ex}$ and output criteria $(1, 1, 1, 1, 1)$ (see Sect. 4.3.3) were given as

Table 4.4 Construction of equations from assembly instructions

Instruction	Equation
`ADD r2 r3`	`carry	r2 = r2 + r3`
`ADC r2 r3`	`carry	r2 = r2 + r3 + carry`
`EOR r2 r3`	`r2 = r2 ⊕ r3`
`AND r2 r3`	`r2 = r2 ∧ r3`
`OR r2 r3`	`r2 = r2 ∨ r3`
`MUL r2 r3`	`r1	r0 = r2 × r3`
`LD/MOV/ST r2 r3`	`r2 = r3`
`ROL r2`	`carry	r2 = r2	carry`
`LSL r2`	`carry	r2 = r2	0`
`LPM r2 Z`	`r2 = TableLookUp(ZH | ZL)`

input to DATAC. The data flow graph $G_{\mathcal{F}_{ex},full}$ for this sample cipher is in Fig. 4.2. DATAC outputs two vulnerable nodes: "`r0` (6)" and "`r1` (7)". The subgraphs with depths 0 and 1, constructed from "`r0` (6)", are shown in Fig. 4.3. As we pointed out in Sect. 4.3.4, the subgraph with depth 1 gives some additional useful information, compared to the one with depth 0.

The equations obtained by using DATAC from the subgraph with depth 1, constructed from "`r0` (6)" (Fig. 4.3b), are as follows:

$$\text{“}\texttt{r0}\text{ (6)”} = \text{“}\texttt{r0}\text{ (4)”} \wedge \text{“}\texttt{0x0F}\text{ (6)”} \tag{4.1}$$

$$\text{“}\texttt{r1}\text{ (7)”} = \text{“}\texttt{r1}\text{ (5)”} \wedge \text{“}\texttt{0xF0}\text{ (7)”} \tag{4.2}$$

$$\text{“}\texttt{r0}\text{ (8)”} = \text{“}\texttt{r0}\text{ (6)”} \vee \text{“}\texttt{r1}\text{ (7)”} \tag{4.3}$$

$$\text{“}\texttt{r2}\text{ (9)”} = \texttt{key2[0]} \tag{4.4}$$

$$\text{“}\texttt{r0}\text{ (11)”} = \text{“}\texttt{r0}\text{ (8)”} \oplus \text{“}\texttt{r2}\text{ (9)”} \tag{4.5}$$

$$\text{“}\texttt{x+}\text{ (13)”} = \text{“}\texttt{r0}\text{ (11)”}. \tag{4.6}$$

Equation (4.1) shows "`r0` (6)" $=$ $\texttt{0000}b_4b_5b_6b_7$ for some $b_j \in \{0, 1\}$ $(j = 4, 5, 6, 7)$. Equation (4.3) shows that if we skip instruction 8, the result of Eq. (4.1) will be used instead of the result of Eq. (4.3) in instruction 11, which corresponds to Eq. (4.5). Together with the information from Eqs. (4.4) and (4.6), the instruction skip attack on instruction 8 would result in the first four bits of `key2[0]` to appear as the first four bits of the faulted ciphertext.

Remark 4.2 The index [0] in the right hand of Eq. (4.4) indicates that the node "`r2` (9)" is the first key word node of `key2`, i.e., the value in "`r2` (9)" is the first byte of the second round key.

4.4 Implementation

DATAC was implemented in Java programming language, to support multi-platform environments and provide efficient running times. In this section, we provide implementation details of DATAC. Section 4.4.1 specifies the input format of the assembly code file. Section 4.4.2 outlines the execution steps of DATAC. Section 4.4.3 provides instructions on the usage of DATAC.

4.4.1 Assembly Code Properties

DATAC assumes the assembly code to be written in a text file, one instruction per line. Some annotations are also required, as well as naming conventions for important variables. These are stated in Table 4.5.

Table 4.5 List of assembly code annotations and variable names required for DATAC

Action/variable name	DATAC format
Loading plaintext	`//load_plaintext`
Start of round *i*	`//round_i`
Storing ciphertext	`//store_ciphertext`
Name of key variable pointer *i*	`keyi`
Name of ciphertext variable pointer	`x`

Another requirement for the code is that it should be unrolled. It would have been possible to add a loop and function dependency analysis into DATAC; however, it was shown before that loops and function calls can be easily disturbed by fault injection (e.g., in [39]) and therefore, it would make more sense to just skip the whole portion of the program instead of trying to analyze it with DFA.

4.4.2 Analysis Steps

In the following, we will outline the steps of the analysis that are being executed each time DATAC is run:
Input: assembly code text file, output criteria, depth.

Step 1: Read the assembly file and construct an array of instructions.

When DATAC reads an instruction, it first recognizes the mnemonics. Based on the type, it reads the operands that follow after the mnemonics. These are stored together with the sequence number.

Some operands are put explicitly, e.g., `LDI r0 0x01` has two operands—input operand is `0x01`, and the output operand is `r0`. In some cases, the operands are implicit, e.g., `MUL r5 r6` has four operands—input operands `r5`, `r6` are explicit and there are two implicit output operands: `r0` and `r1`. The same holds for instructions that use a carry bit, for example, `ADC`, `ROL`—in these cases, the carry is treated as input or/and output operand, since a fault in a carry can cause changes in later values.

Step 2: Construct the data flow graph of the program.

DATAC iterates through the array of instructions and analyzes both the operations and operands. Operands are represented as nodes and operations are represented as edges.

To allow tracing the behavior under fault, cross-dependencies between edges and nodes are also stored—each node has a set of input/output edges, and each edge has its head (output node) and tail (input node).

Some operands involve pointers to memory use pre-/post-increment or decrement operators (`+`/`-`). In this case, an array of consecutive memory values is used by the program, normally either for loading plaintext and key or for storing ciphertext. In the case of storing ciphertext, we treat different entries of the array separately, since it is important to know which word is affected by the fault. Since a fault injection in a memory cell can cause the same data disturbance as a fault on the data bus or register directly, we do not differentiate particular memory cells in case of loading the values. Therefore, such array would appear in the data flow graph as a single variable from which other variables are being loaded.

Step 3: For each node, record the following sets of nodes: Gparent, Gchild, CTGchild, related key words. For each Gchild of the node, calculate Gdistance between the node and this Gchild.

DATAC iterates through each edge. The head of an edge, say v, is a Gchild of the tail of this edge, say u. The Gchildren of v are assigned as Gchildren of u. Furthermore, Gchildren of v are also assigned as Gchildren of the Gparents of u.

Step 4: Select the nodes that satisfy the output criteria, based on the information from the previous step. These are the vulnerable nodes.

Step 5: Generate subgraph for each vulnerable node.

Step 6: Generate one set of equations for each subgraph.

Output: data flow graph, subgraphs, and difference equations for each node that satisfies the output criteria.

4.4.3 Usage of DATAC

Together with the assembly code file, user has to specify the output criteria as an input. We suggest to use relatively tight values of output criteria as a preliminary test to see whether all the key words can be recovered, then loosen the criteria to find possible vulnerable nodes. We would like to point out that it is also possible to automate finding of the optimal values in order to get the desired number of vulnerable nodes. One can start with tight values and then loosen chosen parameters until this number is above some threshold.

Another user-specified parameter is the depth. One can get a good understanding about a usefulness of this parameter by observing Figs. 4.5 and 4.6 from Sect. 4.5. While both vulnerable nodes are related to the same round key, their analysis requires usage of different depths in order to get information about the respective key words. We suggest user to start with higher values of this parameter (e.g., 5) and lower them after each iteration in case there are nodes in the subgraph that are not useful in the analysis.

DATAC is capable of analyzing any microcontroller instruction set, after specifying this set as a subclass of the `Mnemonics` class and specifying instruction properties (linearity, table look-up, etc.) in a subclass of the `MnemonicRecognizer` class.

Also, the relation of the vulnerable node to the key can be adjusted in case some other analysis instead of DFA is required. This can be done in the `analyzeFaultedNodes()` method of the `FaultAnalyzer` class. The class diagram of DATAC is provided in Appendix.

4.5 Case Study

In this section, we will describe a DFA attack on PRESENT that was automatically generated by DATAC. Section 4.5.1 gives the specifications of PRESENT. Section 4.5.2 details the DFA attack we found using DATAC. This DFA attack is new and it requires 16 faults to recover the last round key.

We would like to point out that while all the fault attacks proposed on this cipher so far exploit the differential characteristics of the Sbox (e.g., [13, 21, 31, 33, 34]), our tool was able to find the vulnerable spots in the program that are implementation dependent, easily exploitable, and yet not trivial to find in the assembly code by a manual inspection.

4.5.1 PRESENT Cipher

For the case study, we have chosen a lightweight cipher PRESENT [12]. It is a symmetric block cipher, designed as a substitution-permutation network (SPN). Block length is 64 bits and key length can be either 128 bits or 80 bits (denoted as PRESENT-128 and PRESENT-80, respectively). A round function consists of three operations: `xor` of the state with the round key, followed by a substitution by 4-bit Sbox, and finally, a bitwise permutation. After 31 rounds, there is one more *addRoundKey*, used for post-whitening. The encryption process is depicted in Fig. 4.4. Because of its lightweight character, PRESENT-80 is usually used, therefore we focus on this variant in this section.

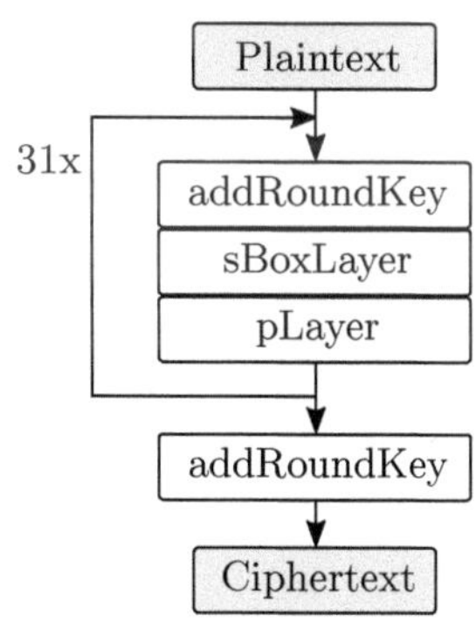

Fig. 4.4 High-level algorithmic overview of PRESENT cipher

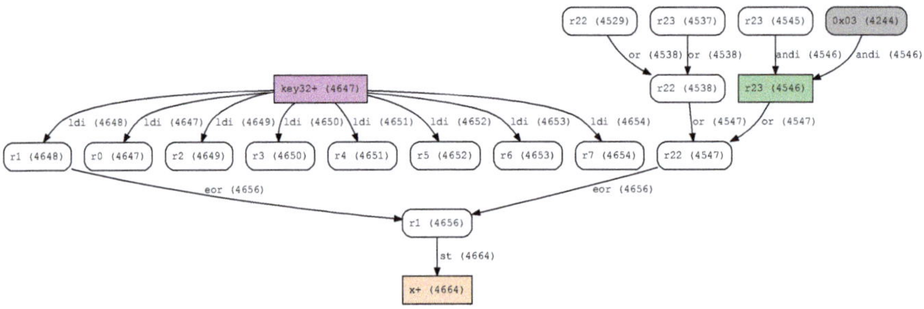

Fig. 4.5 Subgraph with depth 1 generated by DATAC from the assembly implementation of PRESENT, corresponding to vulnerable node "`r23` (4546)"

As a target, we chose a speed-optimized assembly implementation for 8-bit AVR from Verstegen and Papagiannopoulos, publicly available on GitHub.[3] We note that we did not use the key schedule for our analysis, since we were targeting the main encryption routine.

4.5.2 *Fault Analysis*

In order to get the vulnerable nodes from the cipher implementation, we have chosen our output criteria to be (`minAffectedCT`, `minDist`, `maxDist`, `maxKey`, `minKeyWords`)= (1, 1, 1, 1, 1). With this output criteria, DATAC outputs 16 vulnerable nodes, out of the total 4712 nodes. We will explain the fault attack procedure on one of these nodes—"`r23` (4546)". Subgraph for "`r23` (4546)" with depth 1 is stated in Fig. 4.5.

Equations generated for the subgraph with depth 1 from "`r23` (4546)" are as follows:

$$\text{“r22 (4538)”} = \text{“r22 (4529)”} \vee \text{“r23 (4537)”} \tag{4.7}$$

$$\text{“r23 (4546)”} = \text{“r23 (4545)”} \wedge \text{“0x03 (4244)”} \tag{4.8}$$

$$\text{“r22 (4547)”} = \text{“r22 (4538)”} \vee \text{“r23 (4546)”} \tag{4.9}$$

$$\text{“r1 (4648)”} = \texttt{key32[1]} \tag{4.10}$$

$$\text{“r1 (4656)”} = \text{“r1 (4648)”} \oplus \text{“r22 (4547)”} \tag{4.11}$$

$$\text{“x+ (4664)”} = \text{“r1 (4656)”}. \tag{4.12}$$

[3] https://github.com/kostaspap88/PRESENT_speed_implementation.

Equation (4.8) shows "r23 (4545)" $=$ $\texttt{000000}b_6b_7$ for some $b_6, b_7 \in \{0, 1\}$. Together with the other equations we get

$$\text{"x+ (4664)"} = \texttt{key32[1]} \oplus (\text{"r22 (4538)"} \vee \texttt{000000}b_6b_7). \tag{4.13}$$

Consider a bit-flip fault injection with fault mask $\Delta = \texttt{11111100}$ in "r23 (4546)" right before the execution of instruction 4547, which corresponds to Eq. (4.9), then Eq. (4.13) becomes:

$$\text{"x+ (4664)"} = \texttt{key32[1]} \oplus (\text{"r22 (4538)"} \vee \texttt{111111}b_6b_7), \tag{4.14}$$

where "x+ (4664)"$'$ denotes the faulted output. Let $\delta = \delta_0\delta_1\delta_2\delta_3\delta_4\delta_5\delta_6\delta_7 =$ "x+ (4664)"$'$ $\oplus$ "x+ (4664)" and let "r22 (4538)" $= a_0a_1a_2a_3a_4a_5a_6a_7$. Since both $\oplus$ and $\vee$ are bitwise operations, together with Eqs. (4.13) and (4.14) we have

$$\begin{aligned}\delta_0\delta_1\delta_2\delta_3\delta_4\delta_5 &= (a_0a_1a_2a_3a_4a_5 \vee \texttt{000000}) \oplus (a_0a_1a_2a_3a_4a_5 \vee \texttt{111111}) \\ &= a_0a_1a_2a_3a_4a_5 \oplus 111111 \implies a_0a_1a_2a_3a_4a_5 = \delta_0\delta_1\delta_2\delta_3\delta_4\delta_5 \oplus 111111.\end{aligned}$$

Since the value of δ is known and the value of "x+ (4664)" is also known, together with Eq. (4.13), we have

the first 6 bits of key32[1] $=$ first 6 bits of "x+ (4664)"$\oplus\delta_0\delta_1\delta_2\delta_3\delta_4\delta_5\oplus 111111$,

which gives the first 6 bits of the last round key.

With a subgraph constructed from "r22 (4538)" a similar fault analysis helps us to recover the last 2 bits of key32[1]. This is the case where we have to utilize the depth parameter of DATAC to get enough information about the faulted node. The subgraph for the node "r22 (4538)" is stated in Fig. 4.6. As can be seen in this subgraph to see the constants, three layers above the vulnerable node have to be revealed.

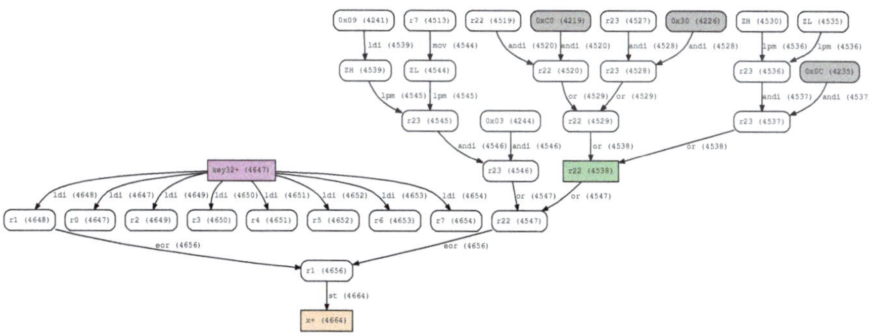

Fig. 4.6 Subgraph with depth 3 generated by DATAC from the assembly implementation of PRESENT, corresponding to vulnerable node "r22 (4538)"

Table 4.6 Assembly code of a table look-up for PRESENT implementation

#	Instruction
0	`LDI ZH 0x06`
1	`MOV ZL r0`
2	`LPM r21 Z`
3	`ANDI r21 0x0C`
4	`LDI ZH 0x07`
5	`MOV ZL r1`
6	`LPM r2 Z`
7	`ANDI r23 0x03`
8	`OR r21 r23`

The same analysis can be carried out for the remaining 14 nodes to get all the bits of the last round key.

To understand the found vulnerability, we examined the assembly code and provide an explanation below on why the cipher implementation contains the exploitable operations output by DATAC. This implementation combines the `pLayer` with the `sBoxLayer` in the form of 5 look-up tables. We will explain how this procedure works on a simple example. Table 4.6 contains the code for two table look-ups, which results into one nibble output. First, a table index is loaded into higher byte of register `Z` (instructions 0 and 4)—this decides which table will be used. Then, the intermediate state is loaded into lower byte of register `Z`—it contains two nibbles of data, therefore, we expect to get 2 bits of data back after the Sbox and the bit permutation is applied. To clear the remaining 6 bits, `ANDI` instruction is used (instructions 3 and 7). Finally, we combine the values of these two look-ups into a nibble with an `OR` instruction. The attack exploits the properties of this combined layer as well as merging of the bits of the intermediate results together into a single register.

4.6 Discussion

This part discusses various aspects of our work. In Sect. 4.6.1, we discuss the output criteria in details, by showing the DATAC analysis results of assembly implementations of SIMON, SPECK, and AES ciphers and compare the results to PRESENT. The scalability of DATAC is discussed in Sect. 4.6.2, where we show analysis results on AES with various number of rounds. Possibilities of extending DATAC to other ciphers and devices are elaborated in Sect. 4.6.3. In Sect. 4.6.4, we discuss possible countermeasures against fault attacks on software implementations and their selection. Finally, in Sect. 4.6.5 we give a remark on security verification of cryptographic software and its relation to our work.

4.6.1 Output Criteria

To test our DATAC tool and give more insight into effect of output criteria, we analyzed two more lightweight ciphers: SIMON and SPECK [10], together with AES [20] as a reference. For SIMON and SPECK, we used assembly implementations for 8-bit AVR from Luo Peng, available from GitHub[4] (the implementation is based on the proposal from the authors of SIMON and SPECK, specified in [9]). More specifically, we tested SIMON 64/128 and SPECK 64/128, both high-throughput implementations. AES implementation, written by Francesco Regazzoni [27], is available from Ecrypt II project website.[5]

Results are shown in Fig. 4.7, which plots numbers of vulnerable nodes for various output criteria values. When looking at SIMON and SPECK, because of the structure of these ciphers, where only half of the state is directly related to one round key, the number of vulnerable nodes is lower compared to PRESENT-80 and AES-128 in most of the cases.

Recall that `maxDist` specifies the maximum number of non-linear operations between the vulnerable node and the ciphertext. To attack nodes in earlier computations, the user needs to set `maxDist` to higher value. Thus, when `maxDist` is set to a bigger value, more nodes would satisfy the output criteria. This can be seen from Fig. 4.7a—when the value of `maxDist` parameter increases, we obtain more vulnerable nodes.

`maxKey` restricts the round keys to be attacked. To attack earlier round keys, the user needs bigger values of `maxKey`. Hence, when this value increases, the number of vulnerable nodes will increase, as shown in Fig. 4.7b. In this case, we set the `maxDist` to a very high number because otherwise the restriction from this parameter would be too tight such that not that many rounds can be involved in the analysis.

Plot depicted in Fig. 4.7c varies the `minKeyWords` parameter. In this case, we set the `maxKey` to 1 to restrict the analysis to the last round key, hence only nodes related to last round key are considered. The graph shows that when the value of `minKeyWords` increases, the number of vulnerable nodes decreases rapidly. For `minKeyWords` value of 1, the number of vulnerable nodes for PRESENT is relatively high. It is because the PRESENT implementation combines `sBoxLayer` and `pLayer`, where multiple operations have to be executed to merge particular bits after this layer, resulting in more nodes in the graph.

In Fig. 4.7d we vary the value of `minAffectedCT` parameter, which is the minimum number of ciphertext words related to vulnerable node. PRESENT and AES have much higher number of nodes per round compared to SPECK and SIMON, therefore the initial number of vulnerable nodes is high for these two ciphers. However, it is uncommon for a node to relate to too many ciphertext nodes, therefore there is a significant drop when the `minAffectedCT` increases from 1.

[4]https://github.com/openluopworld/simon_speck_on_avr/tree/master/AVR.

[5]https://perso.uclouvain.be/fstandae/lightweight_ciphers/source/AES.asm.

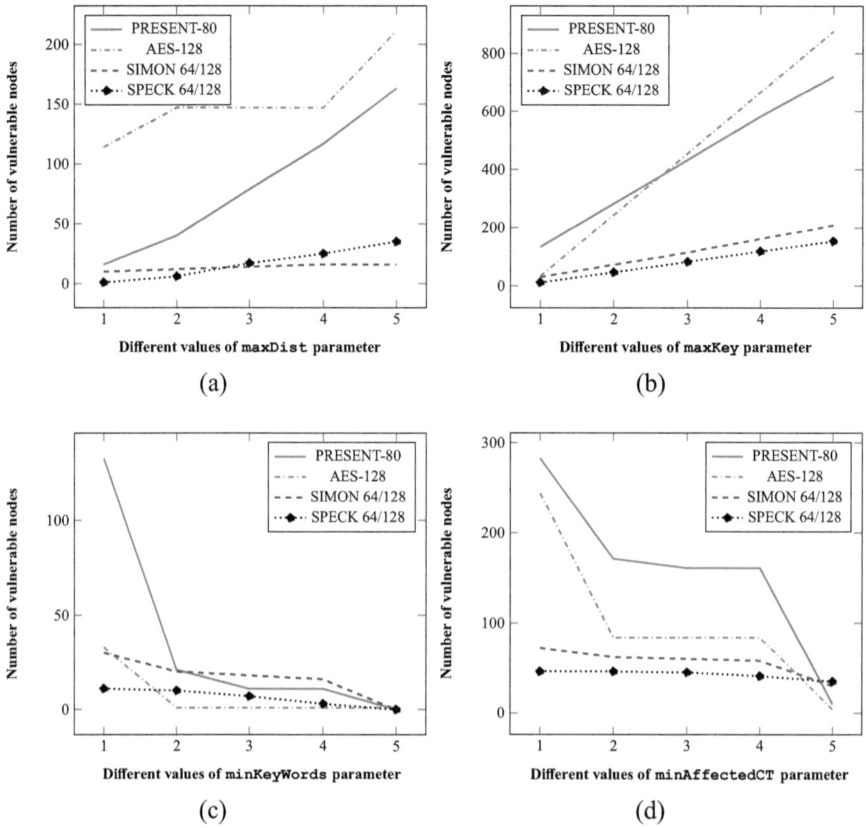

Fig. 4.7 Comparison of different output criteria on different ciphers. Plot (**a**) varies the `maxDist` parameter, while the other parameters are (`minAffectedCT, minDist, maxKey, minKeyWords`) = (1, 1, 2, 1). Plot (**b**) varies the `maxKey` parameter, while the other parameters are (`minAffectedCT, maxDist, minDist, minKeyWords`) = (1, 200, 1, 1). Plot (**c**) varies the `minKeyWords` parameter, while the other parameters are (`minAffectedCT, maxDist, minDist, maxKey`) = (1, 200, 1, 1). Plot (**d**) varies the `minAffectedCT` parameter, while the other parameters are (`maxDist, minDist, maxKey, minKeyWords`) = (200, 1, 2, 1)

The results show that DATAC is capable of finding the vulnerable spots automatically in different implementations, without additional knowledge of the internal cipher structure (only annotations from Sect. 4.4.1 are required). Also, the running times show that our approach is scalable.

4.6.1.1 Vulnerable Nodes and Exploitability

It is not guaranteed that every vulnerable node is exploitable by DFA. For example, for the PRESENT implementation in the case study (Sect. 4.5), the maximum

Table 4.7 Scalability of DATAC tested on AES with different number of rounds

# of rounds of AES	1	10	30	50
# of nodes	281	2060	6300	10,540
# of edges	415	3209	9909	16,609
# of instructions	259	1901	5801	9701
Time (s)	0.07	0.87	5.11	38.89
Average time per round (s)	0.07	0.09	0.17	0.78
Memory (MB)	3	19	170	500

distance between a node used in the last round and its CTGchild is 5. Thus for an extreme example, if we set the output criteria to be (`minAffectedCT, minDist, maxDist, maxKey, minKeyWords`)= $(0, 0, 5, 1, 0)$, we get 150 vulnerable nodes, which involve all the nodes that are related to the last round key. Or for a more reasonable example, if we set output criteria to be (`minAffectedCT, minDist, maxDist, maxKey, minKeyWords`)= $(1, 1, 5, 1, 1)$, we get 133 vulnerable nodes, which involve all the nodes that satisfy the following: affect at least 1 ciphertext node; have at least distance 1 from at least one ciphertext node; are related to the last round key. By an easy analysis, not all such nodes are exploitable. On the other hand, with output criteria (`minAffectedCT, minDist, maxDist, maxKey, minKeyWords`)= $(1, 1, 1, 1, 1)$, we get 16 nodes, which are all exploitable as explained in Sect. 4.5.

As stated in Sect. 4.4.3, we suggest the user to start with relatively tight values for output criteria, analyze the resulting vulnerable nodes. If no exploitable node can be found, then loosen the values of output criteria to get more vulnerable nodes.

4.6.2 *Scalability of DATAC*

Table 4.7 shows the time for analyzing implementations with different number of rounds of AES using DATAC. The results show that DATAC is capable of handling heavy algorithms in reasonable time, while using a laptop computer with average computing power (mobile Intel Haswell family CORE i7 processor with 8 GB RAM). For 50 rounds AES, DATAC needs less than 40 s, with memory consumption of 500 MB. Given the number of instructions in AES, this shows DATAC is capable of evaluating all current block ciphers. The same holds for countermeasures—50 rounds of AES can be considered as a $5\times$ instruction redundancy compared to standard AES-128, while the usual overheads in literature rarely go over $2\times$ the original [36].

We would like to point out that DATAC aims at DFA, where most of the attacks target the last three rounds of the cipher (e.g., [34, 44, 45]). Therefore, for a practical usage of DATAC, it would be sufficient to analyze three rounds of a target block cipher implementation.

4.6.3 Extension to Other Ciphers and Devices

As already mentioned in the introduction, DATAC can be used for analysis of current state-of-the-art block ciphers. We are not aware of any restrictions on the cipher design or structure—since the data propagation is always captured in the assembly implementation, the customized DFG will identify all the necessary information needed for DFA. We have tested DATAC on various cipher designs, including SPN with MDS matrix (AES), SPN with bit permutation (PRESENT), and Feistel with bit shifting (SIMON and SPECK). Round functions of these ciphers encompass different non-linear elements, such as 4-bit and 8-bit Sbox, modular addition, and binary AND.

Our case studies were done on 8-bit AVR assembly code. When it comes to different device architectures, identification of vulnerable nodes, construction of subgraphs, and generation of fault analysis equations do not need any change (see Fig. 4.1). Only the parsing subsystem, responsible for analyzing the assembly code and converting it to DFG, has to be adjusted. More specifically, the `AsmFileReader` class (full class diagram is stated in Fig. 4.8) has to be changed so that nodes are properly identified. For example, for AVR assembly code `AND r0, r1`, the input nodes are `r0` and `r1`, the output node is `r0`. For ARM assembly code `AND r0, r1, r2`, the input nodes are `r1` and `r2`, the output node is `r0`. We would like to point out that we aim at single fault injection. Thus we assume the attacker has an ability to perform a single fault even for pipelined architectures.

4.6.4 Countermeasures

After identifying the vulnerable nodes, a natural question for the implementer would be: How to avoid these vulnerabilities?

Over the time, several approaches to thwart fault attacks in software have emerged. They are often based on instruction duplication/triplication or parity checks [4, 40]. However, as shown in [46], these instruction-level countermeasures can be easily broken by a simple clock-glitch. Especially, the instruction skip protection provided by duplication can be simply overcome by targeting two instructions at once. Therefore, one has to look at proposals offering more guarantees, such as coding theory-based countermeasures [14, 17] or redundancy methods that spread the data across several instructions [38, 43]. Another approach is a technique called *infective countermeasure* that tries to distribute the fault evenly to the whole cipher state so that the attacker does not get any usable information about the secret key [42].

Each of these countermeasures has some assumptions, e.g., fault model, number of faults, fault precision, or number of flipped bits. Since DATAC is capable of analyzing instruction skips and bit-level faults, it can evaluate countermeasures that claim protection against these. However, if breaking the protection technique requires several faults to be injected during a single encryption, such model falls out of the current scope of DATAC. In the future, we would like to enhance the tool capability to include these more sophisticated techniques.

4.6.5 Security Verification

We would like to emphasize that while our approach introduces new methods that can be used in the direction of security verification, it cannot be used to guarantee the security of the underlying implementation directly. To an extent, DATAC works in a similar way to *Sleuth* [8], proposed to verify software implementations against side-channel attacks. *Sleuth*, as a verification tool, provides an analysis of three specific countermeasures against a user-defined leakage model—Boolean masking, arithmetic masking, and random precharging. While it is able to point out the weaknesses of the implementation, it does not guarantee the implementation is secure against all the SCA techniques. In the same way, DATAC is able to detect DFA-vulnerable parts and provide the analyst with the information on weaknesses in the code. However, when no vulnerable nodes can be found, it does not mean the code cannot be broken by utilizing different fault analysis methods, especially since some of them might not be known at the time of analysis.

4.7 Chapter Summary

We have proposed a methodology capable of finding spots vulnerable to DFA in software implementations of cryptographic algorithms. Following our approach, we have created the DATAC tool, which takes an assembly implementation and a user-specified output criteria as an input. It outputs subgraphs for vulnerable nodes in the code, together with equations that can be directly used for DFA on the cipher implementation.

We have presented a detailed DFA attack on PRESENT-80, exploiting implementation weaknesses found by DATAC. Our results show that by using DATAC, it is possible to find fault injection vulnerabilities that are not visible from observing the cipher structure and are hard to find from an assembly code that is normally hundreds to thousands lines long. To further prove its capabilities, we tested another two lightweight cipher implementations, SPECK 64/128 and SIMON 64/128, together with current NIST symmetric key standard, AES-128. All the implementations that we analyzed are publicly available programs. The only adjustment was to add several annotations (and unroll the loops in some cases). The results show that DATAC is scalable and can analyze current algorithms efficiently.

Appendix: Class Diagram of DATAC

See Fig. 4.8.

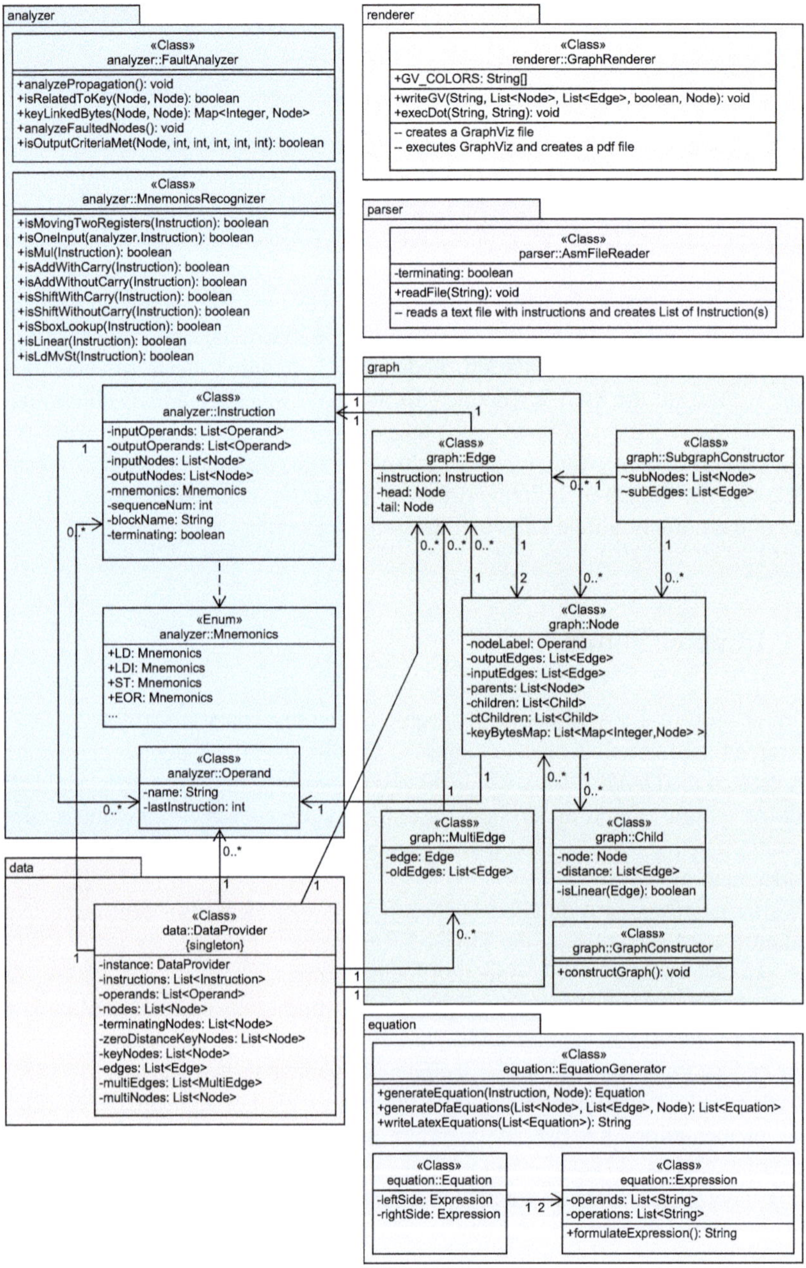

Fig. 4.8 Class diagram of DATAC. Some details were omitted, such as getters/setters and helper methods. Colors represent different packages

References

1. W. Abu-Sufah, D.J. Kuck, D.H. Lawrie, On the performance enhancement of paging systems through program analysis and transformations. IEEE Trans. Comput. **30**(5), 341–356 (1981)
2. G. Agosta, A. Barenghi, G. Pelosi, M. Scandale, Differential fault analysis for block ciphers: an automated conservative analysis, in *Proceedings of the 7th International Conference on Security of Information and Networks (SIN '14)* (ACM, New York, 2014), pp. 137:137–137:144
3. H. Bar-El, H. Choukri, D. Naccache, M. Tunstall, C. Whelan, The Sorcerer's apprentice guide to fault attacks. Proc. IEEE **94**(2), 370–382 (2006)
4. A. Barenghi, G. Pelosi, L. Breveglieri, F. Regazzoni, I. Koren, Low-cost software countermeasures against fault attacks: implementation and performances trade offs, in *Proceedings of the 5th Workshop on Embedded Security (WESS)* (ACM, New York, 2010)
5. A. Barenghi, G.M. Bertoni, L. Breveglieri, M. Pellicioli, G. Pelosi, Injection technologies for fault attacks on microprocessors, in *Fault Analysis in Cryptography* (Springer, Berlin, 2012), pp. 275–293
6. A. Barenghi, L. Breveglieri, I. Koren, D. Naccache, Fault injection attacks on cryptographic devices: theory, practice, and countermeasures. Proc. IEEE **100**(11), 3056–3076 (2012)
7. C. Barrett, "Decision procedures: an algorithmic point of view," by Daniel Kroening and Ofer Strichman, Springer-Verlag, 2008. J. Autom. Reason. **51**(4), 453–456 (2013)
8. A.G. Bayrak, F. Regazzoni, D. Novo, P. Ienne, Sleuth: automated verification of software power analysis countermeasures, in *Proceedings of the 15th International Workshop on Cryptographic Hardware and Embedded Systems (CHES 2013)*, ed. by G. Bertoni, J.-S. Coron (Springer, Berlin, 2013), pp. 293–310
9. R. Beaulieu, D. Shors, J. Smith, S. Treatman-Clark, B. Weeks, L. Wingers, The SIMON and SPECK block ciphers on AVR 8-bit microcontrollers, in *International Workshop on Lightweight Cryptography for Security and Privacy* (Springer, Cham, 2014), pp. 3–20
10. R. Beaulieu, S. Treatman-Clark, D. Shors, B. Weeks, J. Smith, L. Wingers, The SIMON and SPECK lightweight block ciphers, in *2015 52nd ACM/EDAC/IEEE Design Automation Conference (DAC)* (ACM, New York, 2015), pp. 1–6
11. E. Biham, A. Shamir, Differential fault analysis of secret key cryptosystems, in *Advances in Cryptology (CRYPTO '97)*, ed. by B.S. Kaliski Jr. Lecture Notes in Computer Science, vol. 1294 (Springer, Berlin, Heidelberg, 1997), pp. 513–525
12. A. Bogdanov, L.R. Knudsen, G. Leander, C. Paar, A. Poschmann, M.J. Robshaw, Y. Seurin, C. Vikkelsoe, PRESENT: an ultra-lightweight block cipher, in *Proceedings of the 9th International Workshop on Cryptographic Hardware and Embedded Systems (CHES '07)* (Springer, Berlin, 2007), pp. 450–466
13. J. Breier, W. He, Multiple fault attack on present with a hardware trojan implementation in FPGA, in *2015 International Workshop on Secure Internet of Things (SIoT)* (IEEE, Piscataway, 2015), pp. 58–64
14. J. Breier, X. Hou, Feeding two cats with one bowl: on designing a fault and side-channel resistant software encoding scheme, in *Cryptographers' Track at the RSA Conference* (Springer, Cham, 2017), pp. 77–94
15. J. Breier, D. Jap, C.-N. Chen, Laser profiling for the back-side fault attacks: with a practical laser skip instruction attack on AES, in *Proceedings of the 1st ACM Workshop on Cyber-Physical System Security (CPSS '15)* (ACM, New York, 2015), pp. 99–103
16. J. Breier, X. Hou, Y. Liu, Fault attacks made easy: differential fault analysis automation on assembly code. IACR Trans. Cryptogr. Hardw. Embed. Syst. **2018**(2), 96–122 (2018)
17. J. Bringer, C. Carlet, H. Chabanne, S. Guilley, H. Maghrebi, Orthogonal direct sum masking: a smartcard friendly computation paradigm in a code, with builtin protection against side-channel and fault attacks. Cryptology ePrint Archive, Report 2014/665, 2014. http://eprint.iacr.org/2014/665

18. G. Canivet, P. Maistri, R. Leveugle, J. Clédière, F. Valette, M. Renaudin, Glitch and laser fault attacks onto a secure AES implementation on a SRAM-based FPGA. J. Cryptol. **24**(2), 247–268 (2011)
19. A.E. Casavant, D.D. Gajski, D.J. Kuck, Automatic design with dependence graphs, in *17th Design Automation Conference* (ACM, New York, 1980), pp. 506–515
20. J. Daemen, V. Rijmen, *The Design of Rijndael* (Springer, New York, 2002)
21. F. De Santis, O.M. Guillen, E. Sakic, G. Sigl, Ciphertext-only fault attacks on present, in *International Workshop on Lightweight Cryptography for Security and Privacy* (Springer, Cham, 2014), pp. 85–108
22. P. Dey, R.S. Rohit, A. Adhikari, Full key recovery of acorn with a single fault. J. Inf. Secur. Appl. **29**(C), 57–64 (2016)
23. S.V. Dilip Kumar, S. Patranabis, J. Breier, D. Mukhopadhyay, S. Bhasin, A. Chattopadhyay, A. Baksi, A practical fault attack on ARX-like ciphers with a case study on Chacha20, in *2017 Workshop on Fault Diagnosis and Tolerance in Cryptography, FDTC 2017, Taipei, Taiwan, September 25, 2017* (IEEE, Piscataway, 2017), pp. 33–40
24. R. Dreesen, T. Jungeblut, M. Thies, U. Kastens, Dependence analysis of VLIW code for non-interlocked pipelines, in *Proceedings of the 8th Workshop on Optimizations for DSP and Embedded Systems (ODES-8)* (2010)
25. L. Dureuil, M.-L. Potet, P. de Choudens, C. Dumas, J. Clédière, From code review to fault injection attacks: Filling the gap using fault model inference, in *Smart Card Research and Advanced Applications: 14th International Conference, CARDIS 2015, Bochum, Germany, November 4–6, 2015. Revised Selected Papers*, ed. by N. Homma, M. Medwed (Springer, Cham, 2016), pp. 107–124
26. A. Dutta, G. Paul, Deterministic hard fault attack on Trivium, in *Advances in Information and Computer Security: 9th International Workshop on Security, IWSEC 2014, Hirosaki, Japan, August 27–29, 2014. Proceedings*, ed. by M. Yoshida, K. Mouri (Springer, Cham, 2014), pp. 134–145
27. T. Eisenbarth, Z. Gong, T. Güneysu, S. Heyse, S. Indesteege, S. Kerckhof, F. Koeune, T. Nad, T. Plos, F. Regazzoni, F.-X. Standaert, L. van Oldeneel tot Oldenzeel, Compact implementation and performance evaluation of block ciphers in ATtiny devices, in *Progress in Cryptology–AFRICACRYPT 2012: 5th International Conference on Cryptology in Africa, Ifrance, Morocco, July 10–12, 2012. Proceedings*, ed. by A. Mitrokotsa, S. Vaudenay (Springer, Berlin, Heidelberg, 2012), pp. 172–187
28. S. Endo, N. Homma, Y. Hayashi, J. Takahashi, H. Fuji, T. Aoki, A multiple-fault injection attack by adaptive timing control under black-box conditions and a countermeasure, in *Constructive Side-Channel Analysis and Secure Design*, ed. by E. Prouff (Springer, Cham, 2014), pp. 214–228
29. K. Fukushima, R. Xu, S. Kiyomoto, N. Homma, Fault injection attack on Salsa20 and ChaCha and a lightweight countermeasure, in *2017 IEEE Trustcom/BigDataSE/ICESS* (IEEE, Piscataway, 2017), pp. 1032–1037
30. M. Gay, J. Burchard, J. Horacek, A.S.M. Ekossono, T. Schubert, B. Becker, I. Polian, M. Kreuzer, Small scale AES toolbox: algebraic and propositional formulas, circuit implementations and fault equations, FCTRU, 2016, http://hdl.handle.net/2117/99210
31. N.F. Ghalaty, B. Yuce, P. Schaumont, Differential fault intensity analysis on present and led block ciphers, in *International Workshop on Constructive Side-Channel Analysis and Secure Design* (Springer, Cham, 2015), pp. 174–188
32. L. Goubet, K. Heydemann, E. Encrenaz, R. De Keulenaer, Efficient design and evaluation of countermeasures against fault attacks using formal verification, in *Smart Card Research and Advanced Applications: 14th International Conference, CARDIS 2015, Bochum, Germany, November 4–6, 2015. Revised Selected Papers*, ed. by N. Homma, M. Medwed, (Springer, Cham, 2016), pp. 177–192
33. D. Gu, J. Li, S. Li, Z. Ma, Z. Guo, J. Liu, Differential fault analysis on lightweight blockciphers with statistical cryptanalysis techniques, in *2012 Workshop on Fault Diagnosis and Tolerance in Cryptography (FDTC)* (IEEE, Piscataway, 2012), pp. 27–33

34. K. Jeong, Y. Lee, J. Sung, S. Hong, Improved differential fault analysis on present-80/128. Int. J. Comput. Math. **90**(12), 2553–2563 (2013)
35. P. Jovanovic, M. Kreuzer, I. Polian, An algebraic fault attack on the led block cipher. IACR Cryptol. ePrint Arch. **2012**, 400 (2012)
36. D. Karaklajić, J.-M. Schmidt, I. Verbauwhede, Hardware designer's guide to fault attacks, IEEE Trans. Very Large Scale Integr. VLSI Syst. **21**(12), 2295–2306 (2013)
37. P. Khanna, C. Rebeiro, A. Hazra, XFC: a framework for eXploitable fault characterization in block ciphers, in *Proceedings of the 54th Annual Design Automation Conference 2017 (DAC '17)* (ACM, New York, 2017), pp. 8:1–8:6
38. B. Lac, A. Canteaut, J. Fournier, R. Sirdey, Thwarting fault attacks using the internal redundancy countermeasure (IRC), in *International Symposium on Circuits and Systems (ISCAS) 2018* (2018)
39. N. Moro, A. Dehbaoui, K. Heydemann, B. Robisson, E. Encrenaz, Electromagnetic fault injection: towards a fault model on a 32-bit microcontroller, in *2013 Workshop on Fault Diagnosis and Tolerance in Cryptography* (IEEE, Piscataway, 2013), pp. 77–88
40. N. Moro, K. Heydemann, A. Dehbaoui, B. Robisson, E. Encrenaz, Experimental evaluation of two software countermeasures against fault attacks, in *2014 IEEE International Symposium on Hardware-Oriented Security and Trust (HOST)* (IEEE, Piscataway, 2014), pp. 112–117
41. R. Niemann, *Hardware/software co-design for data flow dominated embedded systems* (Springer, Berlin, 1998)
42. S. Patranabis, A. Chakraborty, D. Mukhopadhyay, Fault tolerant infective countermeasure for AES. J. Hardw. Syst. Secur. **1**(1), 3–17 (2017)
43. C. Patrick, B. Yuce, N.F. Ghalaty, P. Schaumont, Lightweight fault attack resistance in software using intra-instruction redundancy, in *International Conference on Selected Areas in Cryptography* (Springer, Cham, 2016), pp. 231–244
44. M. Tunstall, D. Mukhopadhyay, S. Ali, Differential fault analysis of the advanced encryption standard using a single fault, in *IFIP International Workshop on Information Security Theory and Practices* (Springer, Heidelberg, 2011), pp. 224–233
45. H. Tupsamudre, S. Bisht, D. Mukhopadhyay, Differential fault analysis on the families of SIMON and SPECK ciphers, in *2014 Workshop on Fault Diagnosis and Tolerance in Cryptography* (IEEE, Piscataway, 2014), pp. 40–48
46. B. Yuce, N.F. Ghalaty, H. Santapuri, C. Deshpande, C. Patrick, P. Schaumont, Software fault resistance is futile: effective single-glitch attacks, in *2016 Workshop on Fault Diagnosis and Tolerance in Cryptography (FDTC)* (IEEE, Piscataway, 2016), pp. 47–58

Chapter 5
An Automated Framework for Analysis and Evaluation of Algebraic Fault Attacks on Lightweight Block Ciphers

Fan Zhang, Bolin Yang, Shize Guo, Xinjie Zhao, Tao Wang, Francois-Xavier Standaert, and Dawu Gu

5.1 Introduction

This section will make a brief introduction about the background of the lightweight block ciphers, fault attacks, and the algebraic fault attacks. Besides, why this research matters compared to previous works is discussed concisely. Finally, the main work and the organization of this chapter are also summarized.

F. Zhang (✉)
College of Information Science and Electronic Engineering, Zhejiang University, Hangzhou, China

Institute of Cyberspace Research, Zhejiang University, Hangzhou, China

State Key Laboratory of Cryptology, Beijing, China
e-mail: fanzhang@zju.edu.cn

B. Yang
College of Information Science and Electronic Engineering, Zhejiang University, Hangzhou, China
e-mail: yangbolin@zju.edu.cn

S. Guo · X. Zhao
Institute of North Electronic Equipment, Beijing, China

T. Wang
Department of Information Engineering, Ordnance Engineering College, Hebei, China

F.-X. Standaert
UCL Crypto Group, Louvain-la-Neuve, Belgium
e-mail: fstandae@uclouvain.be

D. Gu
Department of Computer Science and Engineering, Shanghai Jiao Tong University, Shanghai, China
e-mail: dwgu@sjtu.edu.cn

J. Breier et al. (eds.), *Automated Methods in Cryptographic Fault Analysis*,
https://doi.org/10.1007/978-3-030-11333-9_5

5.1.1 Background

5.1.1.1 Lightweight Block Ciphers

Data security gets more demanding under resource-constrained environments. Lightweight block ciphers are a cutting-edge technology to provide an efficient and power-saving solution. Frequently used lightweight block ciphers include PRESENT, Piccolo, LED, and LBlock. Most of these ciphers can be implemented with less than 3000 gate equivalents. The complexity of traditional cryptanalysis increases exponentially with the number of rounds. From a theoretical point of view, these ciphers are deemed secure if the number of rounds is sufficiently high.

5.1.1.2 Fault Attacks and Algebraic Fault Attacks

Fault attack can retrieve secret information by actively injecting faults into the cryptosystem. Faults can be generated by changing the power supply voltage, changing the frequency of the external clock, varying the temperature, or exposing the circuits to lasers during the computation [1]. The idea was first reported on RSA-CRT by Biham et al. in 1996 [2]. Later, Biham and Shamir proposed a *differential fault analysis* (DFA) attack on the block cipher DES, which combines a fault attack with differential cryptanalysis [2]. Since then, DFA has been used to break various block ciphers. Traditionally, DFA on block ciphers is mostly conducted through manual analysis. When facing fault injection in a deep round, the fault propagation paths will overlap. The complexity of the analysis among overlapping paths increases exponentially, which is very difficult for the further manual analysis. This also happens when the number of flipped bits is large. A large size is easy for the injections, but it increases the difficulty of the analysis.

To overcome the difficulty of DFA, recent work [8] shows that algebraic cryptanalysis [7] can be combined with fault analysis. A machine solver can be used to automatically recover the secret key. This technique is referred to as *algebraic fault analysis* (AFA). AFA was proposed by Courtois et al. [8] in 2010. They showed that if 24 key bits are known and two bits in the 13th round are altered, DES can be broken with a single fault injection in 0.01 h. The full attack requires about 2^{19} h and works ten times as fast as the brute force attack. Considering their design principles, cryptographic devices with lightweight block ciphers are more vulnerable to fault attacks. Moreover, it is less complicated to solve the algebraic equations for lightweight block ciphers due to their relatively simple structure, making fault exploitations much easier. Zhao et al. [21] and Jovanovic et al. [12] extended AFA to lightweight block ciphers, such as LED. In [21], they used only one fault injection to recover the master key of LED in 1 min with a PC. In 2013, Zhang et al. [19] proposed an improved AFA, which showed that the secret key of Piccolo can be recovered with only one fault injection. Zhao et al. [22] also got a more precise estimation of the LED key search space using AFA.

5.1.2 Why This Research Matters

Previous AFA mostly focused on one particular block cipher. The motivation of this chapter is to standardize the process of AFA and provide a generic framework to analyze fault attacks on different block ciphers, especially on the lightweight ones. In practice, many situations are more challenging. Usually, faults are injected into a state before the linear layer that will bring the diffusion. For example in AES, a fault can be injected into the output of the key addition or substitution, as long as the place for the injection is before the MixColumn layer. However from the adversary's point of view, it is straightforward to ask the question: *where else can I inject a fault during the encryption?* A smart attacker may jump out of the box at a specific state and focus on a local index variable referred to as the *round counter*. Lightweight ciphers have a simple structure for efficiency reasons, but require more rounds to guarantee security. We aim to investigate how fault injections can modify the number of rounds, and how leakages could be used in algebraic fault attacks. The extended case of injecting faults both inside and outside the encryption module therefore requires a thorough study.

5.1.3 Chapter Summary

In this chapter, an algebraic fault attacks framework on block ciphers is comprehensively studied, which can be summarized below:

- A generic description of algebraic fault analysis framework on block ciphers is introduced first. AFA can be described from three levels: the target, the adversary, and the evaluator. At the target level, the design and implementation of cryptographic schemes are considered from three aspects. At the adversary level, the capability of an adversary in four parts is described. At the evaluator level, two metrics are considered: the *approximate information metric* and the *actual security metric*. These metrics can help us to answer two types of questions: for adversaries, *What faults should I inject and how?* For cipher designers and industrial engineers, *How secure is my design?* and *How secure is my implementation?*
- To verify the feasibility of the proposed framework, a comprehensive study of AFA on an ultra-lightweight block cipher called LBlock [18] is performed. LBlock and related fault attacks are described first. Then, this chapter presents how to build the algebraic equation set and provides the strategies on how to solve the equation set. Finally, three scenarios of different applications of AFA to LBlock including fault injection to encryptions, fault injection to key scheduling, and fault injection for round modification are studied in detail, respectively.

5.1.4 *Organization*

Section 5.2 proposes a generic framework for AFA including three levels: the target, the adversary, and the evaluator. Section 5.3 discusses LBlock and its resilience against fault attacks. Section 5.4 evaluates LBlock against fault injections in the encryption procedures. Section 5.5 conducts fault attacks on the key scheduling of LBlock, inspired by previous work [6, 9]. Section 5.6 investigates four cases where faults are injected to a round number or a round counter for round modification. Under each case, our best results show that the key can be recovered with only one fault injection. Section 5.7 concludes the chapter and lists some future work.

5.2 Proposed AFA Framework

In order to overcome the disadvantage of DFA, we propose a generic framework for AFA, which considers three levels: the target, the adversary and the evaluator. The framework tries to standardize the process of AFA and provides a unified solution which could evaluate different targets and adversaries.

5.2.1 *The Target Level*

The target level covers two aspects: *design* and *implementation*. The cryptographic design refers to the *cipher* which utilizes some ideal functions to solve cryptographic problems. For example, LBlock [18] is a cipher. The cryptographic implementation includes two parts: *code* and *device*. The cryptographic *device* refers to the hardware platform to implement the encryption/decryption functions of the *cipher*. For example, a smart card running the LBlock algorithm can be a target device. The cryptographic *code* comprehends the engineering effort of converting the theoretical *cipher* into practical programming code running on the *device*. For example, LBlock has size-optimized and speed-optimized versions in terms of programming code. The target level depicts how a cryptographic code is implemented on a specific device.

Possible targets include block ciphers, stream ciphers, hash functions, message authentication codes (MACs), etc. For this chapter, we carefully choose a block cipher—LBlock [18] as an example, besides, DES [15], PRESENT [3], and Twofish [16] are planned to be researched in our future work. LBlock, DES, and Twofish have a Feistel structure, while PRESENT has an SPN structure. LBlock is quite new but efficient. There is not much work known about it. DES is quite old but well known. Twofish requires some complicated operations such as modulo addition, key-dependent S-Boxes, and the Pseudo-Hadamard Transform (PHT), which makes fault attacks difficult. PRESENT is one of the most famous lightweight block ciphers with an SPN structure. Both LBlock and PRESENT are lightweight. The large number of applications to the LBlock focused in this chapter demonstrates the universality of our framework.

5.2.2 *The Adversary Level*

In our framework, an adversary's capability is characterized by four factors: *the cipher describer*, *the fault model describer*, *the fault injector*, and *the machine solver*. *The cipher describer* refers to its capability of giving the formalizations of the cryptographic codes. *The fault model describer* depicts the attributes of faults to be injected. Both describers are implemented as public interfaces and supported by *equation builders* which automatically transfer those from describers into algebraic equations. *The fault injector* is in charge of injecting the fault into the device [1]. Finally, *the machine solver* takes the equations as inputs and solves them using mathematical automata.

There are three important stages at this level: ① *Fault Injection*, ② *Equation Building*, and ③ *Equation Solving*, which are performed by the fault injector, the describer/builder, and the machine solver in Fig. 5.1, respectively. Figure 5.2 shows the details of how the adversary level works.

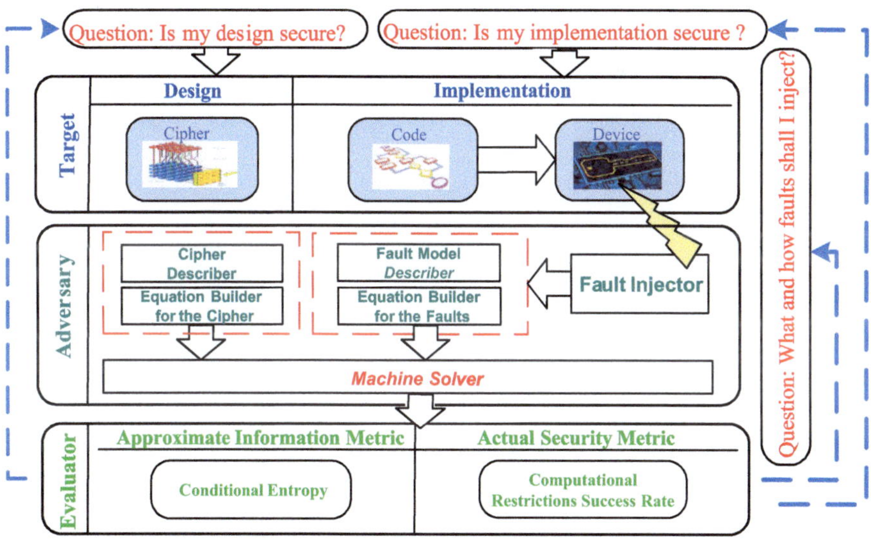

Fig. 5.1 The proposed framework for AFA

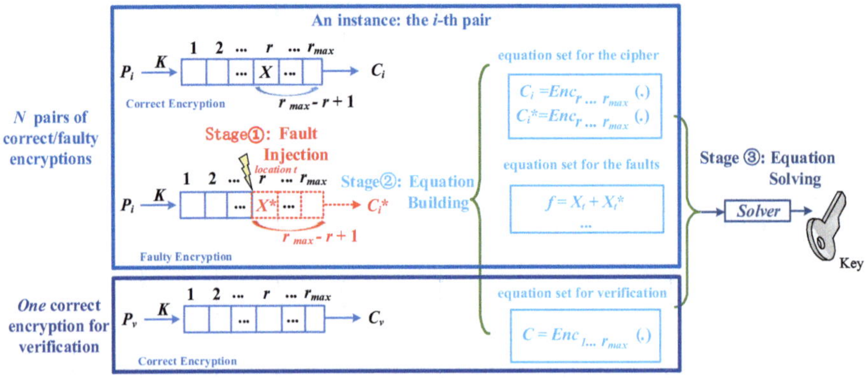

Fig. 5.2 The adversary level of AFA framework

5.2.2.1 The Fault Injector

In Fig. 5.2, Stage ①, i.e., *Fault Injection*, indicates where the fault is injected. Previous work focused on the injections in encryptions. It is possible to extend the scenarios. Inside the encryption, the fault, denoted as f, could be injected into an intermediate state for different linear or nonlinear operations, or a state for storing the total number of rounds, or an instant state called *round counter*. Outside the encryption, f might also be induced to other components such as key scheduling.

There are many practical methods to inject faults, such as optical radiation, clock glitch, critical temperature change, and electromagnetic emission. How to inject faults is discussed in [1], which is out of the scope of this chapter. We focus here on three fault models (bit-based, nibble-based, and byte-based) and conduct injections with simulations.

5.2.2.2 The Fault Model Describer and Its Equation Builder

In Stage ②, i.e., *Equations Building*, the adversary needs to build the equations for the faults.

A formal model F describes what the fault is and how it is related to the cipher. Here, X is an intermediate state. X_i is a unit of X, which determines how X is organized. f is the injected fault. w is the width of f. The fault width w is the maximal number of bits affected by one fault. The value of w might be 1, 4, and 8, which refers to *bit-based*, *nibble-based*, and *byte-based fault models*, respectively. X^* is a faulty state where faults are injected. r is the index for a specific round. r_{max} is the round number, i.e., the total number of rounds. X has different meanings. It can be a state in r-th round of the key scheduling or encryption, thus X is written as X_r^{ks} or X_r^{en}. It can also be a state referred to as the round counter, thus X can be depicted as X_{rc}^{ks} or X_{rc}^{en}, respectively.

Two terms are used throughout this chapter. *Position*, denoted by X, is the state where the fault is situated. It refers to the round in most of the cases. *Location*, denoted by t, is the place where the fault is located inside a state. As in most previous fault attacks [2], we assume that only one unit of X, i.e., X_t, is erroneous with a single fault injection in this chapter. This usually happens in fault attacks to the software implementations of block ciphers. For hardware implementations, multiple units of X might become faulty after a single fault injection. In general, λ, the size of the state, is larger than w. Thus, there are m possible locations for f where $m = \lambda/w$. m denotes the maximal value for the number of possible locations for fault injection. t can be known or unknown depending on the scenarios.

A formal fault model can be described as a tuple of five elements $\mathsf{F}(X, \lambda, w, t, f)$. Basically, it tells us that a fault with value f and width w is injected at location t with respect to a state (or position) X having λ bits.

The injected faults are also represented with algebraic equations. Different parameters such as width w and location t should be considered. The equation set for the faults can be merged with the one for the entire encryption, which can significantly reduce the computation complexity. There is an option to build an additional equation set for verification purposes. It is based on the correct full round encryption of a known plaintext P_v, resulting in a corresponding ciphertext C_v. This equation set enforces the number of solutions to be one.

5.2.2.3 The Cipher Describer and Its Equation Builder

Stage ②, i.e., *Equations Building*, specifies how to construct the equation sets for the cipher. Enc stands for the encryption function. The plaintext, the ciphertext, the master key, and the state are denoted by P, C, K, and X, respectively. On the one hand, the building work has to include *all* the major components in both encryption and key scheduling. On the other hand, it should represent *every* operation. The most difficult part is how to represent the nonlinear operations such as S-Box and modulo addition. More details can be found in [23]. In order to accelerate the solving speed, different strategies can be applied to the solver. For example, as to AFA on block ciphers with SPN structure, it is better to use the pair of correct and faulty ciphertexts to build the equations reversely [22]. In Fig. 5.2, a fault is injected to X in the r-th round. The equation set is built for the last $(r_{max} - r + 1)$ rounds.

5.2.2.4 The Machine Solver

Stage ③, i.e., *Equation Solving*, specifies how to solve the entire equation set. Many automatic tools, such as Gröbner basis-based [19] and SAT-based [18] solver, can be leveraged. The adversary could choose his own according to his skill set.

5.2.3 *The Evaluator Level*

The evaluator level takes the output of machine solvers and evaluates two metrics: the *approximate information metric* and the *actual security metric*. The evaluator answers two types of questions: for adversaries, *What faults should I inject and how?* For cipher designers and industrial engineers, *How secure is my design?* and *How secure is my implementation?*

5.2.3.1 Actual Security Metric

There are two types of security metrics. One is the computational restrictions. The possible criteria of the restrictions can be time complexity (such as the threshold for the timeout and the entire solving time, denote by t_{out} and t_{sol}, respectively), the data complexity (such as the number of fault injections, denoted by N), and the space complexity (such as the memory cost). The other is the success rate (denoted by SR) for extracting the master key. All these objective metrics are either measurable or computable, and thus they can be used to evaluate and compare different factors that may affect algebraic fault attacks.

5.2.3.2 Approximate Information Metric

The *information metric* refers to the conditional entropy of the secret key after N fault injections. It is denoted by $\phi(K)$. In traditional DFAs, the adversary cannot analyze deeper rounds due to the overlap among propagation paths. The full utilization of all faults can be easily done in our AFA framework. The remaining key search space (denoted by $2^{\phi(K)}$) is equivalent to the number of satisfiable solutions if the multiple solution output is supported. Note that if the number of fault injections is small or the fault position is deep, the number of solutions might be too big to search them all. In this case, we can feed κ guessed bits of the secret key into the equation set. As opposed to [17], our information metric actually calculates an approximation to the theoretical complexity of the key search, which can serve as an additional criterion to conduct the evaluations.

5.3 Preliminaries of AFA on LBlock

LBlock [18] is an ultra-lightweight block cipher presented by Wu et al. in CANS 2011. It uses a 32-round Feistel structure with a block size of 64 bits and a key size of 80 bits. The design of LBlock well-balances the trade-off between security and performance. On the one hand, only 1320 gate equivalents and 3955 clock cycles are required for hardware and software implementation respectively, which is outperforming many proposed lightweight block ciphers under mainstream architectures [4, 10]. The good efficiency makes it very suitable for resource-constrained environments. On the other hand, LBlock remains still secure under modern cryptanalysis. It is worth taking a comprehensive investigation to its security features. We are interested in its resilience against fault attacks.

In this section, we first provide the design of LBlock and list those cryptanalyses that are related. Then, the general representations of the equation set for both LBlock and the faults are described.

5.3.1 The Cipher of LBlock

Algorithm 1 shows the encryption of LBlock. Let $P = X_1 \| X_0$ denote the 64-bit plaintext and $C = X_{32} \| X_{33}$ denote the ciphertext, where X_i is 32 bits. $r_{max} = 32$ is the total number of rounds. rc is the round counter.

Algorithm 1: The encryption of LBlock

$r_{max} = 32$;
$P = X_1 \| X_0$;
for $rc = 0; rc < r_{max}; rc{+}{+}$ **do**
  $X_{rc+2} = F(X_{rc+1}, K_{rc+1}) + (X_{rc} <<< 8)$;
$C = X_{32} \| X_{33}$;

The round function F is a nonlinear function with a 32-bit input. It consists of Key Addition (AK), Substitution (SB), and Linear Permutation (PM). $F = PM(SB(AK(X, K_i)))$.

- AK: the leftmost 32 bits of F function input are bitwise exclusive-ORed with a round key.
- SB: the substitution uses every four bits of the exclusive-OR results as index for eight different 4-bit S-Boxes, $s_0, s_1, \ldots, s_7$.
- PM: a permutation of eight 4-bit words Z ($Z = Z_7 || Z_6 || \ldots || Z_0$) to U ($U = U_7 || U_6 || \ldots || U_0$), and it can be illustrated as the following equations:

$$\begin{aligned} &U_7 = Z_6, U_6 = Z_4, U_5 = Z_7, U_4 = Z_5, \\ &U_3 = Z_2, U_2 = Z_0, U_1 = Z_3, U_0 = Z_1 \end{aligned} \tag{5.1}$$

Algorithm 2 shows the key scheduling of LBlock. The master key is denoted by $K = k_{79} || k_{78} || \ldots || k_0$. The leftmost 32 bits of K are used as the first round key K_1. Left32(L) denotes a function to get the leftmost 32 bits of L, where L is a state register of 80 bits. l_i is one bit of L. The other round keys K_{i+1} ($i = 1, 2 \ldots 31$) are generated according to Algorithm 2.

Algorithm 2: The key scheduling of LBlock

$r_{max} = 32$;
$L = K$;
$K_1 = \text{Left32}(L)$;
for $rc = 1; rc < r_{max}; rc{+}{+}$ **do**
    $L <<< 29$;
    $[l_{79} \| l_{78} \| l_{77} \| l_{76}] = s_9[l_{79} \| l_{78} \| l_{77} \| l_{76}]$;
    $[l_{75} \| l_{74} \| l_{73} \| l_{72}] = s_8[l_{75} \| l_{74} \| l_{73} \| l_{72}]$;
    $[k_{50} \| k_{49} \| k_{48} \| k_{47} \| k_{46}] \oplus [rc]$;
    $K_{rc+1} = \text{Left32}(L)$;

LBlock has two software implementations [18]. In the size-optimized implementation, eight 4-bit S-Boxes and 4-bit word permutations are used. In the speed-optimized implementation, the eight S-Boxes and the permutations can be implemented as four 8-bit lookup tables. No additional permutation is required. In the rest of this chapter, we mainly focus on fault attacks on the software implementation of LBlock.

5.3.2 *Related Fault Attacks on LBlock*

Regarding the fault attacks, Zhao et al. [20] proposed the first fault attack on LBlock with DFA. Their best results showed that if a single-bit fault is injected into any round between the 24th and the 31st round, at least eight fault injections are required to extract the master key. In 2013, Jeong et al. [11] presented an improved DFA on LBlock under nibble-based fault model. It requires five fault injections into the left input register of the 29th round, or seven injections into the one of the 30th round. Chen and Fan [5] built eight 8-round integral distinguishers of LBlock and proposed several integral-based fault attacks. When faults are induced into the right part at the end of the 24th round under random nibble fault model, 24 fault injections are required to recover the master key of LBlock. When faults are induced into the right part at the end of the 23rd round under semi-random nibble model, 32 fault injections are required. Li et al. [14] presented the first AFA on LBlock. Under nibble-based fault model in the 27th round, two fault injections are enough to recover the 80-bit master key.

5.3.3 *Building the Equation Set for LBlock*

5.3.3.1 Representing the Overall Encryption

The equations for the overall encryption have already been listed in Algorithm 1 (Line 4) where $0 \le i \le 31$.

$$X_{i+2} = F(X_{i+1}, K_{i+1}) + (X_i <<< 8) \tag{5.2}$$

5.3.3.2 Representing AK

Suppose $X = (x_1, x_2, \ldots, x_{32})$ and $Y = (y_1, y_2, \ldots, y_{32})$ are the two 32-bit inputs to the AK of LBlock. $Z = (z_1, z_2, \ldots, z_{32})$ is the output. AK can be represented as:

$$x_i + y_i + z_i = 0, \ 1 \leq i \leq 32 \tag{5.3}$$

Note that the XOR operation in key scheduling (Line 8 in Algorithm 2) can also be represented with Eq. (5.3). rc can be considered as one input whose value is known.

5.3.3.3 Representing SB

In LBlock, eight S-Boxes $s_0, s_1, \ldots, s_7$ are used in encryption and the other two s_8, s_9 are used in key scheduling. Let the input of S-Box be $(x_1\|x_2\|x_3\|x_4)$ and the output be $(y_1\|y_2\|y_3\|y_4)$. We adopt the method in [13] and represent each S-Box with four equations. For example, the equations for s_0 can be represented as:

$$\begin{aligned} 1 + x_1x_2x_4 + x_1 + x_1x_3 + x_3x_4 + x_2x_4 + y_1 &= 0 \\ 1 + x_1x_2x_4 + x_1x_2x_3 + x_1 + x_4 + x_1x_2 + x_2x_3 + x_2x_4 + x_1x_4 + y_2 &= 0 \\ 1 + x_1 + x_2 + x_4 + x_2x_3 + x_2x_4 + y_3 &= 0 \\ x_1 + x_2 + x_3 + x_4 + x_1x_2 + y_4 &= 0 \end{aligned} \tag{5.4}$$

5.3.3.4 Representing PM

Let the input and output of PM be $(x_1\|x_2\|\ldots\|x_{32})$ and $(y_1\|y_2\|\ldots\|y_{32})$, respectively. The i-th bit of the input can be mapped to the i-th bit of the vector M using Table 5.1.

The PM function can be expressed as:

$$x_i + y_{M[i]} = 0, \ 1 \leq i \leq 32 \tag{5.5}$$

Table 5.1 Permutation sector M

i	1	2	3	4	5	6	7	8
$M[i]$	9	10	11	12	1	2	3	4
i	9	10	11	12	13	14	15	16
$M[i]$	13	14	15	16	5	6	7	8
i	17	18	19	20	21	22	23	24
$M[i]$	25	26	27	28	17	18	19	20
i	25	26	27	28	29	30	31	32
$M[i]$	29	30	31	32	21	22	23	24

5.3.3.5 Representing l-Bit Left Cyclic Shift

Suppose there is an l-bit left cyclic shift to a state register of m bits. LBlock adopts one 8-bit left cyclic shift in encryption ($l = 8, m = 32$) and one 29-bit left cyclic shift in key scheduling ($l = 29, m = 80$). Both can be written as the following equation when the input is $(x_1 \| x_2 \| \ldots \| x_m)$ and the output is $(y_1 \| y_2 \| \ldots \| y_m)$. % stands for a modulo operation.

$$x_{(l+i-1)\ \%\ m\ +\ 1} + y_i = 0,\ \ 1 \leq i \leq m \tag{5.6}$$

Using Eqs. (5.2)–(5.6), each round of key scheduling can be represented with 196 variables and 244 CNF equations, while each round of encryption can be represented with 304 variables and 496 CNF equations. The script size of one full LBlock encryption is 449 KB.

5.3.4 *Building the Equation Set for Faults*

Let X denote the λ-bit correct data unit of LBlock. $X = x_1 \| x_2 \| \ldots \| x_\lambda$. X might represent a 32-bit left state register in the encryption ($\lambda = 32$), or an 80-bit key register in the key scheduling ($\lambda = 80$). Let Y denote the faulty value of X. $Y = y_1 \| y_2 \| \ldots \| y_\lambda$. There are m possible locations for the injected faults where $m = \lambda / w$. Let Z denote the fault difference of X and Y:

$$Z = z_1 \| z_2 \| \cdots \| z_\lambda,\ z_i = x_i + y_i,\ \ 1 \leq i \leq \lambda \tag{5.7}$$

Then, Z can be divided into m parts: $Z_1 \| Z_2 \| \ldots \| Z_m$

$$Z_i = z_{w\times(i-1)+1} \| z_{w\times(i-1)+2} \| \cdots \| z_{w\times i},\ 1 \leq i \leq m \tag{5.8}$$

According to whether the adversary knows the exact location t ($1 \leq t \leq m$) or not, the algebraic equation representation of Z may have different formats.

5.3.4.1 Representing the Fault with Known t

Suppose t is known. Then, Z can be denoted as:

$$Z_i = 0,\ \ 1 \leq i \leq m, i \neq t \tag{5.9}$$

Z_t has a nonzero value of w-bits. We introduce a single bit variable u_t to represent that Z_t is faulty.

$$u_t = (1 \oplus z_{w\times(t-1)+1})(1 \oplus z_{w\times(t-1)+2}) \cdots (1 \oplus z_{w\times t}) = 0 \tag{5.10}$$

Using Eqs. (5.9) and (5.10), Z can be represented with $w+1$ variables and $w(m+1)+2$ CNF equations.

5.3.4.2 Representing the Fault with Unknown t

In practical attacks, the fault location t may be unknown. We introduce a variable u_i of m bits to represent whether Z_i is faulty or not.

$$\begin{aligned} u_i = &\left(1 \oplus z_{w\times(i-1)+1}\right)\left(1 \oplus z_{w\times(i-1)+2}\right)\cdots \\ &\left(1 \oplus z_{w\times i}\right),\ 1 \le i \le m \end{aligned} \tag{5.11}$$

If $u_i = 0$, Z_i will be the variable that is associated with the w-bit fault. Assuming that one and only one fault is injected, there should be only one zero among $u_1, u_2, \ldots, u_m$. This constraint can be represented as:

$$\begin{aligned} &(1-u_1) \vee (1-u_2) \vee \cdots \vee (1-u_m) = 1, \\ &u_i \vee u_j = 1,\ 1 \le i < j \le m \end{aligned} \tag{5.12}$$

Using Eqs. (5.11) and (5.12), Z can be represented with $m(w+2)$ variables and $m(2w+0.5m+1.5)+1$ CNF equations. These equations can also be represented when different values of w, m, and λ are given.

5.3.5 *Equation Solving Strategies*

In this chapter, we choose CryptoMiniSAT v2.9.6 as our equation solver. It has two modes. Mode A works with a pair of known plaintext P_v and corresponding ciphertext C_v, which enforces the number of solutions to be one all the time. The purpose of this mode is to get the statistics of different solving times with different numbers of fault injections, which is one type of the actual security metrics mentioned in Sect. 5.2.3.1. Mode B works without (P_v, C_v). The solver is running a multiple solution mode to estimate $\phi(K)$, the remaining entropy of the master key. It is the approximate information metric mentioned in Sect. 5.2.3.2.

Next, we describe how to use CryptoMiniSAT to roughly estimate $\phi(K)$ given N fault injections under Mode B. Let len denote the key length and κ denote the number of guessed secret bits fed into the solver. To estimate $\phi(K)$, κ is usually chosen from a larger value to a smaller one. Let $\eta(\kappa)$ denote the number of solutions for given κ. When the number of solution for one AFA is larger than 2^{18}, it is difficult for CryptoMiniSAT to find out all possible solutions within affordable time. In this case, a threshold τ for the maximal number of solutions can be set as $\tau = 2^{18}$. The detailed algorithm is shown in Algorithm 3.

Algorithm 3: Estimate $\phi(K)$ under Mode B

```
Input  : len, N, τ
Output: φ(K)
GenerateAFAES(N);
GenKnownKeySet (S_k);
for κ=len; κ >-1; κ − − do
    FeedRandKeyBits(S_k) ;
    RemoveRandKeyBit(S_k);
    RunAFAModeB();
    CalcSolutionCount(η(κ));
    if η(κ) ≥ τ and κ > 0 then
        φ(K)=κ + log2(η(κ));
        break;
    if η(κ) ≤ τ and κ==0 then
        φ(K)=log2(η(κ));
```

In Algorithm 3, `GenerateAFAES` generates the equation set of the last few rounds after the fault is injected. `GenKnownKeySet` generates the value of the known key bits into set S_k. S_k is initialized to len (80 for LBlock) bits of the secret key. `FedRandKeyBits` feeds the value of κ key bits in S_k to the equation set. `RemoveRandKeyBit` removes one random key bit from S_k. `RunAFAModeB` means using CryptoMiniSAT to solve for all possible solutions. `CalcSolutionCount` represents counting the solutions of the secret key from the output file of CryptoMiniSAT. From Algorithm 3, we can see that when $\eta(\kappa) \geq \tau$ and $\kappa > 0$, $\phi(K)$ can be roughly estimated as $\kappa + log_2\eta(\kappa)$. If $\kappa = 0$ and $\eta(\kappa) \leq \tau$, the accurate value of $\phi(K)$ is $log_2\eta(\kappa)$.

5.4 Application to LBlock: Fault Injection to Encryption (Scenario 1)

This section will make a thorough inquiry on one of the scenarios: fault injection to the process of encryption of lightweight block ciphers.

5.4.1 Fault Model

In this scenario, the fault f is injected into X_{rc+1} in LBlock encryption which is marked with a red double box in Algorithm 4. The fault model can be described as $\mathsf{F}(X_r^{\text{en}}, \lambda, w, t, f)$. More specifically, a fault is injected into the left 32-bit register of the encryption ($\lambda = 32$), whose value f is unknown. We consider three cases for the fault width ($w = 1, 4, 8$) and two cases for the location (t is known or unknown).

Algorithm 4: Fault injection to encryption

$r_{max} = 32$;
$P = X_1 \| X_0$;
for $rc = 0; rc < r_{max}; rc{+}{+}$ **do**
    $X_{rc+2} = F(\boxed{f \rightsquigarrow X_{rc+1}}, K_{rc+1}) + (X_{rc} <<< 8)$;
$C = X_{32} \| X_{33}$;

5.4.2 AFA Procedure

The attack is described in Algorithm 5. Both Mode A and Mode B of CryptoMiniSAT are considered. We define an *instance* as one run of our algorithm under one specific fault model. In one *instance*, the algorithm may be repeated many times, each of which requires one pair of plaintext and ciphertext, and one fault injection. We define N as the number of fault injections. For these N fault injections, the fault model F is the same. As for the inputs of Algorithm 5, b_t is a flag to indicate whether the location t is known or not. r is the specific round of X_r^{en}. If the fault is induced into a deeper round, the value of r is smaller. If the solver is under Mode A, the output is the solving time t_{sol} if it is successful. Otherwise, the algorithm stops at a time out t_{out}. We define a success rate SR for extracting the master key, which is the number of instances with a successful solving within t_{out} over the number of all instances. If it is under Mode B, the output is the remaining key entropy $\phi(K)$.

In Algorithm 5, $\mathbb{P}$ and $\mathbb{K}$ denote the sets for plaintexts and round keys, respectively. `KS` and `Enc` denote the key scheduling and encryption function, respectively. `RandomPT` generates one or more random plaintexts. `InjectFault` induces one fault. A function in Algorithm 5 will generate an equation set if its name is prefixed with `Gen` and suffixed with `ES`. The attack can be described as follows. The adversary A generates N pairs of plaintext/ciphertext and starts constructing equations. First, he builds the equations for key scheduling (`GenKSRdES` in Line #1). For each P_i, he will build the equation set for the correct encryption (R_r to R_{32}) using C_i (`GenEnRdES` in Line #2). For each injection, he needs to build the equation set for the faulty encryption (R_r to R_{32}) using C_i^* (in Line #3) together with the one for the fault itself (`GenFaultyES` in Line #4). Besides that, A has to generate the equation set for a full round encryption (in Line #5). The equation set based on a pair of (P_v, C_v) in Line #6 is for the verification purpose under Mode A. Finally, these combined equation sets are fed into the solver for key recovery (`RunAFA` in Line #7).

Algorithm 5: The AFA procedure of scenario 1

Input : N, r, w, b_t
Output: t_{sol} in Mode A, $\phi(K)$ in Mode B

RandomPT($\mathbb{P}$) ;
$\mathbb{K}$=KS(K, L) ;
for $rc = 1; rc < r_{max}; rc$++ **do**
    GenKSRdES(rc, K_{rc+1}) ; // #1
for $i = 0; i < N; i$++ **do**
    C_i=Enc(P_i, K) ;
    for $rc = r - 1; rc < r_{max}; rc$++ **do**
        GenEnRdES($X_{rc+1}, X_{rc}, K_{rc+1}$) ; // #2
    GenInputES(C_i) ;
    C_i^*=InjectFault(Enc(P_i, K), X_r);
    for $rc = r - 1; rc < r_{max}; rc$++ **do**
        GenEnRdES($X_{rc+1}, X_{rc}, K_{rc+1}$) ; // #3
    GenInputES(C_i^*) ;
    GenFaultyES($f = X_r + X_r^*$) ; // #4
RandomPT(P_v) ;
C_v=Enc(P_v, K) ;
for $rc = 0; rc < r_{max}; rc$++ **do**
    GenEnRdES($X_{rc+1}, X_{rc}, K_{rc+1}$) ; // #5
GenInputES(P_v, C_v) ; // #6
$(T_{sol}, \phi(K))$ = RunAFA() ; // #7

5.4.3 Case Study 1: Bit-Based Fault Model

Under the bit-based fault model, the different fault positions and known/unknown locations are considered.

5.4.3.1 The Location *t* Is Unknown

For a specific state X_r^{en}, we decrease r from 30 to 24. For each r, 100 instances of AFA are conducted under Mode A. For each instance, there are N fault injections. The statistics of different values of (r, N) are shown in Fig. 5.3. The horizontal axis is the solving time in seconds. The vertical axis is the percentage.

In Fig. 5.3, the statistics seem to follow an exponential distribution. N can be reduced when r is smaller, which means that an injection to a deeper round could reduce the number of faults that are required. When $(r, N) = (30, 10)$ or $(29, 5)$, the 80-bit master key of LBlock can be recovered within 5 min, $SR = 100\%$. If $(r, N) = (28, 3)$ or $(27, 2)$, it can be extracted in 1 min, $SR = 100\%$.

Note that the single-bit fault model in [20] can be converted to our fault model in this chapter. The work in [20] assumed that a single-bit fault is randomly injected into the internal state at the end of the $(r - 1)$-th round. This is equivalent to our

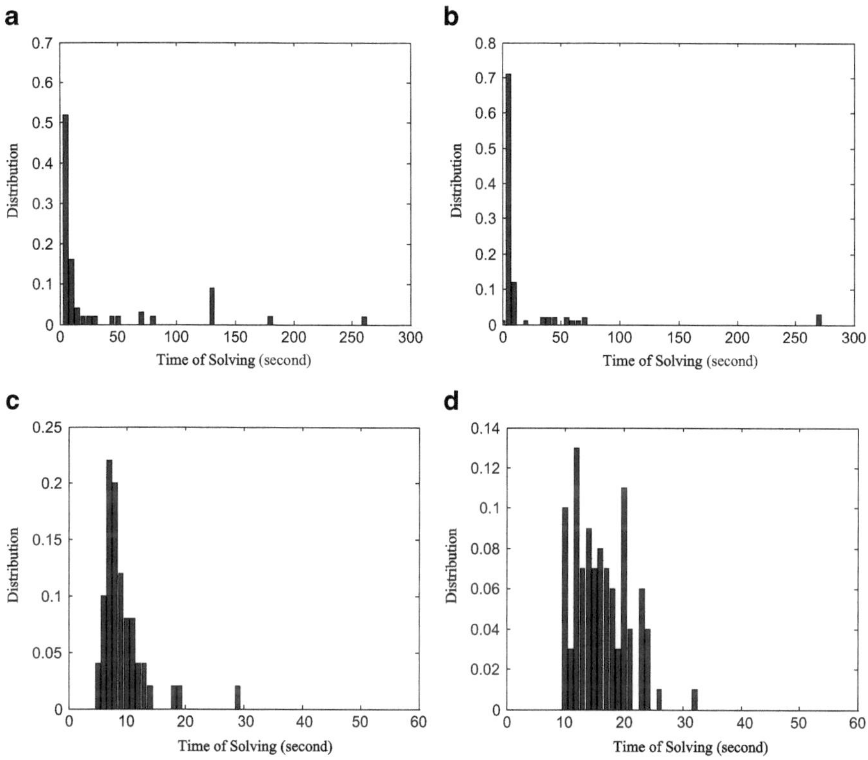

Fig. 5.3 Distribution of solving time under bit-based fault model, t is unknown (Mode A). (**a**) $r = 30$, $N = 10$, (**b**) $r = 29$, $N = 5$, (**c**) $r = 28$, $N = 3$, (**d**) $r = 27$, $N = 2$

Table 5.2 The number of injections in comparison with previous work under random bit-based fault model

r	DFA in [20]	AFA in this chapter
30	24	10
29	24	5
28	24	3
27	12	3
26	8	2
25	24	5

bit-based fault model in the r-th round, where single-bit fault is randomly injected into the left input register of the r-th round. The comparison with [20] is shown in Table 5.2. With our framework, we first verify the result in previous work for specific rounds. In contrast, the efficiency and effectiveness of our work are demonstrated when the fault is injected into the same round. Our attack requires only a few injections. For example, two injections are enough for our AFA in R_{27}, while about eight injections or more are required for most cases in [20].

Table 5.3 Bit-based fault model, t is known (Mode A)

r	N	t_{sol} (s)	Success rate
30	10	7	100%
29	4	15	100%
28	3	6	100%
27	2	10	100%
26	2	15	100%
26	1	1997	92%
25	2	221	91%
24	5	321	85%
23	50	654	65%

5.4.3.2 The Location t Is Known

If t is known, we first conduct 100 AFA instances for each r under Mode A, $t_{out} = 3600$ s. The results in Table 5.3 show that N becomes smaller compared to that for the same r when t is unknown. For example, $(r, N) = (29, 5)$ in Fig. 5.3, while $(r, N) = (29, 4)$ in Table 5.3. Moreover, fault injections in deeper rounds can help retrieve the key. For instance, when r is decreased from 28 to 27, N can also be reduced from 3 to 2.

In particular, when a single bit fault is injected into the left register in R_{26}, it might be possible to recover the master key. In this special case, we first try to solve for the secret key directly under Mode A. When $t_{out} = 2$ h, SR is only 18% for most instances, which indicates that it is difficult for CryptoMiniSAT to find the solution. To overcome this, we guess an 8-bit value of the master key and feed this value into the solver. The attack stops when the solver finds out a satisfiable solution for one key guess. Since there are 256 possible values, we can conduct at least 1, at most 256 (on average 128) guesses for each instance. When more guessed key variables are fed into the solver, CryptoMiniSAT can either find a satisfiable solution or output "unsatisfiable." The statistics of the solving time of 100 AFA instances are listed in Fig. 5.4. The master key can be recovered within 1997 s on average and $SR = 92\%$ when $t_{out} = 2$ h. *To the best of our knowledge, this is the first time LBlock has been attacked with only one injection under bit-based fault model.*

To interpret the results in Table 5.4, we evaluate $\phi(K)$ for one fault injection $(N = 1)$ under Mode B. Let ψ denote the number of the faulty nibbles in the ciphertext for one injection. Let $\bar{\psi}$ denote the average of ψ where 10,000 random instances are collected. Results of ψ, $\bar{\psi}$, and $\phi(K)$ are listed in Table 5.4.

From Table 5.4, we can see that when $28 \leq r \leq 30$, only a few nibbles in the ciphertext become faulty. Since $r = 26$, all 16 nibbles in the ciphertext are faulty. When $(r, \psi) = (26, 16)$, our best result of AFA shows that $\phi(K)$ can be reduced to 17.3. Note that in Table 5.4, $N = 1$. When $(r, N) = (30, 1)$, $\phi(K)$ can be reduced to about 70.2, which means that 9.8 key bits can be recovered with a single injection. Then, when $(r, N) = (30, 10)$, $\phi(K)$ can be reduced to a smaller value. This can also explain why CryptoMiniSAT can output the correct solution within a

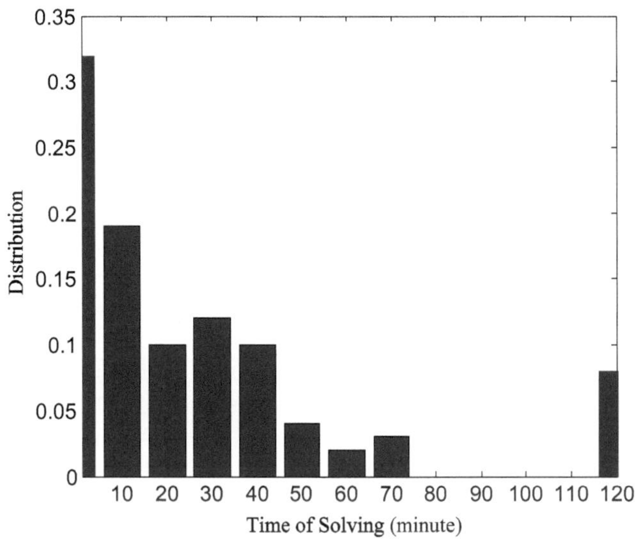

Fig. 5.4 Distribution of solving time with one injection to R_{26} under bit-based fault model (Mode A)

Table 5.4 ψ, $\bar{\psi}$, and $\phi(K)$ under bit-based fault model

r	ψ	$\bar{\psi}$	Best $\phi(K)$
30	5	5	70.2
29	8	8	61.6
28	$11 \leq \psi \leq 12$	11.81	50.4
27	$12 \leq \psi \leq 15$	14.57	32.6
26	$10 \leq \psi \leq 16$	15.21	17.3
25	$9 \leq \psi \leq 16$	14.99	18.5
24	$9 \leq \psi \leq 16$	14.99	$\leq$24
23	$8 \leq \psi \leq 16$	15.00	$\leq$40

few seconds for $(r, N) = (30, 10)$ in Table 5.3. In particular, when $(r, N) = (26, 1)$, $\phi(K)$ can be reduced to about 17.3 in Table 5.4. It explains why CryptoMiniSAT can find the secret key within affordable time under Mode A in Table 5.3.

5.4.4 *Case Study 2: Nibble-Based Fault Model*

In [5, 11], the adversary has to build the distinguishers manually and deduce the fault position. Specific algorithms must be customized for each fault position. We conduct AFA under nibble-based fault model as in [5, 11]. However, with our framework, the solver can automatically deduce the fault position and solve for the key. The workload for customizations can be saved.

Table 5.5 Comparison with previous work under nibble-based fault model

	[11]		[5]		This chapter	
r	N	$\phi(K)$	N	$\phi(K)$	N	$\phi(K)$
30	7	30	–	–	5	23
29	5	25	–	–	5	13
28	–	–	–	–	3	7.56
27	–	–	–	–	2	4.6
26	–	–	–	–	2	0.1
25	–	–	24	0	3	0
24	–	–	32	0	5	0

We extend the faults into deeper rounds and calculate $\phi(K)$ for a given amount of fault injections. The comparison with previous work [5, 11] under Mode B is shown in Table 5.5. Under the same fault model, our AFA can use less injections. For example, when $r = 30$, we can reduce N from 7 to 5 as compared to [11].

Our AFA can further reduce $\phi(K)$. In [11], $\phi(K)$ is 30 and 25 when $(r, N) = (30, 7)$ and $(29, 5)$, respectively. As for $(r, N) = (29, 5)$, $\phi(K)$ is 13 in our AFA, compared to 25 in [11]. The estimation on $\phi(K)$ in [11] might not be accurate. This is because the manual analysis may miss some faulty states in the propagation path, while the solver fully utilizes all the faults along all paths. Each faulty state can contribute his own entropy to reducing $\phi(K)$. As a result, our AFA can achieve better efficiency.

Significant enhancements are achieved when the injections are to R_{24} or R_{25}. In Table 5.5, our attack requires only 3 and 5 injections, compared with 24 injections for R_{25} and 32 injections for R_{24} in [5], respectively.

5.4.5 *Case Study 3: Byte-Based Fault Model*

Previous fault attacks on LBlock [5, 11, 20] are mainly under bit-based or nibble-based model. As aforementioned in Sect. 5.3, LBlock usually adopts the size-optimized or speed-optimized implementation on 8-bit microcontrollers. For speed-optimized implementation, the fault width is one byte. Under byte-based fault model, the fault propagation becomes more complicated.

We are concentrating on challenging AFA on LBlock under byte-based fault model. We implement the speed-optimized version of LBlock. One single byte fault is injected into the input of the big S-Box. The results under Mode B are listed in Table 5.6 where our AFA can still reduce $\phi(K)$ to a smaller value. For example, when $(r, N) = (26, 2)$, $\phi(K)$ can be further reduced to 0.

Table 5.6 AFA under random byte-based fault model

	t is unknown		t is known	
r	N	$\phi(K)$	N	$\phi(K)$
30	5	20.8	5	16
29	3	20.4	3	13.5
28	3	15.6	2	12.4
27	–	-	2	2.3
26	–	-	2	0
25	–	-	3	0

5.4.6 *Comparisons with Previous Work*

Compared with previous fault attacks on LBlock [5, 11, 14, 20], our work demonstrates that the data complexity of previous work is not optimal and AFA can work at much deeper rounds. Meanwhile, under different fault models, AFA can automatically evaluate the remaining key search space. For the first time, only one fault injection is required to recover the master key. To the best of our knowledge, this is the best result for fault attacks on LBlock in terms of data complexity.

5.5 Application to LBlock: Fault Injection to Key Scheduling (Scenario 2)

This section will inquire on another scenario: fault injection to the process of key scheduling of lightweight block ciphers.

5.5.1 *Fault Model*

In this scenario, a state register for round keys is altered due to the injected fault. The fault will be propagated to the remaining rounds of key scheduling. This case is equivalent to injecting multiple faults simultaneously into multiple rounds. The manual analysis is difficult due to the complexity. In contrast, the automatic analysis by CryptoMiniSAT is expected to be much more efficient. This is because the more equations that are generated, the more entropies are utilized in the same problem solving.

In this model $\mathsf{F}(X_r^{\text{ks}}, \lambda, w, t, f)$, a fault is injected into the left 32-bit of the 80-bit key register L in the r-th round key scheduling ($\lambda = 32$), as shown in Algorithm 6. The round key $K_r, K_{r+1}, \ldots, K_{32}$ are faulty. We consider three cases for the fault width ($w = 1, 4, 8$) and the location t is known.

Algorithm 6: Fault injection to key scheduling

$r_{max} = 32$;
$L = K$;
$K_1 = \text{Left32}(L)$;
for $rc = 1; rc < r_{max}; rc{+}{+}$ **do**
  $\boxed{f \rightsquigarrow L} <<< 29$;
  $[l_{79} \| l_{78} \| l_{77} \| l_{76}] = s_9[l_{79} \| l_{78} \| l_{77} \| l_{76}]$;
  $[l_{75} \| l_{74} \| l_{73} \| l_{72}] = s_8[l_{75} \| l_{74} \| l_{73} \| l_{72}]$;
  $[k_{50} \| k_{49} \| k_{48} \| k_{47} \| k_{46}] \oplus [rc]$;
  $K_{rc+1} = \text{Left32}(L)$;

5.5.2 AFA Procedure

The detailed procedure is depicted by Algorithm 7 where there are only two slight differences with Algorithm 5. In Line #3, the adversary has to build the equation set for the faulty key scheduling (R_r to R_{31}). In Line #4, he has to build the equation set for the faulty encryption (R_r to R_{32}) using the faulty round keys.

5.5.3 Case Study 1: Bit-Based Fault Model

First, we evaluate $\phi(K)$ for different r under bit-based fault model under Mode B. ψ and $\bar{\psi}$ are collected from 10,000 instances with single fault injection. $\phi(K)$ is calculated from 100 full AFA attacks. Results of ψ, $\bar{\psi}$, and $\phi(K)$ are listed in Table 5.7.

From Table 5.7, we can see that when $27 \le r \le 30$, only a few nibbles in the ciphertext become faulty. Since $r = 25$, all 16 nibbles in the ciphertext are faulty. When $(r, \psi) = (24, 16)$, our best result of AFA shows that $\phi(K)$ can be reduced to 16.

It is interesting to see that if $r \ge 23$, $\phi(K)$ increases when r decreases. For instance, $\phi(K)$ changes from 16 to less than 30 if the injection changes from R_{24} to R_{23} in key scheduling. Meanwhile, $\bar{\psi}$ is approximately 15 for $r = 23$, which is even slightly smaller than $\bar{\psi} = 15.09$ for $r = 24$. The reason behind is the overlap of the faults in the last few rounds.

Note that Table 5.7 can be used to determine the optimal round position for the injection and estimate the total number of injections that is required. From Table 5.7, we can deduce that it is into R_{24} where we should inject a bit-based fault in order to minimize $\phi(K)$.

In our attack, when $r = 24, 25, 26$, two single-bit fault injections ($N = 2$) can reduce $\phi(K)$ to 0 under Mode B. In particular, we also conducted AFA with only one single-bit fault injection under Mode A for $r = 24, 25$. As in Sect. 5.4.3, we guess an 8-bit value of the master key and feed this value into the solver. The results show that the full key can be recovered within 2 h where SR is about 85%.

Algorithm 7: The AFA procedure of scenario 2

```
Input : N, r, w, b_t
Output: t_sol in Mode A, φ(K) in Mode B
RandomPT(ℙ);
𝕂=KS(K, L);
for rc = 1; rc < r_max; rc++ do
    GenKSRdES(rc, K_{rc+1});                              // #1
for i = 0; i < N; i++ do
    C_i=Enc(P_i, K);
    for rc = r − 1; rc < r_max; rc++ do
        GenEnRdES(X_{rc+1}, X_{rc}, K_{rc+1});            // #2
    GenInputES(C_i);
    𝕂*=InjectFault(KS(K, L));
    for rc = r; rc < r_max; rc++ do
        GenKSRdES(rc, K*_{rc+1});                         // #3
    C*_i=Enc(P_i, 𝕂*);
    for rc = r; rc < r_max; rc++ do
        GenEnRdES(X_{rc+1}, X_{rc}, K*_{rc+1});           // #4
    GenInputES(C*_i);
    GenFaultyES(f = L + L*);                              // #5
RandomPT(P_v);
C_v=Enc(P_v, K);
for rc = 0; rc < r_max; rc++ do
    GenEnRdES(X_{rc+1}, X_{rc}, K_{rc+1});                // #6
GenInputES(P_v, C_v);                                     // #7
(T_sol, φ(K)) = RunAFA();                                 // #8
```

Table 5.7 $\bar{\psi}$ and $\phi(K)$ under bit-based fault model

r	ψ	$\bar{\psi}$	Best $\phi(K)$
30	$2 \le \psi \le 3$	2.10	74
29	$3 \le \psi \le 4$	3.40	71
28	$4 \le \psi \le 8$	6.17	62
27	$5 \le \psi \le 12$	9.52	42
26	$8 \le \psi \le 15$	12.89	28
25	$9 \le \psi \le 16$	14.83	20
24	$9 \le \psi \le 16$	15.09	16
23	$10 \le \psi \le 16$	15.00	$\le$30
22	$10 \le \psi \le 16$	15.00	$\le$42

5.5.4 Case Study 2: Nibble-Based Fault Model

The results under nibble-based fault model are shown in Table 5.8, where 10,000 random instances are collected. We can see that the fault propagation is faster under this model than under bit-based model. For example, for the same $r = 27$, $\bar{\psi} = 10.09$ in Table 5.8, while $\bar{\psi} = 9.52$ in Table 5.7. Note that $\phi(K) \le 40$ when

Table 5.8 $\bar{\psi}$ and $\phi(K)$ under nibble-based fault model

r	ψ	$\overline{\psi}$	Best $\phi(K)$
30	$2 \le \psi \le 3$	2.11	75
29	$3 \le \psi \le 5$	3.60	66
28	$3 \le \psi \le 10$	6.70	62
27	$4 \le \psi \le 13$	10.09	38
26	$5 \le \psi \le 15$	13.24	32
25	$8 \le \psi \le 16$	14.92	20
24	$10 \le \psi \le 16$	15.06	$\le$32
23	$10 \le \psi \le 16$	15.00	$\le$40
22	$10 \le \psi \le 16$	15.00	$\le$60

Table 5.9 $\phi(K)$ and ψ under byte-based fault model

r	ψ	$\overline{\psi}$	Best $\phi(K)$
30	$2 \le \psi \le 5$	3.98	75
29	$3 \le \psi \le 9$	3.65	60
28	$4 \le \psi \le 14$	10.45	39
27	$6 \le \psi \le 15$	13.32	35
26	$8 \le \psi \le 16$	14.92	28
25	$8 \le \psi \le 16$	15.02	26
24	$10 \le \psi \le 16$	15.01	16
23	$10 \le \psi \le 16$	15.00	$\le$45
22	$10 \le \psi \le 16$	15.00	$\le$65

$23 \le r \le 27$. Our best results show that two fault injections can recover the master key of LBlock when $24 \le r \le 26$. Similarly, it is in R_{25} where we should inject a nibble-based fault in order to minimize $\phi(K)$, which could be used as an empirical parameter to guide the physical injections if possible.

5.5.5 *Case Study 3: Byte-Based Fault Model*

The results under byte-based fault model are shown in Table 5.9. We can observe that the fault propagation under byte-based model is very fast. $\bar{\psi}$ is close to 4 when $r = 30$. $\phi(K) \le 40$ when $23 \le r \le 28$. Our best results show that when $24 \le r \le 28$, two fault injections can recover the full key of LBlock.

5.6 Application to LBlock: Fault Injection for Round Modification (Scenario 3)

In this section, a scenario that fault injection is for the purpose of round modification is considered comprehensively.

5.6.1 *Fault Model*

During a typical implementation, *round number*, denoted by r_{max}, is the total number of rounds to be executed. *round counter*, denoted by rc, is a variable that specifies which round it is executing. In this section, we evaluate the security of LBlock against *round modification attack* (RMA). RMA can induce the misbehavior of round operations by fault injections. A fault could be injected either into r_{max} or rc. The new values are denoted by r'_{max} or rc'. The change in the execution of LBlock can facilitate subsequent cryptanalysis.

In LBlock, there are 31 rounds in key scheduling. The round keys generated from key scheduling will be further utilized in the 32-round encryption. Two round counters are actually used for key scheduling and encryption. $r_{max} = 32$ before the fault injection. Due to page limitation, we mainly discuss the scenario when a fault is injected to modify the round during encryptions.

In this model $\mathsf{F}(X^{\text{en}}_{rc}, \lambda, w, t, f)$, a fault is injected into X^{en}_{rc} in encryption. As in previous RMA work [6, 9], we assume that both the fault value f and the fault location t are known. $\lambda = w = 8$. We consider two cases for the fault position, as shown in Algorithm 8.

Algorithm 8: Fault Injection to r_{max} or rc

$P = X_1 \| X_0$;

$\boxed{f \rightsquigarrow r_{max}} = 32$;

for $\boxed{f \rightsquigarrow rc} = 0$ *to* $r_{max} - 1$ **do**

  $X_{rc+2} = F(X_{rc+1}, K_{rc+1}) + (X_{rc} <<< 8)$;

$C = X_{32} \| X_{33}$;

5.6.2 *AFA Procedure*

The detailed procedure is depicted by Algorithm 9, where there are only three slight differences with Algorithm 5. Line #3 and Line #4 show how the adversary can build the equation set for the faulty encryption (R_r to R_{31}) if the fault is injected into r_{max} or rc (determined by b), respectively. Line #4 in Algorithm 5 is discarded here.

Algorithm 9: The AFA procedure of scenario 3

```
Input  : N, b, r, rc', r'_max
Output: t_sol in Mode A, φ(K) in Mode B
RandomPT(ℙ) ;
𝕂=KS(K, L) ;
for rc = 1; rc < r_max; rc++ do
    GenKSRdES(rc, K_{rc+1}) ;                                   // #1
for i = 0; i < N; i++ do
    C_i=Enc(P_i, 𝕂) ;
    for rc = 0; rc < r_max; rc++ do
        GenEnRdES(X_{rc+1}, X_rc, K_{rc+1}) ;                    // #2
    GenInputES(C_i) ;
    switch b do
        case 0 do
            C*_i=InjectFault(r'_max, Enc(P_i, 𝕂)) // #3 for rc = 0;
            rc < r'_max; rc++ do
                GenEnRdES(X_{rc+1}, X_rc, K_{rc+1})
            GenInputES(C*_i) ;
        case 1 do
            C*_i=InjectFault(r,rc',Enc(P_i, 𝕂)) // #4 b_tag = 0 ;
            for rc = 0; rc < r_max; rc++ do
                GenEnRdES(X_{rc+1}, X_rc, K_{rc+1});
                if rc = r − 2 and b_tag = 0 then
                    b_tag ++;
                    rc = rc';
                    if rc > 31 then
                        break;
            GenInputES(C*_i) ;
RandomPT(P_v) ;
C_v=Enc(P_v, 𝕂) ;
for rc = 0; rc < r_max; rc++ do
    GenEnRdES(X_{rc+1}, X_rc, K_{rc+1}) ;                        // #5
GenInputES(P_v,C_v) ;                                            // #6
(T_sol, φ(K)) = RunAFA() ;                                       // #7
```

5.6.3 *Case Study 1: Injecting Faults to Modify r_{max}*

In this case, a fault is injected into r_{max} in Line 2 of Algorithm 8. r_{max} could be accessed at the beginning of each instance where the fault may cause an increase or decrease in the total number of rounds.

5.6.3.1 Case 1: $r'_{max} \geq 32$

In this case, LBlock will proceed $(r'_{max} - 32)$ additional rounds after the normal encryption. These extra rounds use invalid values of round keys (for instance, four 0xcc bytes observed from physical experiments) which are known to the adversary. This case does not provide the adversary with any useful information.

5.6.3.2 Case 2: $r'_{max} < 32$

In this case, LBlock will only proceed with the first r'_{max} rounds and skip the remaining $(32 - r'_{max})$ rounds. As for the adversary, the key recovery is a reduced $(32 - r'_{max})$ round cryptanalysis. We are interested in the cases $r'_{max} = 28$ or 29 which are difficult for previous work.

We first run 100 random AFA instances under Mode A. Time statistics for $r'_{max} = 28$ and $r'_{max} = 29$ are shown in Fig. 5.5. The solver can output the correct solution within 1 min for $r'_{max} = 28$ and 2 min for $r'_{max} = 29$. Under Mode B, we also run 100 random AFA instances and calculate $\phi(K)$ for $r'_{max} = 28$ and 29. The results show that $\phi(K)$ can be reduced to $16 \sim 17$ which could be done with a brute force. This can also explain why the solver can recover the master key within a limited time under Mode A.

Meanwhile, we conduct AFA on LBlock for $r'_{max} = 3$ or 4. Under unknown plaintext scenario, since the key recovery is equivalent to analyzing the $(32 - r'_{max})$ round LBlock, it is difficult for the solver to recover the secret key within limited time. However, under known plaintext/ciphertext scenario, it can be converted into the algebraic analysis of a reduced r'_{max} round LBlock. Under Mode A, the solver can always solve the problem within 1 min.

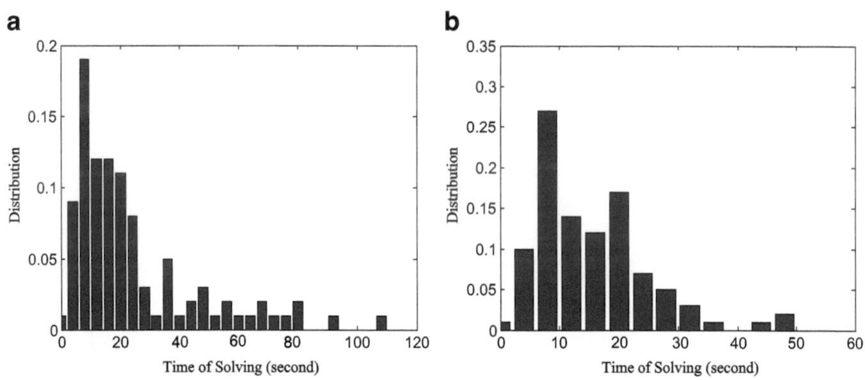

Fig. 5.5 Distribution of solving time for AFA when modifying r_{max}. (**a**) $r'_{max} = 28$ and (**b**) $r'_{max} = 29$

5.6.4 *Case Study 2: Injecting Faults to Modify* rc

In this case, a fault is injected to rc in Line 3 of Algorithm 8 at the beginning of R_r, the r-th round. Depending on the instant value of rc and the faulty value rc', various changes may occur during encryption, such as adding, reducing, or even repetitively executing several rounds.

5.6.4.1 Case 1: $rc' < rc < r_{max}$

In this case, $(rc - rc')$ intermediate encryption rounds can be repeated. We illustrate a simple case where rc $=$ 30 and $rc' = 29$. The sequence of rounds during encryption is as shown below:

$$R_1, R_2, \ldots, R_{29}, R_{30}, R_{30}, R_{31}, R_{32} \tag{5.13}$$

We can see that R_{30} is repeated twice. During the key recovery, two types of equation sets are built: those for $R_1, \ldots, R_{29}, R_{30}, R_{31}, R_{32}$ with a correct ciphertext, and those for $R_1, \ldots, R_{29}, R_{30}, R_{30}, R_{31}, R_{32}$ with a faulty ciphertext.

Under known ciphertext scenario, we conduct 100 AFA instances. The results show that under Mode A, the solver can finish in 2 min with 100% success rate; under Mode B, $\phi(K)$ can be reduced to $16 \sim 17$.

5.6.4.2 Case 2: $rc < rc' < r_{max}$

In this case, $(rc' - rc)$ intermediate encryption rounds can be skipped. We investigate the case when $rc = 29$ and $rc' = 31$. The sequence of those rounds during encryption is as shown below. R_{30} and R_{31} are skipped. The total number of rounds actually executed is 30.

$$R_1, R_2, \ldots, R_{29}, R_{32} \tag{5.14}$$

Then, the key recovery is converted into the algebraic analysis with two equation sets: one for $R_1, R_2, \ldots, R_{29}, R_{30}, R_{31}, R_{32}$ with a correct ciphertext, and one for $R_1, R_2, \ldots, R_{29}, R_{32}$ with a faulty ciphertext. Results achieved are similar to the ones in Case 1. One fault injection is enough to recover the master key of LBlock within 2 min.

5.6.4.3 Case 3: $rc < r_{max} < rc'$

In this case, $(33 - rc)$ intermediate encryption rounds can be skipped. One more example can be given for $rc = 30$ and $rc' = 35$. The sequence is $R_1, R_2, \ldots, R_{29}$. Note that R_{30}, R_{31}, R_{32} are skipped. This case is equivalent to our Case Study 1 when $r_{max} = 29$. The result is similar to Case 1. One fault injection is enough to recover the full key within 1 min.

It should be noted that AFA can also be used to recover the master key when a fault is injected to modify the round during key scheduling. Since only the number of rounds in key scheduling has been modified and that in the encryption is always 32, the equation sets to be built are slightly different from those in this section. Our experiment results show that, if a single fault could be injected into either r_{max} or rc in key scheduling of LBlock, $\phi(K)$ can also be reduced to $16 \sim 17$.

5.7 Conclusion and Future Work

This chapter proposes a generic framework for algebraic fault analysis on block ciphers. The framework could be used to analyze the efficiency of different fault attacks, to compare different scenarios, and to evaluate the factors that may determine the solving time and the success rate.

First, we highlight a conceptual overview of the framework. The important levels and roles are clarified, and four functional parts and three workflow stages are depicted. Then, we select LBlock as a start point to illustrate how our framework can work on a block cipher, especially a lightweight one. To demonstrate the flexibility of the framework, three scenarios are exploited, which include injecting a fault to encryption, to key scheduling, or to modify the rounds.

Future work can be derived in different directions. One possible area is to further improve the efficiency of the framework. The current version still meets some difficulties in AFA on deep round of extremely complicated ciphers. With an enhanced solver, more compact equation builders, and other advanced techniques, the AFA framework might work with more rounds of those ciphers. In addition, the framework proposed in this chapter can be extended to other well-known lightweight block ciphers, such as DES, PRESENT, Twofish, and so on.

Acknowledgements This work was supported in part by the National Natural Science Foundation of China under the grants 61472357 and 61571063, the Zhejiang University Fundamental Research Funds for the Central Universities under the grant 2018QNA5005, the Open Fund of State Key Laboratory of Cryptology under the grand MMKFKT201805, the Alibaba-Zhejiang University Joint Institute of Frontier Technologies, the Major Scientific Research Project of Zhejiang Lab under the grant 2018FD0ZX01, and the European Commission through the ERC project 280141 (acronym CRASH).

References

1. H. Bar-El, H. Choukri, D. Naccache, M. Tunstall, C. Whelan, The Sorcerer's apprentice guide to fault attacks. Proc. IEEE **94**(2), 370–382 (2006)
2. E. Biham, A. Shamir, Differential fault analysis of secret key cryptosystems, in B.S. Kaliski Jr. (eds.), *Advances in Cryptology - CRYPTO '97*. Lecture Notes in Computer Science, vol. 1294 (Springer, Berlin, 1997), pp. 513–525

3. A. Bogdanov, L.R. Knudsen, G. Leander, C. Paar, A. Poschmann, M.J. Robshaw, Y. Seurin, C. Vikkelsoe, PRESENT: an ultra-lightweight block cipher, in *Proceedings of the 9th International Workshop on Cryptographic Hardware and Embedded Systems, CHES '07* (Springer, Berlin, 2007), pp. 450–466
4. M. Cazorla, K. Marquet, M. Minier, Survey and benchmark of lightweight block ciphers for wireless sensor networks, in *SECRYPT* (2013), pp. 543–548
5. H. Chen, L. Fan, Integral based fault attack on LBlock, in *ICISC* (2014), pp. 227–240
6. H. Choukri, M. Tunstall, Round reduction using faults, in *FDTC* (2015), pp. 13–24
7. N.T. Courtois, J. Pieprzyk, Cryptanalysis of block ciphers with overdefined systems of equations, in *Proceedings of the 8th International Conference on the Theory and Application of Cryptology and Information Security (ASIACRYPT)*, Queenstown, December 2002 (Springer, Berlin, 2002), pp. 267–287
8. N.T. Courtois, K. Jackson, D. Ware, Fault-algebraic attacks on inner rounds of des, in *e-Smart'10 Proceedings: The Future of Digital Security Technologies* (Strategies Telecom and Multimedia, Montreuil, 2010)
9. A. Dehbaoui, A.-P. Mirbaha, N. Moro, J.-M. Dutertre, A. Tria, Electromagnetic glitch on the AES round counter, in *International Workshop on Constructive Side-Channel Analysis and Secure Design* (Springer, Berlin, 2013), pp. 17–31
10. D. Dinu, Y. Le Corre, D. Khovratovich, L. Perrin, J. Großschädl, A. Biryukov, Triathlon of lightweight block ciphers for the internet of things. J. Cryptogr. Eng. **209**, 1–20 (2015)
11. K. Jeong, C. Lee, J.I. Lim, Improved differential fault analysis on lightweight block cipher LBlock for wireless sensor networks. EURASIP J. Wirel. Commun. Netw. **2013**(151), 1–9 (2013)
12. P. Jovanovic, M. Kreuzer, I. Polian, An algebraic fault attack on the LED block cipher. IACR Cryptol. ePrint Archive **2012**, 400 (2012)
13. L. Knudsen, C. Miolane, Counting equations in algebraic attacks on block ciphers. Int. J. Inf. Secur. **9**(2), 127–135 (2010)
14. W. Li, J. Zhao, X. Zhao, J. Zhu, Algebraic fault analysis on LBlock under nibble-based fault model, in *IMCCC* (2013), pp. 1525–1529
15. NIST, Data encryption standard. Federal Information Processing Standards Publications, May 1977
16. B. Schneier, J. Kelsey, D. Whiting, D. Wagner, C. Hall, N. Ferguson, Twofish: A 128-bit block cipher. NIST AES Proposal, 15 June 1998
17. F.-X. Standaert, T. Malkin, M. Yung, A unified framework for the analysis of side-channel key recovery attacks, in *EUROCRYPT* (2009), pp. 443–461
18. W. Wu, L. Zhang, LBlock: a lightweight block cipher, in *ACNS* (2011), pp. 327–344
19. F. Zhang, X. Zhao, S. Guo, T. Wang, Z. Shi, Improved algebraic fault analysis: a case study on piccolo and applications to other lightweight block ciphers, in *Proceedings of the 4th International Workshop Constructive Side-Channel Analysis Secure Design (COSADE)*, Paris, March 2013 (Springer, Berlin, 2013), pp. 62–79
20. L. Zhao, T. Nishide, K. Sakurai, Differential fault analysis of full LBlock, in *COSADE* (2012), pp. 135–150
21. X. Zhao, S. Guo, F. Zhang, T. Wang, Z. Shi, K. Ji, Algebraic differential fault attacks on LED using a single fault injection, in *IACR Cryptology ePrint Archive* (2012)
22. X. Zhao, S. Guo, F. Zhang, Z. Shi, C. Ma, T. Wang, Improving and evaluating differential fault analysis on LED with algebraic techniques, in *Proceedings of the 10th IEEE Workshop on Fault Diagnosis and Tolerance in Cryptography (FDTC)*, Santa Barbara, August 2013 (IEEE, Los Alamitos, 2013), pp. 41–51
23. X. Zhao, S. Guo, F. Zhang, T. Wang, Z. Shi, D. Gu, C. Ma, Algebraic fault analysis on GOST for key recovery and reverse engineering, in *FDTC* (2014), pp. 29–39

Chapter 6
Automatic Construction of Fault Attacks on Cryptographic Hardware Implementations

Ilia Polian, Mael Gay, Tobias Paxian, Matthias Sauer, and Bernd Becker

6.1 Introduction

Security features of today's electronic systems are often realized by dedicated cryptographic circuits. For example, the exponential growth in network traffic, together with a stronger demand for encrypted and/or authenticated transmission, may soon render software-only solution insufficient. Moreover, cryptographic modules are increasingly integrated into systems-on-a-chip (SoCs) used in a variety of critical applications, from driving-assistance to mobile payment. System security is determined by its weakest link, and therefore different threats must be accounted for during system design. Historically, such threats included the use of insecure (weak, flawed, or outdated) cryptographic algorithms, communication of data through inadequately protected channels, their storage in insufficiently secure memories, software-level manipulations (e.g., buffer overflow), and social engineering (e.g., extracting passwords from conversations in social networks, or phishing attacks).

More recently, the system hardware components shifted into the focus of attackers and, consequently, system designers and security architects. Fault-injection attacks [3], also called fault attacks, are one prominent class of hardware-related physical attacks. This chapter focuses on fault-injection attacks on cryptographic circuits. These attacks consist in imposing a physical disturbance (fault) during the execution of the circuit, observing the circuit's fault-affected behavior, and

I. Polian (✉) · M. Gay
University of Stuttgart, Stuttgart, Germany
e-mail: ilia.polian@informatik.uni-stuttgart.de; mael.gay@informatik.uni-stuttgart.de

T. Paxian · M. Sauer · B. Becker
University of Freiburg, Freiburg, Germany
e-mail: paxiant@informatik.uni-freiburg.de; sauerm@informatik.uni-freiburg.de; becker@informatik.uni-freiburg.de

J. Breier et al. (eds.), *Automated Methods in Cryptographic Fault Analysis*,
https://doi.org/10.1007/978-3-030-11333-9_6

exploiting this information for deducing sensitive information, such as the secret key or the plaintext.

It has to be noted that fault attacks are neither restricted to cryptographic functions nor to pure hardware implementations. One attack outside the cryptographic domain is manipulating a status bit in a microprocessor's memory-management unit. If successful, the attacker would gain access to data in the protected regions of the system memory. Another example not (necessarily) involving cryptographic functions would be manipulating the register that stores the outcome of a password check, granting the adversary authorizations even when the check had failed. With regard to software implementations, a cryptographic algorithm running on a microprocessor stores data in the processor's registers, loads them from and stores them to caches and memories, and uses the processor's control path for decisions. All these components are as vulnerable to fault attacks as application-specific circuitry.

The significance and relevance of fault-injection attacks are strengthened by the current technology trends. On the one hand, more applications become cyber-physical and are now found in an environment where their users have physical access to them, making it easier for attackers to apply physical disturbances. For example, electronic payment functions are now found in mobile phones, in cars, but also in (publicly available) infrastructure , e.g., integrated parking systems. On the other hand, the capabilities of fault-injection equipment improve and its cost decreases. A decade ago, overview articles on fault attacks focused on low-cost injection techniques such as underpowering or overclocking [2]. Today, public research institutes have access to dual-beam laser systems capable of injecting a fault in an encryption circuit while simultaneously disabling that circuit's error-detecting features [23].

After the seminal paper by Boneh et al. [5] has pointed out the vulnerability of cryptographic procedures to fault attacks, a large number of successful attacks have been reported [1, 6, 11, 13, 14, 26, 27]. Most of these attacks targeted a specific cipher and made use of its (often intricate) cryptographic properties. Given a new cipher, previous attacks were not easily transferable, and the construction of a new attack required nontrivial contributions by a cryptographer. In the last few years, there is a trend towards more generic approaches that aim at automatic construction of attacks for a broader class of cryptographic functions (see Chaps. 2 and 5 of this book). These approaches started with a functional description of the attacked algorithm and created a set of formulae or equations to represent the effect of the fault injection and its propagation during the execution of the algorithm. The equations incorporated information of interest for the adversary (typically, bits of the secret key), and solving the equations yielded these bits or restricted the set of key candidates such that brute-force search became practical.

This chapter focuses on the question whether it is possible to automatically construct a fault attack on a given cryptographic circuit, specified on gate or register transfer level (RTL) in a hardware description language such as Verilog, and how this construction works. In contrast to other frameworks for automatic or semiautomatic fault attack construction [15, 21, 29], the methodology does not need an algebraic description of the cipher's functionality (this information is derived directly from the circuit's description). It has the following benefits:

- The task to create a system of algebraic or Boolean equations for describing the cipher under attack, as required by previous approaches, and to verify its correctness, is feasible but difficult in practice. Automatic construction of an attack directly from a circuit description is far easier, since a verified circuit implementation is readily available.
- The descriptions are more detailed than purely functional models used in traditional algebraic fault analysis. Not only the cipher's state bits but also all internal signal lines are explicitly represented, and faults can be defined on these lines. This is useful if the attacker has the capability to target individual gates (e.g., by high-resolution optical or electromagnetic techniques), and the effect of the resulting fault cannot be easily mapped to failing state bits.
- The technique benefits from advances in both SAT solving (this is the case for other algebraic attacks as well) and in SAT-based modeling techniques. For example, recent research demonstrated efficient incorporation of a detailed circuit timing into SAT formulae [22], and such extensions can be integrated with the fault attack construction.
- The approach is easy to (re-)apply when a small modification to the circuit is done (e.g., an SBox is replaced by one with a lower power consumption or better resistance against side-channel analysis). Such seemingly trivial modification may give rise to fault attacks which did not exist in the original design. The same applies to modifications that aim at reducing the cipher's cost, such as reducing its state size or number of rounds. If the modification is done in the hardware description, the procedure simply needs to be rerun, and no modification of the algebraic model is required.
- It is easily possible to integrate certain types of fault attack countermeasures. For example, if certain locations in the circuit are connected to on-chip attack detectors (e.g., power droop sensors [4]), it is possible to suppress fault scenarios that affect these locations, because the sensor will detect the attack and counteract by, for instance, rekeying [17].

The procedure described in this chapter roughly follows the principles proposed in [7] and incorporates later extensions, e.g., the support of multiple fault injections. In contrast to [7], this chapter focuses on how to use a circuit implementation together with fault injections to finally gain a formula readable (and hopefully solvable) by a SAT solver. It provides underlying concepts such as the Tseitin transform to obtain a suitable Boolean formula out of the circuit and the time-frame expansion of sequential circuits. We avoid discussion of performance of the presented approach in comparison to its alternatives, because: (1) performance is a moving target that depends on solver technology and algorithmic details of the approach not in the focus of this chapter and (2) its main distinguished feature is the hardware-oriented modeling rather than speed. The procedure in [7] was able to break the LED-64 block cipher using a single fault, and its extension to multiple fault injections (described in Sect. 6.3.4.3) scales to AES-128.

The remainder of this chapter is organized as follows. The next section explains how to map combinational and sequential circuits to Boolean formulae using the Tseitin transform. This approach is being used in formal verification [12] and SAT-based test pattern generation [8] and is not restricted to cryptographic circuits. Section 6.3 starts with the generalization of the modeling introduced so far to cryptographic circuits which process secret information. Models of injected faults are discussed next, before several variants of attack construction are explained. Section 6.4 provides an overview of alternative approaches to automatic (or semiautomatic) fault attack construction; none of them is explicitly optimized for circuit implementations. Section 6.5 summarizes this chapter.

6.2 Modeling of Circuits by Boolean Formulae

In this section, we will first describe the mapping of combinational circuits to Boolean formulae that can be processed by a SAT solver using the Tseitin transform. We will then extend this approach to sequential circuits via time-frame expansion. These steps are necessary to construct the model of the cryptographic circuit under attack, which will be described in the next section.

6.2.1 Combinational Circuits

A *combinational circuit* is a directed, acyclic graph with *logic gates* (inverters, AND, XOR gates, etc.) as nodes and signal lines between the gates as edges. Every input of a gate is either connected to an output of a different gate or is a *primary input* of the circuit. Similarly, an output of a gate can drive a gate input and/or be a *primary output* of the circuit. It is possible to describe combinational circuits hierarchically. For example, a combinational circuit realization of AES-128[1] can be broken into ten (almost) identical submodules corresponding to ten encryption rounds and a further submodule for the key schedule (see Fig. 6.1). The round submodule can be further divided into four parts according to the four AES steps; this spans a hierarchy AES-128 $\rightarrow$ round $\rightarrow$ step $\rightarrow$ logic gate.

All submodules at any hierarchy depth, including individual logic gates, are combinational circuits themselves. A combinational circuit C with n inputs and m outputs implements a *Boolean function* $f_C : \mathbb{B}^n \rightarrow \mathbb{B}^m$. This function can be represented by its function table (with 2^n rows and $n + m$ columns), a compressed data structure such as a Boolean Decision Diagram, or by a Boolean expression [12]. The Boolean expression of a function derived from a circuit may or may not follow the circuit's structure; for example, the Boolean function of a circuit consisting of an

[1] 128-bit version of the Advanced Encryption Standard [9].

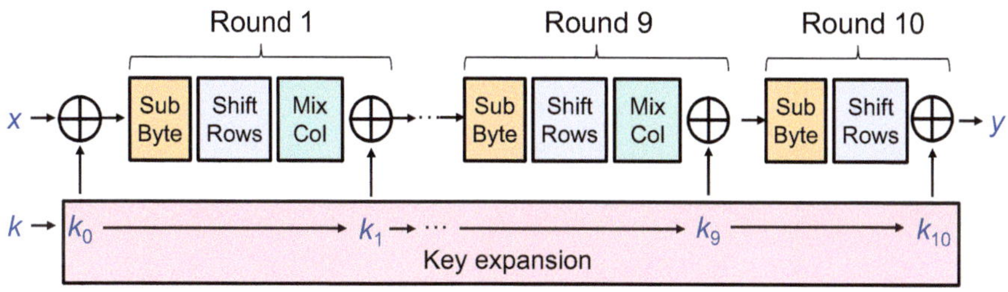

Fig. 6.1 Diagram of combinational circuit realization of AES-128

XOR gate with inputs a, b and output c can be written as $c = a \oplus b$ or, equivalently, as $c = (a \vee \neg b) \wedge (\neg a \vee b)$.

A Boolean expression is in *conjunctive normal form* (CNF, also known as POS or product-of-sums) if it is a conjunction (AND) of *clauses*, where clauses are disjunctions (OR) of *literals* and literals are variables or negated variables. The *Boolean satisfiability problem* (SAT) aims, given a Boolean formula $f : \mathbb{B}^n \to \mathbb{B}$ in CNF, at either finding an assignment of logic-0 and logic-1 values to its n variables for which f assumes value 1 or to prove that no such assignment exists. SAT is an NP-complete problem [12], and yet modern SAT solvers work efficiently for many circuit-related problem instances.

Given an n-input, m-output combinational circuit C, its function $f_C : \mathbb{B}^n \to \mathbb{B}^m$ can be equivalently written as m functions $f_{C,1}, \ldots, f_{C,m} : \mathbb{B}^n \to \mathbb{B}$ (each representing one primary output of the circuit). It is then possible to formulate the SAT problem for each of the function $f_{C,j}$ (i.e., either to find an input combination $i \in \mathbb{B}^n$ with $f_{C,j}(i) = 1$ or to prove that no such input combination exists. However, the derivation of Boolean expressions in CNF from the circuits is, while always possible, in general inefficient. For example, the n-input, 1-output circuit which calculates the XOR of all its n inputs has only exponential-length CNFs. A remedy for this problem is to use the *Tseitin transform* [25], which guarantees linear CNF length at the expense of additional variables in the formula. This transform is based on the notion of a characteristic function.

Let G be a logic gate with n_G inputs $i_1, \ldots, i_{n_G}$ and one output o (the extension to multi-output gates is straightforward). Let f_G be the Boolean function of G: $f_G(i_1, \ldots, i_{n_G}) = o$ iff G with values $i_1, \ldots, i_{n_G}$ applied to its inputs produces o at its output. The *characteristic function* of G is $\chi_G : \mathbb{B}^{n_G+1} \to \mathbb{B}$ with

$$\chi_G(i_1, \ldots, i_{n_G}, o) = \begin{cases} 1, \text{ if } f_G(i_1, \ldots, i_{n_G}) = o \\ 0, \text{ otherwise} \end{cases}$$

Table 6.1 contains characteristic functions of several basic logic gates. For example, it holds $\chi_{\text{AND}}(1, 1, 1) = 1$ because $\text{AND}(1, 1) = 1$; $\chi_{\text{AND}}(1, 0, 0) = 1$ because $\text{AND}(1, 0) = 0$; but $\chi_{\text{AND}}(1, 0, 1) = 0$ because $\text{AND}(1, 0) \neq 1$.

Table 6.1 Characteristic functions of basic logic gates

Logic gate	Characteristic function
$o = \text{INV}(i_1) = \neg i_1$	$\chi_{\text{INV}}(i_1, o) = (i_1 \vee \neg o) \wedge (\neg i_1 \vee o)$
$o = \text{AND}(i_1, i_2) = i_1 \wedge i_2$	$\chi_{\text{AND}}(i_1, i_2, o) = (i_1 \vee \neg o) \wedge (i_2 \vee \neg o) \wedge (\neg i_1 \vee \neg i_2 \vee o)$
$o = \text{OR}(i_1, i_2) = i_1 \vee i_2$	$\chi_{\text{OR}}(i_1, i_2, o) = (\neg i_1 \vee o) \wedge (\neg i_2 \vee o) \wedge (i_1 \vee i_2 \vee \neg o)$
$o = \text{XOR}(i_1, i_2) = i_1 \oplus i_2$	$\chi_{\text{XOR}}(i_1, i_2, o) = (i_1 \vee i_2 \vee \neg o) \wedge (i_1 \vee \neg i_2 \vee o) \wedge (\neg i_1 \vee i_2 \vee o) \wedge(\neg i_1 \vee \neg i_2 \vee \neg o)$

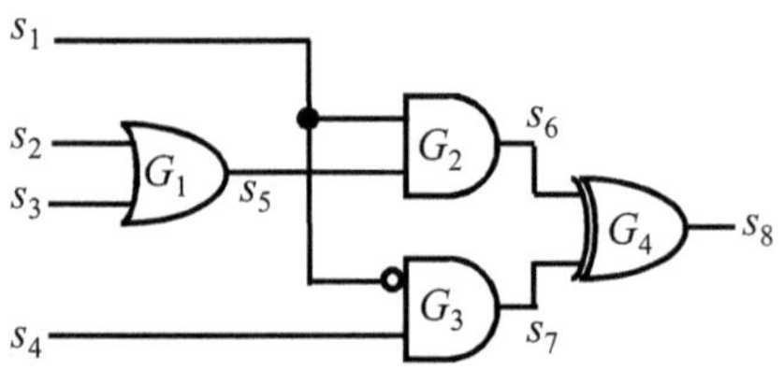

$$
\begin{aligned}
&\chi_{\text{OR}}(s_2, s_3, s_5) \wedge \chi_{\text{AND}}(s_1, s_5, s_6) \wedge \\
&\chi_{\text{AND}}(\neg s_1, s_4, s_7) \wedge \chi_{\text{XOR}}(s_6, s_7, s_8) = \\
&(\neg s_2 \vee s_5) \wedge (\neg s_3 \vee s_5) \wedge (s_2 \vee s_3 \vee \neg s_5) \wedge \\
&(s_1 \vee \neg s_6) \wedge (s_5 \vee \neg s_6) \wedge (\neg s_1 \vee \neg s_5 \vee s_6) \wedge \\
&(\neg s_1 \vee \neg s_7) \wedge (s_4 \vee \neg s_7) \wedge (s_1 \vee \neg s_4 \vee s_7) \wedge \\
&(s_6 \vee s_7 \vee \neg s_8) \wedge (s_6 \vee \neg s_7 \vee s_8) \wedge \\
&(\neg s_6 \vee s_7 \vee s_8) \wedge (\neg s_6 \vee \neg s_7 \vee \neg s_8)
\end{aligned}
$$

Fig. 6.2 Illustration of Tseitin transform

The *Tseitin transform* of a combinational circuit C is a Boolean formula, in CNF, which has one variable for each signal line of the circuit and is the conjunction of characteristic functions of all the gates which are part of C. Figure 6.2 illustrates the Tseitin transform for a circuit with four gates and eight signals, including four inputs, three internal signals, and one output. Note that since the characteristic function of each gate is in CNF, their conjunction is in CNF as well. Moreover, note that the negated input of gate G_3 is incorporated into the formula without adding an inverter. Every gate of a certain type is represented by a fixed number of clauses (3 for a two-input AND/OR gate, and 4 for a two-input XOR gate) and literals (7 for a two-input AND/OR gate, and 12 for a two-input XOR gate), and therefore the length of the formula is linear in the size of the circuit. Note that the function realized by the circuit has four variables $s_1, \ldots, s_4$, whereas the Tseitin transform is defined over eight variables.

Once the Tseitin transform is available, it can be used for solving justification problems, i.e., finding consistent assignments to the circuit's signal lines. It is possible to *force* any signal line s_i to the logical value of 1 by adding the clause (s_i) to the CNF (to force s_i to logic-0, clause $(\neg s_j)$ has to be added). For example, one may be interested whether the lines in the circuit from Fig. 6.2 can assume the values $s_5 = 1$ and $s_6 = 0$ at the same time. This task is solved by adding clauses (s_5) and $(\neg s_6)$ to the formula and handing it over to a SAT solver. If it finds a solution, the assignments to $s_1, \ldots, s_4$ give the input vector which justifies the desired values, and further variables correspond to the values in the circuit under this input. It is also possible to formulate more complex conditions, e.g., that at least two out of three signals s_1, s_6, and s_8 equal 1, by representing such conditions by one or multiple clauses and adding these clauses to the CNF. This feature will be essential for modeling fault injections and deducing values of internal signals that are consistent with the observed fault-affected behavior.

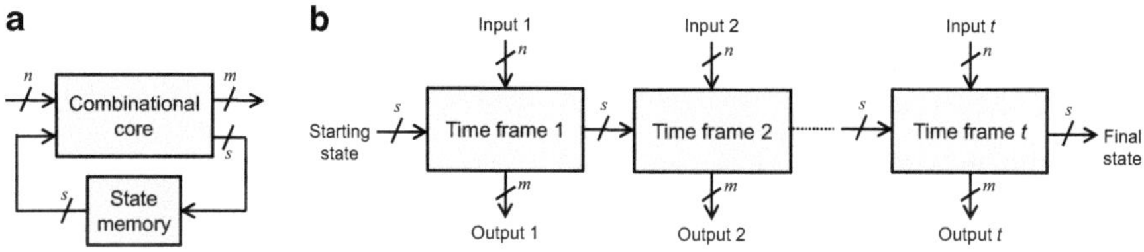

Fig. 6.3 Sequential circuit (**a**) and its time-frame expansion (**b**)

6.2.2 Sequential Circuits

A (synchronous) sequential circuit[2] consists of a combinational part and a clocked memory that holds the circuit's *state*. The combinational part of a circuit with n primary inputs, m primary outputs, and s state bits computes, in each clock cycle, two Boolean functions. The *state-transition function* $\delta : \mathbb{B}^{n+s} \rightarrow \mathbb{B}^s$ determines the state assumed by the circuit based on its current state and values applied to its primary inputs. The *output function* $\lambda : \mathbb{B}^{n+s} \rightarrow \mathbb{B}^m$ defines the values produced on the circuit's primary outputs based on the same data.

Assume that the *input sequence* applied to a sequential circuit (i.e., the number t of clock cycles and the values on the primary inputs in each cycle) and the circuit's initial state are known ahead of time. These assumptions are typically fulfilled in cryptographic applications where the circuit accepts the plaintext in the first clock cycle, sets the initial state based on this plaintext, and then performs, e.g., one round of encryption per clock cycle. Under these assumptions, it is possible to construct a combinational circuit that is functionally equivalent to the sequential circuit by a process known as "unrolling" or "time-frame expansion." For each of the t clock cycles, one copy of the circuit's combinational part is created; the i-th such copy is called the i-th *time frame*. The outputs of time frame i which have fed the state memory in the sequential circuit are connected to the inputs of time frame $i + 1$. Figure 6.3 illustrates the construction of time-frame expansion.

Time frame i in the unrolled circuit models the i-th cycle of execution in the original sequential circuit. Its state-transition logic would have written some state S_i into the state memory, and the circuit would read the same state S_i in cycle $i + 1$; in the time-frame expanded version, there is a direct connection skipping the state memory. Consider the sequential version of AES-128 where one round of encryption takes one clock cycle (one such round is indicated in Fig. 6.4a). Figure 6.1 can be regarded as the time-frame expansion of that figure. For example, the logical value at the least-significant bit of the ShiftRows operation in the 9-th clock cycle of the circuit from Fig. 6.4a is identical to the least-significant bit of the ShiftRows operation in "Round 9" part of Fig. 6.1.

[2]Asynchronous circuits have been suggested for cryptographic implementations [16] but they are not in scope of this chapter.

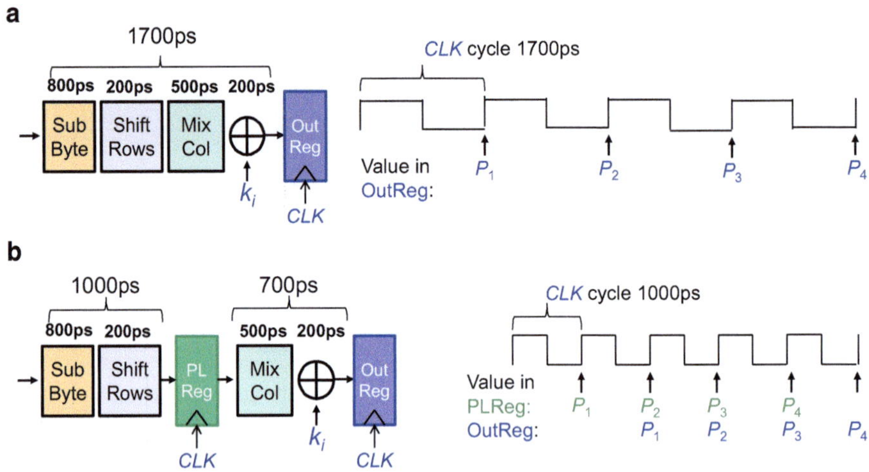

Fig. 6.4 Sequential realization of AES-128: diagram of one round (**a**) without and (**b**) with pipelining

One important design optimization which results in sequential behavior is *pipelining*. Figure 6.4b shows a pipelined version of one AES round from Fig. 6.4a, along with a diagram comparing the execution time for processing four plaintexts. Applying time-frame expansion to pipelined circuitry is straightforward: the pipeline registers are simply removed from the circuit model.

Once the time-frame expansion of a sequential circuit has been constructed, it can be treated as a combinational circuit. In particular, it is possible to apply Tseitin transform to this circuit and solve justification problems in the same manner as described in the previous section. Note that every signal of the original circuit is present t times in the time-frame expansion, and therefore will be mapped to t Boolean variables, one for its value in each considered clock cycle. The justification problems can stretch over multiple cycles. For example, it is possible to set the circuit's starting state to the all-0 value and the primary output of the last (t-th) time frame to a combination of values Q by adding single-literal clauses to the Tseitin transform of the time-frame expansion. For instance, if the circuit has four outputs represented by variables $s_{11}, s_{12}, s_{15}, s_{20}$ and $Q = (1, 1, 0, 1)$, then the added clauses will be (s_{11}), (s_{12}), $(\neg s_{15})$, and (s_{20}). If the SAT solver can find a satisfying assignment to the resulting formula's variables, the values on the inputs of time frames 1 through t give the sequence of t input vectors which must be applied to the circuit to generate the output combination $Q = (1, 1, 0, 1)$ after t cycles, given that the circuit was initially reset into the all-0 state [19].

We now have introduced most of the concepts needed to automatically find fault attacks on a given cryptographic implementation. In the next section, we will discuss the modeling of secret key bits, of single and multiple fault injections, and how to combine all these models into an integrated framework for attack construction.

6.3 SAT Models for Fault Attacks

In this section, we will first describe the extensions of models from the last section to cryptographic circuits which process secret information, and discuss modeling of faults injected by adversaries. Then, we focus on the construction of the actual attack, starting with a basic scenario. A circuit model is generated, converted into a Boolean formula using techniques from the last section, and passed, along with information observed during physical fault injection, to the SAT solver. Finally, extensions of this basic scenario are discussed.

6.3.1 *Modeling of Cryptographic Circuits*

SAT-based circuit analysis usually assumes that the circuit's complete functionality is known; for instance, it is possible to predict the values on the circuit's outputs if its input values are given. This assumption does not hold in a cryptographic implementation which process deterministic, but unknown, secret key bits. The secret key can either reside on the circuit in a (protected) nonvolatile memory, be produced by a physical unclonable function (PUF), be stored in a one-time memory (typically, an array of fuses blown after fabrication), or be (securely) transmitted from outside the circuit and stored in a volatile memory. However, from the perspective of the combinational circuit, secret key bits are fed on the circuit's additional inputs, similar to its primary inputs and state memory outputs discussed further above.

A cryptographic function can be (functionally) described by an equation which includes the secret key K. For example, encryption *enc* can be written as $C = enc(P, K)$, where P is the plaintext being encrypted and C is the ciphertext. If P and C are known (e.g., when the adversary has access to a circuit computing *enc* and can apply P to the circuit's inputs and observe C on its outputs), the secret key is the solution of equation $C = enc(P, K)$ after the unknown variable K. Solving this equation in the absence of further information should not be practically feasible; otherwise, the encryption is not sufficiently strong.

A typical fault attack consists in a repeated application of encryption (or a different cryptographic function) by running the circuit implementing the function with the same input in the presence and in the absence of a fault. It is assumed that the circuit uses the same secret key K during both applications. For example, suppose that a specific fault modifies the encryption function *enc* into enc'. Then, executing the circuit with input P results in output $C = enc(P, K)$ in the absence of faults and in $C' = enc'(P, K)$ in the presence of the abovementioned faults. It is possible to perform multiple fault injections, resulting in a number of fault-affected outputs: $C'_1 = enc'_1(P, K)$, $C'_2 = enc'_2(P, K)$, ... Fault attacks deduct knowledge about (parts of) K by analyzing the relationship between the difference of observed outputs $C \oplus C'_j$ and the effect of the fault within the circuit. To do so, it is necessary to establish a model of fault injection and to express it by Boolean functions.

6.3.2 Modeling of Faults

A variety of fault-injection techniques have been proposed, and these techniques result in different effects within the circuit [18]. In the simplest case, the result of a fault injection is a signal line being set to a specific logical value (0 or 1), known as *(single) stuck-at fault*. For example, illuminating the PMOS transistor in an inverter with focused laser light [20] may create parasitic currents that will switch on the transistor and force the logic-1 value on the inverter's output. Another useful fault model is the *bit-flip fault*, where an existing value on a line is replaced by its complement. One physical mechanism with this effect is applying an electromagnetic pulse [10] to a memory cell, perturbing its value. Stuck-at and bit-flip faults can affect multiple signal lines, e.g., all outputs of a 4-bit register.

Some fault-injection techniques may lead to effects which are not captured by simple fault models. For example, lowering the circuit's clock cycle duration (by injecting a glitch on the circuit's clock line) will create fault effects on a subset of the circuit's outputs. In more detail, outputs driven by sensitized paths with a longer delay will be affected and outputs driven by shorter paths will be not. Note that the same path can be sensitized under some circuit inputs and not sensitized under others. This is problematic when considering a cryptographic circuit where path sensitization depends on the secret key and that key is unknown; for such a circuit, it is impossible to predict precisely which outputs will fail. A further source of uncertainty is the variability in timing stemming from the fabrication process, environmental conditions (temperature), noise (e.g., jitter), and inaccuracies of the fault-injection equipment.

As a consequence, most published fault attacks employ fault models that allow uncertainty to some extent. A typical model is a *byte fault* where any subset of bits within a byte can be flipped, and the adversary does not know which bits were flipped but assumes that no bits outside this byte were affected by the fault. Analogous models can be defined on 4-bit nibbles or further objects of certain bit multiplicity.

6.3.3 Basic Fault Attack Construction

Let the cryptographic function under attack have N_p regular (controllable) inputs $p_1, \ldots, p_{N_p}$, N_k secret key inputs $k_1, \ldots, k_{N_k}$, N_c outputs $c_1, \ldots, c_{N_c}$, and N_s internal signals $s_1, \ldots, s_{N_s}$. (We use p_j as in "plaintext" and c_j as in "ciphertext", which corresponds to an encryption, but the construction is compatible with decryption and other cryptographic functions.) Moreover, let all circuit locations affected by faults (i.e., inputs, outputs, or signal lines where fault can be injected) be $z_1, \ldots, z_{N_Z}$. If the circuit is sequential, its time-frame expansion is used. Note that it is possible to model faults during different clock cycles as multiple faults in the unrolled version.

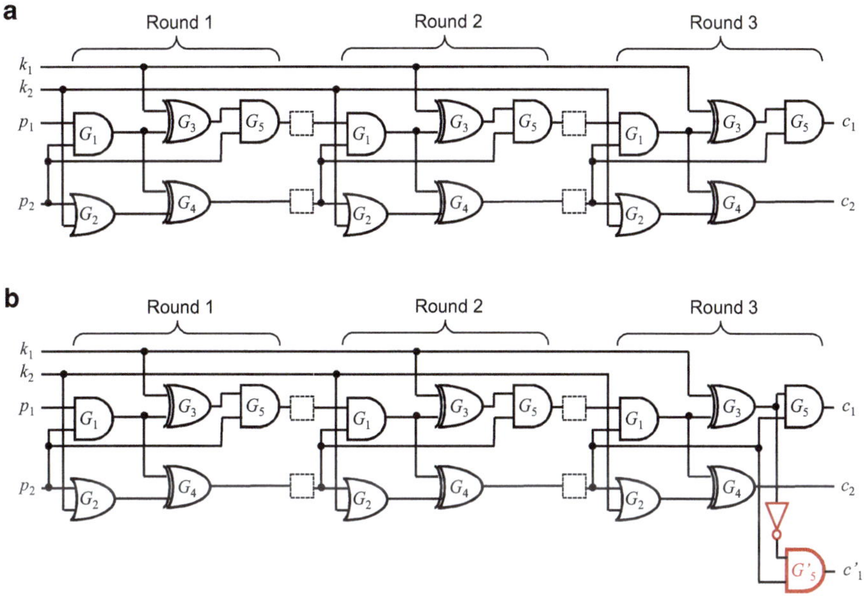

Fig. 6.5 (**a**) An example circuit, and (**b**) miter circuit for fault attack using a bit-flip fault in round 3

Figure 6.5a shows a (hypothetical) example circuit with two controllable inputs p_1, p_2, two secret key bits k_1, k_2, and two outputs c_1, c_2. It has been obtained by a time-frame expansion over three rounds, and the eliminated state memories between rounds are indicated by dashed boxes.

To mount the attack, a *miter circuit* is created and translated into a Boolean formula in CNF, using the Tseitin transform. After the physical fault attack, the adversary knows the values of P, C, and C' and feeds them, along with the obtained CNF, to a SAT solver. Miter circuits are often used in applications like equivalence checking or automatic test pattern generation. They represent the fault-unaffected and the fault-affected operation in the same circuit model. We will now describe the creation of the miter circuit assuming a single bit-flip fault, using Fig. 6.5 for illustration (we will extend the construction to other fault models in the next section).

First, reproduce the original circuit, with all its inputs, outputs, and internal signal lines. Mark the fault-injection location f and *duplicate all logic gates of the circuit in f's transient fanout*, i.e., for each gate G which is accessible from f via a path through the circuit, add a new gate G'. Connect an inverter to f with output f'; this line represents the immediate effect of the fault injection. For every added gate G', connect its inputs as follows:

- If an input of the original G was driven by f, connect that input of G' to f' (the output of the new inverter).

- If an input of G was driven by an output of gate H and H has been duplicated, connect that input of G' to the output of the duplicated gate H'.
- If an input of G was driven by an output of gate H that has *not* been duplicated (i.e., it is not affected by the fault), connect that input of G' to the output of H.

Figure 6.5b depicts the miter circuit for the fault injection on the output of G_3 in the last round. The inverter modeling the single bit-flip fault is shown in red. Only one gate, G_5 from the third round, is in the transitive fanout of the fault-injection location, and therefore only one gate duplicate, G_5' shown in red, is added to the circuit. This implies that the miter circuit has only one extra output c_1'. One would expect the second output c_2' to be added as well, but skipping it is correct because c_2 is not affected by the fault on the output of G_3 in round 3. If an adversary runs the attack and records the values $p_1, p_2, c_1, c_2, c_1', c_2'$, the ciphertext bits c_2 and c_2' must be equal; otherwise, the fault was not injected on the line assumed. Now, the attacker can apply Tseitin transform to the miter circuit to obtain a CNF, add single-literal clauses for p_1, p_2, c_1, c_2, c_1', and feed the formula to a SAT solver; the solution for variables representing lines k_1, k_2 gives a secret key candidate.

In general, a SAT solver can produce one of the following outcomes: generation of a solution (assignments of Boolean values to all variables of the model), report of unsatisfiability, or timeout. If a solution has been generated, the variables describing the secret key (k_j) are evaluated, and the *secret key candidate* k is produced as the concatenation of their values. This key candidate should be simulated with a known plaintext–ciphertext pair in order to verify that it is indeed the correct key. If this is the case, the attack was successful. The verification step is essential, because k can turn out to *not* be the correct key, in particular when the physically injected fault was inconsistent with the fault modeled in the CNF formula.[3] The same inconsistency can lead to unsatisfiability of the formula. In such a situation, the adversary can either rerun the analysis with a less restrictive model assumption or attempt a new physical fault injection to obtain a new fault-affected ciphertext C''.

6.3.4 Extensions

6.3.4.1 Reduced-Round Miter Circuit

To control the complexity of the analysis in a round-based cryptographic circuit, it is possible to use the **reduced-round miter circuit** where all rounds before the fault injection are eliminated. For example, some attacks on AES inject faults in round 8, and in this case, the first seven rounds can be excluded from the circuit. The

[3]In most cases, a cryptosystem's input and output (e.g., the plaintext and the ciphertext of an encryption) determine the secret key uniquely and there is only one consistent solution. If the circuit under attack implements a cryptosystem where this property does not hold, it can become necessary to search for different key candidates among the formula's solutions (cf. Sect. 6.3.4.1).

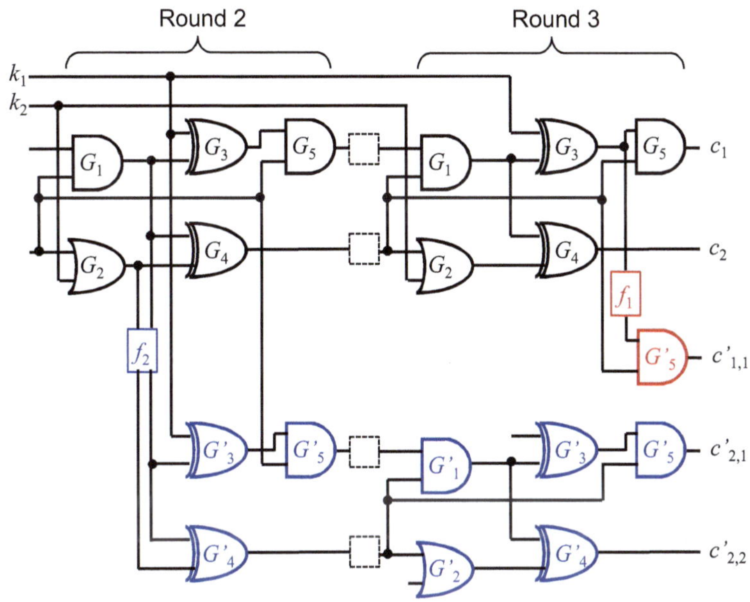

Fig. 6.6 Reduced-round miter circuit for a double fault attack using a bit-flip fault f_1 in round 3 and an uncertain fault f_2 (one or both of the outputs of G_1 and G_2 flip) in round 2

model becomes smaller and potentially better tractable for the SAT solver, but the knowledge of input P can no longer be used, since the new circuit's primary inputs correspond to an intermediate state that cannot be derived from P without knowing the secret key. Figure 6.6 illustrates this optimization when two fault injections are considered: one in round 3 and one in round 2, and therefore round 1 can be eliminated. Note that the controllable inputs of the circuit are no longer marked p_1, p_2 because they represent the circuit's intermediate state after round 1 (whereas k_1, k_2 still represent the secret key).

The reduced-round model is far less restricted than the full-circuit model and tends to have a large number of solutions. In the ideal case, the SAT solver is capable of generating all found solutions, and if their number is reasonable (less than around 2^{40}), then the adversary can apply brute-force search, i.e., try all key candidates sequentially. If the SAT solver does not have such a feature and produced a key candidate k that turned out to not be the correct key, it is possible to add "conflict clauses" to the formula that suppress regeneration of candidate k and attempt to solve the extended formula. This can be repeated, excluding more and more unsuitable key candidates.

6.3.4.2 More Generic Fault Models

The circuit in the previous section assumed a single bit-flip fault represented by an inverter. It is straightforward to model multiple bit-flips using several inverters, and to model stuck-at faults by constant values in the duplicated circuit. However, the flexibility of SAT modeling allows the adversary to express far more complex fault patterns and to represent uncertainty to some extent. Such models do not need to be mapped to logic gates but can be arbitrary Boolean expressions. For example, consider a *byte fault* on signals $l_1 \dots l_8$ in the original circuit. This means that at least one (but possibly more, up to eight) of the signals l_j can flip. Introduce the corresponding signals $l'_1 \dots l'_8$ and duplicate all gates in the transitive fanout of $l_1 \dots l_8$. Connect the duplicated gates as explained in the last section for the case of the inverter output f'. However, instead of modeling an inverter, add the condition $(l_1 \oplus l'_1) \vee \cdots \vee (l_8 \oplus l'_8)$. This condition must be converted into CNF and added to the overall Boolean formula obtained by Tseitin transform from the miter circuit.

6.3.4.3 Multiple Fault Injections

Some ciphers require **multiple fault injections** for successful cryptanalysis. This is achieved by replicating the fault-affected part of the circuit multiple times, according to the number of fault injections. This strategy is particularly useful for the reduced-circuit model where combining conditions from different fault injections can vastly reduce the number of possible key candidates and therefore the complexity of the brute-force search. It is important to distinguish multiple fault injections (i.e., separate experiments where more than one faulty ciphertext is recorded) from faults affecting multiple circuit locations during one fault injection. For example, one fault injected into a sequential circuit may flip the same location during several consecutive clock cycles; this corresponds to a multiple bit-flip fault.

Figure 6.6 illustrates the application of two fault injections on the circuit from Fig. 6.5a. The first injection (f_1) is the same single bit-flip fault as in Fig. 6.5b, and the second injection takes place in round 2 and affects (flips) either the output of G_1, the output of G_2, or both at the same time. Instead of introducing an inverter, f_1 and f_2 are functional models in CNF (the exact equations are not shown). The resulting miter circuit models two fault injections via a "parallel construction." It has three sets of outputs: fault-free (c_1, c_2), affected by f_1 (only $c'_{1,1}$, as output c_2 is not affected by that fault), and affected by f_2 ($c'_{2,1}, c'_{2,2}$). Note the different number of gates replicated for the two modeled fault injections.

6.4 Alternative Approaches

A few automated fault attack frameworks have been proposed in recent years. In this section, we will review some of the current approaches concerning the automatic construction of fault attack. All of them start with an algebraic description of the cryptosystem under attack, i.e., a system of equations that specify the system's functionality. This is in contrast to the procedure described in the previous sections, which takes the circuit description as its input. We will first explain how to find suitable scenarios for the realization of fault attacks, which is the first step into an automated construction, by examining the two frameworks from [15] and [21] (cf. Chap. 2 of this book). These frameworks were specifically designed to this extent. Further below, we will delve into algebraic fault analysis (AFA) frameworks such as the ones presented in [28–30], which take similar but yet different approach to the problem of automated fault attacks.

6.4.1 Fault Characterization and Key Space Evaluation

One of the first stages in fault attacks is to identify the proper position (in most cases, the round) and location (nibble/byte) for a fault injection. This, of course, depends on the chosen fault model, and we will discuss this choice when necessary. One of the first approaches towards the automation of fault attack is therefore to extract the fault-injection parameters that are the most likely to yield to a successful attack. The XFC framework from [15] proposed a coloring-based procedure, restricted to block ciphers, capable of characterizing such faults.

The XFC framework takes as input a block cipher specification and a fault model. The block cipher specification focuses on its composition of linear and nonlinear functions. By separating linear from nonlinear functions, according to the different inputs involved, XFC generates a color-based cipher description which traces the fault propagation (according to the chosen fault model) throughout the cipher. When a fault is injected, still according to the chosen fault model, XFC assigns a new color to the affected parts of the block cipher. Then, depending on the input of each function and whether it is linear or not, XFC assigns a new color at each step of the encryption through which the fault is propagated. The output of this step is then fed to the next stage of XFC, which is the estimation of the key space.

XFC uses the estimated size of the key space as an indication of the attack complexity. In order to estimate the key space size, XFC progresses backwards and refers to the previously generated colored cipher. Each color refers to a variable and XFC looks for specific equations related to those variables, which would lead to a recovery of some related portions of the key. While this process is only semiautomatic, as it needs some additional inputs, it allows for a good estimate of the complexity of the attack. The authors applied XFC to a few block ciphers, including AES. The offline complexity of the AES attack found by XFC (by injecting a fault in one byte at the 8th round) is the same as for common differential fault attacks (DFA) on AES, which supports the correctness of the framework. The whole XFC flow replicates what is manually done for DFA and as such XFC manages to automate this step and provides an estimation of the attack's complexity according to a specified fault model.

The color-based approach of XFC, while functional, has some limitations. For instance, the authors of [21] pointed out that an impossible differential fault analysis (IDFA) cannot be processed by XFC. As an example of this, they show how XFC is unable to find a suitable attack scenario for a fault injection at the beginning of round 7 in the case of the AES. To solve this problem, the authors proposed a similar framework to XFC. Compared with XFC, it supports more fault models and fully automates the evaluation of the key space. The framework uses a data mining-like approach instead of the color-based one.

Three steps are involved in the proposed framework. First, and similar to XFC, a distinguisher for the attack needs to be identified. This is once again the key step for any DFA. Then comes the divide and conquer stage. In this stage, the found distinguisher is evaluated in order to verify its computability. Finally, an automated evaluation of the key space is performed, in order to output the feasibility of the attack. In more detail, the distinguisher identification is performed through computation of state entropies, related to a chosen fault model, and compared to the maximum state entropy. The measure of the entropy allows to carry information related to the fault propagation path and the studied distinguisher. A differential state is a distinguisher if its entropy is inferior to the maximum entropy of the same state. The identification steps return several possible distinguishers that go through the second stage.

In order to proceed to a divide and conquer strategy, the cipher description is broken down into a graph, composed of subgraphs for each operation. This graph, called cipher dependency graph, allows to evaluate the computability of the chosen distinguisher. By searching through the graph, it is possible to identify related key bits that can be computed from the differential distinguisher and to know which remaining portion of the key needs to be guessed. Finally, all the distinguisher properties deduced from the previous step are fed into the algorithm that evaluates the key space reduction, estimating the key space size and the number of required fault injections.

The authors evaluated their framework on the AES and PRESENT, and, for instance, found that the key space size for a fault injection at the beginning of round 7 in the AES was roughly 2^{32}–2^{26}. Even though the authors do not provide an estimation of the number of fault injections required in this specific case, it should be noted that such an attack scenario was found with this framework, which was not possible through the use of XFC. The previous framework takes, similarly to XFC, a functional description of the cipher and a fault model as input. But, contrary to XFC, the fault model is not limited and can be easily extended to different kinds of faults. While similar to XFC, it is more versatile and gives a better estimation of the key space.

Both frameworks constitute a first step towards the automated construction of fault attack, but they do not proceed to actually implement an actual attack and only provide a complexity estimation of the attack at the found position and location.

6.4.2 Algebraic Fault Analysis Frameworks

Algebraic fault analysis (AFA) combines fault attacks and algebraic fault analysis. Such attacks take as inputs a cipher description and a fault model, both converted to algebraic equations. One of the advantages of AFA is that they do not strictly require manual analysis of the cipher, and can therefore be automated by using different types of solvers, such as SAT solvers or Gröbner basis solvers.

The first step of any AFA is to provide the correct inputs. As previously mentioned, AFA needs a set of equations and a fault model as input. The authors from [28, 30] and [29] assume functional description of the cipher as equations. Deriving such equations from the description of the cipher is feasible but can be difficult and error prone in practice, especially for nonlinear functions such as SBoxes or complicated operations such as MixColumns. While [28] assumes a specific fault model, which could be extended, [30] and [29] are not restricted to specific fault models.

In [28], authors propose an AFA framework for the Piccolo cipher that can be extended to some other lightweight ciphers. The functional description of Piccolo is converted to formulae in algebraic normal form (ANF). The chosen fault model, a nibble-based fault in the 23rd round of Piccolo with known plaintext and ciphertext, is also translated into ANF formulas. In order to do so, the author proposed a method to represent the faults. They introduce new variables for each fault injection (expressed as XORs) and variables representing the presence of the aforementioned faults. Once all equations are available, they are fed to a SAT solver, in this case to CryptoMiniSat [24]. The SAT solver handles the solving and returns the correct key. The authors evaluated their framework on both the encryption and the decryption of Piccolo. While the framework was unable to solve the algebraic equations for the encryption, it recovered the key by attacking the decryption in 5 h on average, with one fault, and 2 h with two faults. The authors also discuss the possible extension of their framework to different ciphers (for which they also provide benchmarking) and fault models.

In summary, this framework takes as input a set of ANF formulas describing the cipher and fully automates the solving. While the equation creation step is partially automated in [28], thanks to their method of conversion, it is still cipher and fault model dependent. Such a method also allows for partial automation of the equation creation as a script can automate this process for different fault positions and/or locations.

The authors of [30] (cf. Chap. 5 of this book) suggest to use, in addition to AFA equations, differential fault equations derived by dedicated cryptanalysis, to improve the solving time. The improved algebraic differential fault analysis (ADFA) proposed in [30] is focused on the cipher LED but could be extended to different ciphers. The process is similar to the work presented in [28] but with the addition of equations based on a DFA approach. Similar equations are created from the cipher description but also for reverse operations, such as inverse SBoxes, since they are needed for differential fault equations. The solving will handle going backwards through a portion of the cipher, similarly to DFA or XFC. Finally, the fault difference is also expressed as ANF formulas, and all sets of equations are sent as input to the SAT solver (again CryptoMiniSat). For LED, the authors verified the effectiveness of their framework and obtained an average of 97.2% successful key retrievals within 10 min, assuming a nibble-based fault in the 30th round.

While the experimental results show good performance compared to DFA and support the possibility to add additional cryptanalytic input for better performance, the authors also compared the key space reduction of their approach to DFA. ADFA approach improves the reduction of the key space by almost ten orders of magnitude, which further supports ADFA automation as an efficient method.

In a more recent work, the authors of [29] expand on the proposed framework of [28], making it more versatile and not cipher restricted. The approach is similar to the previous two frameworks but the authors consider various ciphers and fault models, showcasing the versatility of their framework. The first step is similar to other approaches. Equations are created both from the cipher description and the fault model which will then be fed to a SAT solver (CryptoMiniSat). An interesting feature of this framework is the two solving modes. Similarly to [28], the framework has a mode A that can be used to solve the set of equations to retrieve the key, but it also has a mode B that automatically estimates the key space of the considered attack. In the second mode, no plaintext/ciphertext pair is provided and a modified version of the SAT solver evaluates all possible solutions and returns the key space size. As shown in [15] and [21], this is an important feature as it allows to evaluate the feasibility of an attack with the chosen fault model.

Another important feature of the framework is the large number of possible different inputs. Previous frameworks were limited to either few ciphers, few fault models, or in general few different inputs, but the authors of [29] validated their framework with several different inputs. They considered a fault injection during the encryption, the key schedule, or even in the round counter itself. They also considered bit-based, nibble-based, and byte-based fault models, as well as several different ciphers (LBlock, DES, PRESENT, and Twofish). In all reported cases, the proposed framework was capable of solving different instances with faults at different positions and locations. This provides evidence that AFA frameworks are extremely versatile and can support numerous ciphers and fault models, and as such are suitable for automated construction of fault attacks.

For all AFA approaches discussed in this section, the SAT solver can be swapped for any other solver as long as the equation input is of the correct format. Furthermore, it is also important to note that the practicability of automating such attacks is tied to the performance of aforementioned solvers. For instance, as SAT solver become more and more efficient, AFA frameworks also become more efficient.

6.5 Chapter Summary

Hardware-implemented cryptographic functions are a natural target for physical attacks, and their vulnerability to such attacks should be assessed and, ideally, eliminated. This chapter explained in detail how to automatically construct attacks starting with a cryptographic circuit description. The construction leverages concepts like time-frame expansion or Tseitin transform that have been in use in hardware domain and thus allows their combination with recent improvements in attack construction and SAT technology. While the overall objective of the described procedure is comparable to methods from other chapters of this book, its focus on models derived directly from circuit descriptions simplifies their construction and provides a foundation for direct integration of complex fault models. With respect to performance, the described flow is still under development. The version presented in [7] could handle lightweight 64-bit ciphers, whereas our current implementation is applicable to 128-bit AES; this appears to be comparable with most alternative approaches.

Acknowledgements This work was partially supported by the DFG project "Algebraic Fault Attacks" (funding ID PO 1220/7-2, BE 1176/20-2).

References

1. S. Banik, S. Maitra, S. Sarkar, A differential fault attack on the grain family of stream ciphers, in *International Workshop on Cryptographic Hardware and Embedded Systems*. Lecture Notes in Computer Science, vol. 7428 (Springer, Berlin, 2012), pp. 122–139
2. H. Bar-El, H. Choukri, D. Naccache, M. Tunstall, C. Whelan, The sorcerer's apprentice guide to fault attacks. Proc. IEEE **94**(2), 370–382 (2006)
3. A. Barenghi, L. Breveglieri, I. Koren, D. Naccache, Fault injection attacks on cryptographic devices: theory, practice, and countermeasures. Proc. IEEE **100**(11), 3056–3076 (2012)
4. R.P. Bastos, F.S. Torres, J.-M. Dutertre, M.-L. Flottes, G. Di Natale, B. Rouzeyre, A bulk built-in sensor for detection of fault attacks, in *2013 IEEE International Symposium on Hardware-Oriented Security and Trust (HOST)* (IEEE Computer Society, Silver Spring, 2013), pp. 51–54
5. D. Boneh, R.A. DeMillo, R.J. Lipton, On the importance of checking cryptographic protocols for faults (extended abstract), in *Proceedings of the International Conference on the Theory and Application of Cryptographic Techniques – Advances in Cryptology – EUROCRYPT '97* (Springer, Berlin, 1997), pp. 37–51
6. E. Brier, D. Naccache, P.Q. Nguyen, M. Tibouchi, Modulus fault attacks against RSA-CRT signatures. J. Cryptogr. Eng. **1**(3), 243–253 (2011)
7. J. Burchard, M. Gay, A.S.M. Ekossono, J. Horácek, B. Becker, T. Schubert, M. Kreuzer, I. Polian, Autofault: towards automatic construction of algebraic fault attacks, in *2017 Workshop on Fault Diagnosis and Tolerance in Cryptography (FDTC)* (IEEE Computer Society, Silver Spring, 2017), pp. 65–72
8. A. Czutro, M. Sauer, T. Schubert, I. Polian, B. Becker, SAT-ATPG using preferences for improved detection of complex defect mechanisms, in *2012 IEEE 30th VLSI Test Symposium (VTS)* (IEEE Computer Society, Silver Spring, 2012), pp. 170–175
9. J. Daemen, V. Rijmen, *The Design of Rijndael* (Springer, New York, 2002)
10. A. Dehbaoui, J.-M. Dutertre, B. Robisson, A. Tria, Electromagnetic transient faults injection on a hardware and a software implementations of AES, in *2012 Workshop on Fault Diagnosis and Tolerance in Cryptography (FDTC)* (IEEE, Piscataway, 2012), pp. 7–15
11. W. Fischer, C.A. Reuter, Differential fault analysis on Grøstl, in *2012 Workshop on Fault Diagnosis and Tolerance in Cryptography* (IEEE Computer Society, Silver Spring, 2012), pp. 44–54
12. G.D. Hachtel, F. Somenzi, *Logic Synthesis and Verification Algorithms* (Springer, Berlin, 1996)
13. P. Jovanovic, I. Polian, Fault-based attacks on the Bel-T block cipher family, in *2015 Design, Automation & Test in Europe Conference & Exhibition (DATE)* (ACM, New York, 2015), pp. 601–604

14. P. Jovanovic, M. Kreuzer, I. Polian, A fault attack on the LED block cipher, in *Proceedings of the 3rd International Workshop on Constructive Side-Channel Analysis and Secure Design* (Springer, Berlin, 2012), pp. 120–134
15. P. Khanna, C. Rebeiro, A. Hazra, XFC: a framework for eXploitable fault characterization in block ciphers, in *Proceedings of the 54th Annual Design Automation Conference 2017 (DAC '17)* (ACM, New York, 2017), pp. 8:1–8:6
16. I. Levi, A. Fish, O. Keren, Low-cost pseudoasynchronous circuit design style with reduced exploitable side information. IEEE Trans. Very Large Scale Integr. VLSI Syst. **26**(1), 82–95 (2018)
17. M. Medwed, F.-X. Standaert, J. Großschädl, F. Regazzoni, Fresh re-keying: security against side-channel and fault attacks for low-cost devices, in *International Conference on Cryptology in Africa* (Springer, Berlin, 2010), pp. 279–296
18. F. Regazzoni, I. Polian, Securing the hardware of cyber-physical systems, in *2017 22nd Asia and South Pacific Design Automation Conference (ASP-DAC)* (IEEE, Piscataway, 2017), pp. 194–199
19. S. Reimer, M. Sauer, T. Schubert, B. Becker, Using MaxBMC for Pareto-optimal circuit initialization, in *2014 Design, Automation & Test in Europe Conference & Exhibition (DATE)* (European Design and Automation Association, 2014), pp. 1–6
20. C. Roscian, J.-M. Dutertre, A. Tria, Frontside laser fault injection on cryptosystems—application to the AES'last round, in *2013 IEEE International Symposium on Hardware-Oriented Security and Trust (HOST)* (IEEE, Piscataway, 2013), pp. 119–124
21. S. Saha, U. Kumar, D. Mukhopadhyay, P. Dasgupta, Differential fault analysis automation. IACR Cryptology ePrint Arch. **2017**, 673 (2017)
22. M. Sauer, B. Becker, I. Polian, PHAETON: a SAT-based framework for timing-aware path sensitization. IEEE Trans. Comput. **65**(6), 1869–1881 (2016)
23. B. Selmke, J. Heyszl, G. Sigl, Attack on a DFA protected AES by simultaneous laser fault injections, in *2016 Workshop on Fault Diagnosis and Tolerance in Cryptography (FDTC)* (IEEE, Piscataway, 2016), pp. 36–46
24. M. Soos, K. Nohl, C. Castelluccia, Extending SAT solvers to cryptographic problems, in *Proceedings of the 12th International Conference on Theory Applications Satisfiability Testing (SAT)* (Springer, Berlin, 2009), pp. 244–257
25. G. Tseitin, *On the Complexity of Derivation in Propositional Calculus*. Studies in Constructive Mathematics and Mathematical Logic, 1968
26. M. Tunstall, D. Mukhopadhyay, S. Ali, Differential fault analysis of the advanced encryption standard using a single fault, in *IFIP International Workshop on Information Security Theory and Practices* (Springer, Berlin, 2011), pp. 224–233
27. H. Tupsamudre, S. Bisht, D. Mukhopadhyay. Differential fault analysis on the families of SIMON and SPECK ciphers, in *2014 Workshop on Fault Diagnosis and Tolerance in Cryptography* (IEEE, Piscataway, 2014), pp. 40–48
28. F. Zhang, X. Zhao, S. Guo, T. Wang, Z. Shi, Improved algebraic fault analysis: a case study on piccolo and applications to other lightweight block ciphers, in *Proceedings of the 4th International Workshop Constructive Side-Channel Analysis Secure Design (COSADE)* (Springer, Berlin, 2013), pp. 62–79
29. F. Zhang, S. Guo, X. Zhao, T. Wang, J. Yang, F.-X. Standaert, D. Gu, A framework for the analysis and evaluation of algebraic fault attacks on lightweight block ciphers. IEEE Trans. Inf. Forensics Secur. **11**(5), 1039–1054 (2016)
30. X. Zhao, S. Guo, F. Zhang, Z. Shi, C. Ma, Tao Wang, Improving and evaluating differential fault analysis on LED with algebraic techniques, in *Proceedings of the 10th IEEE Workshop on Fault Diagnosis Tolerance Cryptography (FDTC)* (IEEE, Piscataway, 2013), pp. 41–51

Part II
Automated Design and Deployment of Fault Countermeasures

Chapter 7
Automated Deployment of Software Encoding Countermeasure

Jakub Breier and Xiaolu Hou

7.1 Introduction

As it was shown before [15], fault countermeasures often lower the implementation resistance against side-channel attacks. Therefore, it is necessary to consider these two classes of physical attacks together when protecting the algorithm. In this chapter, we show how to automatically construct a code-based countermeasure that significantly reduces the success of a fault injection attack while keeping low information leakage via side-channels.

There are two main countermeasure classes to protect implementations against side-channel attacks. *Masking* [9] is a software-level countermeasure which tries to "mask" the relationship between the intermediate values and power leakage. *Hiding* [20] tries to reduce the signal and increase noise by utilizing various techniques—it "hides" the operations performed by the device. While masking can make fault attacks more challenging, it does not help to prevent them. On the other hand, some hiding techniques, such as dual-rail precharge logic (DPL), help in preventing fault attacks by detecting faults [18].

In 2011, DPL was extended to software by Hoogvorst et al. [10], by using balanced encoding schemes. Since then, there were several other proposals

---

This research was conducted when author "J. Breier" was with Temasek Laboratories, NTU.
This research was conducted when author "X. Hou" was with Nanyang Technological University.

J. Breier
Underwriters Laboratories, Singapore, Singapore
e-mail: jbreier@jbreier.com

X. Hou (✉)
Acronis, Singapore, Singapore
e-mail: ho0001lu@e.ntu.edu.sg

J. Breier et al. (eds.), *Automated Methods in Cryptographic Fault Analysis*,
https://doi.org/10.1007/978-3-030-11333-9_7

[6, 13, 14, 19], all of them using various coding techniques to prevent side-channel leakage. However, it was shown that unlike hardware DPL representation, its software counterpart is not fault resistant by default [4]. Therefore, to prevent both attack techniques, it is necessary to design the coding scheme from the beginning with this goal in mind.

In this chapter, we focus on approach presented in [2]. We first explain the theoretical background necessary for designing software hiding countermeasures that are resistant to both side-channel and fault attacks. We provide an algorithm to automatically find optimal codes for various code distances and number of codewords with given code length. We also provide evaluation of the codes—by using detection and correction probabilities and by an automated fault simulator. This simulation is done by using a general-purpose microcontroller implementation and an instruction set simulator that is capable of injecting different fault models into any instruction of the code. Our evaluation shows that the codes generated by our algorithm provide a high security level with respect to both side-channel and fault attacks.

The rest of the chapter is organized as follows: Section 7.2 provides an overview of the related work in this field, together with necessary background on coding theory. Section 7.3 defines the properties of codes with respect to fault attacks. Section 7.4 details our algorithm and provides estimated and simulated results on chosen codes. These results are further discussed in Sect. 7.5. Finally, Sect. 7.6 summarizes this chapter.

7.2 General Background

In this section we provide a necessary background on software encoding-based side-channel countermeasures and on coding theory necessary for developing a combined countermeasure. Section 7.2.1 overviews the related work in the field. Section 7.2.2 provides basic definitions that are used later in this chapter.

7.2.1 Related Work

After the paper by Hoogvorst et al. [10], who presented a method to extend the DPL to software implementations, several works were published in the area of software hiding schemes.

Rauzy et al. [14] developed a scheme that encodes the data by using bit-slicing, where only one bit of information is processed at a time. They claim that this kind of protection is 250 times more resistant to power analysis attacks compared to the unprotected implementation, while being 3 times slower. For testing, they used PRESENT cipher, running on an 8-bit microcontroller.

Chen et al. [6] proposed an encoding scheme that adds a complementary bit to each bit of the processed data, resulting in a constant Hamming weight code. Their countermeasure was implemented on a Prince cipher, using an 8-bit microcontroller.

Servant et al. [19] introduced a constant weight implementation for AES, by using a (3,6) code. To improve the performance, they split 8-bit variables into two 4 bit words and encode them separately. This implementation was also capable of detecting faults with 93.75% probability. Their implementation used a 16-bit microcontroller.

Maghrebi et al. [13] proposed an encoding scheme that differs from the previous proposals. For their case, they did not assume the Hamming weight leakage model for register bits; therefore, they concluded that balanced codes might not be the optimal ones to use generally. In their method, they first obtain the profile of a device to get a vector of register bit leakages. Then they estimate leakage values for each codeword and build a code by using codewords with the lowest leakage. Their algorithm selects the optimal code by ranking the codes based on the difference in power consumption between the codewords and on the power consumption variance. Our algorithm extends this idea by adding the variance of register bits in order to achieve better leakage characteristics and by adding conditions for error detection and correction.

In general, none of the previous schemes has been designed for fault resistance. Schemes proposed in [6, 14] have been analyzed with respect to fault attacks by Breier et al. [4], concluding that without additional modifications to assembly code, the probability of a successful fault attack is non-negligible. Therefore, in this chapter we focus on design and automated generation of fault tolerant and side-channel resistant coding schemes.

When it comes to combined countermeasures, in [17], Schneider et al. proposed a hardware countermeasure based on combining threshold implementation with linear codes. As stated in the paper, their proposal is not considered for software targets. In the execution process, there are multiple checking steps that protect the implementation against faults. However, in software, it would be easy to overcome such checks by multiple fault injections [21]. Also, it would be possible to inject faults that are impossible with hardware implementations, such as instruction skips [3].

This chapter provides the reader with the following information:

- We specify theoretical bounds for encoding schemes with respect to fault attacks that are necessary to be taken into account when designing a fault resistant scheme.
- We show how to automatically design a code that is capable of protecting the implementation against side-channel and fault attacks and we show trade-offs between these two resistances.
- We adopt the ranking algorithm proposed in [13] and show how to improve it for constructing side-channel resistant codes with better properties—by ranking the codes according to the codeword with the highest leakage, and by calculating

the register bit variance. We add the conditions for selecting the codes with the desired error-detection/correction capabilities in an automated way.
- We analyze the codes constructed by the code generation algorithm—we calculate leakages, fault detection, and correction probabilities, and we simulate the assembly code implementing the codes on a general-purpose microcontroller.

7.2.2 Coding Theory Background

A *binary code*, denoted by C, is a subset of the n-dimensional vector space over $\mathbb{F}_2$-$\mathbb{F}_2^n$, where n is called the *length* of the code C. Each element $\boldsymbol{c} \in C$ is called a *codeword* in C and each element $\boldsymbol{x} \in \mathbb{F}_2^n$ is called a *word* [11, p. 6]. Take two codewords $\boldsymbol{c}, \boldsymbol{c}' \in C$, the *Hamming distance* between $\boldsymbol{c}$ and $\boldsymbol{c}'$, denoted by $\text{dis}\left(\boldsymbol{c}, \boldsymbol{c}'\right)$, is defined to be the number of places at which $\boldsymbol{c}$ and $\boldsymbol{c}'$ differ [11, p. 9]. More precisely, if $\boldsymbol{c} = c_1c_2\ldots c_n$ and $\boldsymbol{c}' = c_1'c_2'\ldots c_n'$, then

$$\text{dis}\left(\boldsymbol{c}, \boldsymbol{c}'\right) = \sum_{i=1}^{n} \text{dis}\left(c_i, c_i'\right),$$

where c_i and c_i' are treated as binary words of length 1 and hence

$$\text{dis}\left(c_i, c_i'\right) = \begin{cases} 1 & \text{if } c_i \neq c_i' \\ 0 & \text{if } c_i = c_i' \end{cases}.$$

Furthermore, for a binary code C, the *(minimum) distance* of C, denoted by $\text{dis}\,(C)$, is [11, p. 11]

$$\text{dis}\,(C) = \min\{\text{dis}\left(\boldsymbol{c}, \boldsymbol{c}'\right) : \boldsymbol{c}, \boldsymbol{c}' \in C, \boldsymbol{c} \neq \boldsymbol{c}'\}.$$

Definition 7.1 ([7, p. 75]) For a binary code C of length n, $\text{dis}\,(C) = d$, let $M = |C|$ denote the number of codewords in C. Then C is called an (n, M, d)-binary code.

This minimum distance of a binary code is closely related to the error-detection and error-correction capabilities of C.

Definition 7.2 ([11, p. 12]) Let u be a positive integer. C is said to be *u-error-detecting* if, whenever there is at least one but at most u errors that occur in a codeword in C, the resulting word is not in C.

From the definition, it is easy to prove that C is u-error-detecting if and only if $\text{dis}\,(C) \geq u + 1$ [11, p. 12]. A common decoding method that is used is *nearest neighbor decoding*, which decodes a word $\boldsymbol{x} \in \mathbb{F}_2^n$ to the codeword $\boldsymbol{c_x}$ such that

$$\text{dis}\left(\boldsymbol{x}, \boldsymbol{c_x}\right) = \min_{c \in C} \text{dis}\left(\boldsymbol{x}, \boldsymbol{c}\right). \tag{7.1}$$

When there are more codewords $\boldsymbol{c_x}$ satisfies (7.1), the *incomplete decoding rule* requires a retransmission [11, p. 10].

Definition 7.3 ([11, p. 13]) Let v be a positive integer. C is *v-error correcting* if minimum distance decoding with incomplete decoding rule is applied, v or fewer errors can be corrected.

Remark 7.1 C is v-error correcting if and only if $\text{dis}\,(C) \geq 2v + 1$ [11, p. 13].

Definition 7.4 ([8]) An (n, M, d)-binary code C is called an *equidistant code* if $\forall \boldsymbol{c}, \boldsymbol{c}' \in C,\ \text{dis}\left(\boldsymbol{c}, \boldsymbol{c}'\right) = \text{dis}\,(C)$.

For our purpose, we will use binary code for protecting the underlying implementation.

We propose two choices of look-up tables:

1. Correction Table: This table will treat a word $\boldsymbol{x} \in \mathbb{F}_2^n$ the same as the codeword $\boldsymbol{c_x} \in C$ which satisfies $\text{dis}\,(\boldsymbol{c_x}, \boldsymbol{x}) \leq \lfloor \frac{d-1}{2} \rfloor$, where d is the distance of C. Note that this is equivalent to using *bounded distance decoding* [12, p. 36] and taking the bounded distance to be $\lfloor \frac{d-1}{2} \rfloor$. To use this table we require that $\text{dis}\,(C) \geq 3$.
2. Detection Table: This is a normal look-up table that returns a null value when $\boldsymbol{x} \notin C$ is accessed.

We will give a theoretical criterion to measure the bit flip fault resistant capability of a binary code when it is used as an encoding countermeasure against fault injection attacks in Sect. 7.3. Afterwards we propose three coding schemes. The encoding scheme will be simulated (and implemented) and evaluated in Sect. 7.4.

Let m be a positive integer such that $1 \leq m \leq n$, where n is the code length.

Definition 7.5 An *m-bit fault* is a fault injected in the codeword that flips exactly m bits. We assume each bit has equal probability to be flipped.

Definition 7.6 When the fault is analyzed, we adopt the following terminologies:

- *Corrected*: Fault is detected and corrected.
- *Null*: Fault is detected and results into zero output.
- *Invalid*: Fault is detected and results into an output that is not a codeword.
- *Valid*: Fault is not detected and fault injection is successful, i.e., it results in the output of a valid but incorrect codeword.

7.3 Theoretical Analysis

In this section we will first give the theoretical analysis for the fault resistant capabilities of binary code in general. Then we propose two different coding schemes and analyze their fault resistant probabilities.

7.3.1 Correction Table

Definition 7.7 For an (n, M, d)-binary code C such that $d \geq 3$, let

$$F_{c,m} := \left\{ x \in \mathbb{F}_2^n : \text{dis}\,(c, x) = m \text{ and } \exists c' \in C \text{ such that dis}\left(x, c'\right) \leq \left\lfloor \frac{d-1}{2} \right\rfloor \right\}.$$

Then

$$p_{m,(e)} := \begin{cases} 1 & m \leq \lfloor \frac{d-1}{2} \rfloor \\ 1 - \frac{1}{M\binom{n}{m}} \sum_{c \in C} |F_{c,m}| & m > \lfloor \frac{d-1}{2} \rfloor \end{cases} \tag{7.2}$$

is called the *m-bit fault resistance probability with error correction* for C.

As mentioned earlier, when a Correction Table is used, it is equivalent to using bounded distance decoding. When $m \leq \lfloor \frac{d-1}{2} \rfloor$ bits are flipped, by Remark 7.1, the error will be corrected and hence $p_{m,(e)} = 1$. When $m > \lfloor \frac{d-1}{2} \rfloor$ bits are flipped, the fault will be valid if the resulting word is at distance at most $\lfloor \frac{d-1}{2} \rfloor$ from any codeword. Thus by Definition 7.6, $1 - p_{m,(e)}$ gives the theoretical probability of a *Valid* fault and the bigger the $p_{m,(e)}$ is, the more resistant the binary code to m-bit fault. Furthermore, when $m = 1$, the fault will be corrected and most of the cases are expected to return *Corrected*.

Another interesting fault model is random fault, i.e., assuming there is an equal probability for m-bits fault to occur $\forall 1 \leq m \leq n$. Taking this into account, we define the following.

Definition 7.8 For an (n, M, d)-binary code C such that $d \geq 3$, let $p_{m,(e)}$ be its m-bit fault resistance probability with error for $1 \leq m \leq n$, then

$$p_{\text{rand},(e)} := \sum_{m=1}^{n} \frac{1}{n} p_{m,(e)}$$

is called the *overall resistance index with error correction* for C.

As suggested by the name, the bigger the $p_{\text{rand},(e)}$ is, the more resistant the code C is to random faults.

7.3.2 Detection Table

Now we consider Detection Table.

Definition 7.9 For an (n, M, d)-binary code C such that $d \geq 2$, let

$$S_m := \sum_{c \in C} |\{c' \in C : \text{dis}\left(c', c\right) = m\}|.$$

Then

$$p_m := 1 - \frac{S_m}{M\binom{n}{m}} \tag{7.3}$$

is called the *m-bit fault resistance probability* for C.

When an m-bit fault is injected in the codeword, if the resulting word is not a codeword then the value will be set to *Null*. The only case when the fault is valid is when after m bits are flipped, the resulting word is still a codeword. Thus by Definition 7.6, $1 - p_m$ gives the theoretical probability of a *Valid* fault. Hence, the bigger the p_m, the better the m-fault resistance of the binary code.

Remark 7.2 When $m \leq d$, no codeword is at distance m from each other and hence $p_m = 1$.

Note that if $S_n = M$, i.e., for each codeword $\boldsymbol{c} \in C$, there exists a $\boldsymbol{c}' \in C$ such that $\text{dis}\left(\boldsymbol{c}, \boldsymbol{c}'\right) = n$, then we have

$$p_n = 1 - \frac{M}{M\binom{n}{n}} = 1 - 1 = 0.$$

That means, for this code, n-bit fault will always be injected successfully. In view of this, we exclude these kind of codes from our selection (see Algorithm 1). In practice, n and M are the fixed known values, from Eq. (7.3), to get bigger p_m the goal of choosing the code C is to make S_m small. There are several ways of achieving this depending on the preference of the user:

1. For small values of m, make $p_m = 0$: Choose code with a bigger minimum distance d, then p_m will be 1 for more values of m. Of course, there is a limit for the minimum distance that can be achieved (see Table 7.1). This particular scheme will be discussed in Sect. 7.3.3, where it is called Detection Scheme.
2. A certain m_0-bit fault resistance is desired: Choose code such that $S_{m_0} = 0$.
3. Sacrificing one m_0-bit fault resistance to achieve m-bit fault resistance for all other values of $m \neq m_0$: This is possible by using equidistant codes. That is, take code such that $|S_{m_0}| = M$. This particular scheme will be discussed in Sect. 7.3.3, where it is called Equidistant Detection Scheme.
4. Making all p_m almost equally large: Choose C such that S_m are similar for all $m > d$. Note that

$$\sum_{m=d+1}^{n} S_m = 2M$$

is always true.

Similar to last subsection, considering random fault, we define the following.

Algorithm 1: Ranking algorithm that chooses the code with the optimal leakage properties

Input : n: the codeword bit-length, M: number of codewords, d: minimum distance of the code, α_i: the leakage bit weights of the register, where i in $[\![1, n]\!]$

Output: An (n, M, d) binary code

for *Every set S of M words* **do**

  for $x == 0; x < |S|; x{+}{+}$ **do**

    for $y == x + 1; y < |S|; y{+}{+}$ **do**

      Calculate the distance dis $(S[x], S[y])$;

      if dis $(S[x], S[y]) < d$ *(or* dis $(S[x], S[y])! = d$*, depends on equidistance condition)* **then**

        continue with a different set S;

      if dis $(S[x], S[y]) == n$ **then**

        $n_{distance}{+}{+}$

    if $n_{distance} == n$ **then**

      continue with a different set S;

    Compute the estimated power consumption for codeword $S[x]$ and store the result in table A: $A[S[x]] = \Sigma_{i=1}^{n} \alpha_i S[x][i]$;

    Compute the estimated variance for bit leakages in $S[x]$ and store the result in table B: $B[S[x]] = \Sigma_{i=1}^{n}((\alpha_i S[x][i]) - \mu_{S[x]})^2$;

    Compute the bit with the highest bit leakage in $S[x]$ and store the result in table C: $C[S[x]] = max(\alpha_i S[x][i])$;

  Compute the register leakage variance for codewords in S and store the result in table D: $D[S] = \Sigma_{S[x]=1}^{|S|}(A[S[x]] - \mu_S)^2$;

  Choose the highest variance for register bit leakages for codewords in S and store the result in table E: $E[S] = max(B)$;

  Choose the value of the highest register bit leakage among the codewords in S and store the result in table F: $F[S] = max(C)$;

Get the optimal candidate using the following criteria:

1. Choose the candidates with the lowest register variances from $D[S]$;
2. From this set, choose the candidates with the lowest value of the highest leakage according to $F[S]$;
3. Finally, choose from the previous set, take the candidate with the lowest bit leakage variance according to $E[S]$;

return M codewords in case all the conditions are met, or an empty set otherwise

Table 7.1 Possible (n, M, d)-binary codes for $n = 8, 9, 10$, $M = 16$ and $n = 8$, $M = 4$

n	M	d
8	4	2, 3, 4, 5
8	16	2, 3, 4
9	16	2, 3, 4
10	16	2, 3, 4

Definition 7.10 For an (n, M, d)-binary code C such that $d \geq 2$, let p_m be its m-bit fault resistance probability for $1 \leq m \leq n$, then

$$p_{\text{rand}} := \sum_{m=1}^{n} \frac{1}{n} p_m$$

is called the *overall resistance index* for C.

Note that the bigger the p_{rand} is, the more resistant the code C is to random faults.

Lemma 7.1 *For an (n, M, d)-binary code C, if it is equidistant, then*

$$p_m = \begin{cases} 1 & m \neq d \\ 1 - \frac{M-1}{\binom{n}{d}} & m = d \end{cases}, \quad \text{and} \quad p_{\text{rand}} = 1 - \frac{M-1}{\binom{n}{d} n}.$$

7.3.3 Coding Schemes

Here we propose two different coding schemes:

1. Detection Scheme: Using binary code which has minimum distance at least 2.
2. Correction Scheme: Using binary code which has minimum distance at least 3 with error correction enabled look-up table.

Furthermore, as will be seen from the rest of this chapter, equidistant codes have different behaviors than codes that are not equidistant. Hence when equidistant codes are used, we emphasize the usage by referring to the schemes as "Equidistant detection scheme" and "Equidistant correction scheme," respectively.

We will analyze the m-bit fault resistant probability (with error) as well as overall resistance index (with error) for each of them using (n, M, d) binary codes for $n = 8, 9, 10$ and $M = 4, 16$. We chose $M = 4$ because it is easy to analyze and explain, and $M = 16$ because it can encode one nibble of the data; therefore, it is usable in a practical scenario. To illustrate the usage of the schemes we refer the reader to Appendix 2 for calculations of the probabilities for some specific codes as examples.

First, we discuss the possible values of the minimum distance d. As is well known in coding theory, fixing the length of the code n and minimum distance d, M is upper bounded by certain value. This upper bound is tight for small values n and d and still open for a lot of other values [7, p. 247]. In particular, for $n = 8, 9, 10$ and different values of d we know the exact possible values of M. In return, the possible values of d are known when n, M are fixed. In Table 7.1 we list the possible minimum distances that can be achieved for $n = 8, 9, 10$ and $M = 4$ or 16. Note that the values are taken from [7, p. 247, 248] and [5].

For equidistant binary code, we have the following constraint on d.

Lemma 7.2 *Let C be an (n, M, d) equidistant binary code such that $M \geq 3$, then d is even.*

Proof Recall the *Hamming weight* of a word $\boldsymbol{x} \in \mathbb{F}_2^n$ denoted by wt($\boldsymbol{x}$) is defined to be the number of nonzero coordinates in $\boldsymbol{x}$ [11, p. 46]. And we have the following relation (see [11, Corollary 4.3.4 and Lemma 4.3.5]):

$$wt(\boldsymbol{x}) + wt(\boldsymbol{y}) \equiv \text{dis}(\boldsymbol{x}, \boldsymbol{y}) \mod 2.$$

Take an (n, M, d) equidistant binary code C and any three distinct codewords $\boldsymbol{x}, \boldsymbol{y}, \boldsymbol{z} \in C$, we have

$$\text{dis}(\boldsymbol{x}, \boldsymbol{y}) + \text{dis}(\boldsymbol{y}, \boldsymbol{z}) + \text{dis}(\boldsymbol{z}, \boldsymbol{x}) \equiv 2wt(\boldsymbol{x}) + 2wt(\boldsymbol{y}) + 2wt(\boldsymbol{z}) \equiv 0 \mod 2.$$

Hence, d cannot be odd.

Furthermore we have $M \leq n + 1$[8]. Thus we will only consider $(8, 4, 2)$ and $(8, 4, 4)$ equidistant binary codes. The fact that such codes exist can be derived from [8].

7.4 Automated Generation and Evaluation of Codes

In this section, we will utilize the findings stated in Sect. 7.3 to design the algorithm that automatically generates codes with the optimal side-channel and fault detection properties for a given code length. First, we present the algorithm that finds the codes based on searching criteria in Sect. 7.4.1. Then we show properties of the codes that were produced by the algorithm in Sect. 7.4.2. To verify our theoretical results, we simulate fault injections into these codes, by using an automated fault simulator which will be explained in Sect. 7.4.3. Finally, we present and discuss the simulation results in Sect. 7.4.4.

7.4.1 Code Generation and Ranking Algorithm

When it comes to device leakage, it normally depends on the processed intermediate values. In [13], they proposed the first encoding scheme that assumed a stochastic leakage model over the Hamming weight model. In such model, leakage is formulated as follows:

$$T(x) = L(x) + \epsilon, \tag{7.4}$$

where L is the leakage function mapping the deterministic intermediate value (x) processed in the register to its side-channel leakage, and ϵ is the (assumed) mean-free Gaussian noise. For 8-bit microcontroller case, we can specify this function as $L(x) = \alpha_0 + \alpha_1 x_1 + \cdots \alpha_8 x_8$, where x_i is the ith bit of the intermediate value, and α_i is the ith bit weight leakage for specific register [16]. The α_i values can be obtained by using the following equation:

$$\alpha = (\mathbf{A}^T\mathbf{A})^{-1}\mathbf{A}^T\mathbf{T}, \tag{7.5}$$

where **A** is a matrix of intermediate values and **T** is a set of traces. After the device profiling which obtains the α values, we can use our ranking algorithm to select the optimal code with given inputs (Algorithm 1). Note that one can still use the Hamming weight model—for that case, α has to be defined as unity. In the following, we will explain how the algorithm works.

First, the inputs have to be specified—length (n), number of the codewords (M), minimum distance (d), and leakages of the register bits (α_i). Depending on these values, the algorithm analyzes every possible set of M codewords that can be a potential code candidate. Lines 2–3 iterate over every combination of two codewords. Lines 4–6 test if the minimum distance condition is fulfilled. Then, lines 7–10 check, whether for each codeword there exists another codeword which is at distance n from it—if yes, we skip this set. This condition is necessary in order to get a code resistant against n-bit flip (we will detail such case in Sect. 7.5). Lines 11–13 compute the 3 values that are used in order to calculate the values for the whole code in the later phase: estimated power consumption for the codeword, stored in table A, estimated variance for bit leakages in the codeword, stored in table B, and the highest bit leakage value, stored in table C. Next, the codeword value is stored in the index table I.

Lines 14–16 use the values from tables A, B, C to compute the register leakage variance ($\mu_{S[x]}$ denotes the mean leakage for a word $S[x]$), highest variance for bit leakages within registers, and value of the highest bit leakage within registers for the set S. These values are stored in tables D, E, F, respectively, and are used in the final evaluation.

The final evaluation is the last phase of the algorithm. First, it takes a subset of D with the best register leakage variance (μ_S denotes the mean leakage for codewords in S). It narrows this subset to candidate codes with the lowest value of the highest bit leakage according to set E. From these, it chooses the code with the lowest bit leakage variance using table F.

7.4.2 Properties of Generated Codes

Codes with the best side-channel and fault resistance properties according to Algorithm 1 with 4 codewords and length 8 can be found in Table 7.2. Their detailed properties are stated in Table 7.3. More codes with cardinality 16 and various distances can be found in Appendix 1.

Table 7.2 Codes used in evaluation

Code	Distance	Denoted by
`0x3D, 0x9D, 0xAD, 0xBC`	$= 2$	$C_{8,4,eq2}$
`0x0B, 0x19, 0x35, 0xA6`	$>= 2$	$C_{8,4,min2}$
`0x19, 0x35, 0x8A, 0xA6`	$>= 3$	$C_{8,4,min3}$
`0x55, 0x93, 0xA5, 0xC6`	$= 4$	$C_{8,4,eq4}$
`0x19, 0x27, 0x8A, 0xB4`	$>= 4$	$C_{8,4,min4}$
`0x19, 0x6A, 0x87, 0xF4`	$>= 5$	$C_{8,4,min5}$

For calculating the register variance, we follow the similar methodology as used in [13], together with their generated α values, but we improved their ranking algorithm by calculating the bit variances inside registers and by selecting the code which has the lowest leakage value for the highest leaking codeword. First part of Table 7.3 shows these three values, with the order of preference according to our ranking algorithm. Second part of the table shows bit fault resistance probabilities, denoted by p_m for m-bit flips in the codeword, as well as overall resistance index, denoted by p_{rand} for the code. The last part of the table shows the fault resistance probabilities with error correction, denoted by $p_{m,(e)}$, as well as overall resistance index with error correction, which is denoted by $p_{\text{rand},(e)}$. We do not consider codes with distance 1 because such codes do not provide protection against 1-bit flips and therefore the fault protection would be very low. However, such codes can still be used for minimizing the side-channel leakage.

In general, if we aim for higher distance values, we get better detection and correction capabilities, but the side-channel leakage is higher as well. That is because if the distance is higher, it is more likely that the variance of leakage among the codewords is bigger. Also, we can see that equidistant codes have a constant detection probability of 1 except the case when number of bit flips is the same as the code distance. Moreover, if we sum up the probabilities of all the bit flip faults for non-equidistant codes, the overall detection probability is lower. However, the side-channel leakage of equidistant codes is more than 10 times higher compared to non-equidistant codes.

7.4.3 Automated Fault Simulation

The fault simulator we used was customized for the purpose of evaluating a microcontroller assembly table look-up implementation of the encoding schemes presented in this chapter. More details on this simulator are provided in [1]. This simulator helps us to extend the theoretical results to real-world results, where one has to use capabilities of microprocessors for computing the results.

Table 7.3 Side-channel and fault properties of the codes

α = [0.613331, 0.644584, 0.602531, 0.190986, 0.586268,0.890951, 1.838814, 1.257943, 0.899922, 0.614699]

Code	$C_{8,4,eq2}$	$C_{8,4,min2}$	$C_{8,4,min3}$	$C_{8,4,eq4}$	$C_{8,4,min4}$	$C_{8,4,min5}$
Codeword variance	0.0158	1.150×10^{-5}	9.800×10^{-6}	0.0021	6.440×10^{-5}	6.743×10^{-3}
Highest leakage	4.9003	3.3445	3.3413	2.9514	3.3377	3.3445
Bit variance	0.2492	0.2748	0.2776	0.1535	0.2776	0.3702
p_1	1	1	1	1	1	1
p_2	0.8929	0.9821	1	1	1	1
p_3	1	0.9911	0.9821	1	1	1
p_4	1	0.9929	0.9857	0.9571	0.9857	1
p_5	1	0.9821	1	1	0.9643	0.9643
p_6	1	1	1	1	1	0.9643
p_7	1	0.9375	0.8750	1	1	1
p_8	1	1	1	1	1	1
p_{rand}	0.9866	0.9857	0.9804	0.9946	0.9938	0.9911
$p_{1,(e)}$	–	–	1	1	1	1
$p_{2,(e)}$	–	–	0.8929	1	1	1
$p_{3,(e)}$	–	–	0.9107	0.7857	0.9286	1
$p_{4,(e)}$	–	–	0.9143	0.9571	0.8429	0.8571
$p_{5,(e)}$	–	–	0.9286	0.7857	0.8929	0.8571
$p_{6,(e)}$	–	–	0.7500	1	0.7857	0.75
$p_{7,(e)}$	–	–	0.8750	1	1	0.75
$p_{8,(e)}$	–	–	0	1	1	1
$p_{\text{rand},(e)}$	–	–	0.7839	0.9411	0.9313	0.9018

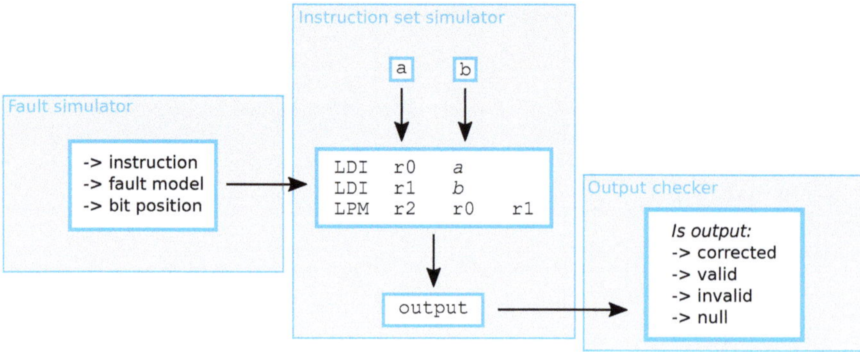

Fig. 7.1 Fault simulator operation overview

A high-level overview is given in Fig. 7.1. There are three instructions in total—the first two `LDI` load the two operands into registers `r0` and `r1`. Both of the operands are already encoded according to one of the coding schemes. The `LPM` instruction loads the data from the look-up table stored in the memory by using the values in `r0` and `r1`, and the result is stored to register `r2`. This part works as a standard instruction set simulator. During each execution, a fault is injected into the code. For each type of fault, we test all the possible combinations of codewords, and we disturbed all the instructions in our code. We have tested the following fault models:

- **Bit faults:** In this fault model, one to n bits in the destination register change its value to a complementary one.
- **Random byte faults:** The *random byte fault* model changes random number of bits in the destination register.
- **Instruction skip:** Instruction skip is a very powerful model that is capable of removing some countermeasures completely. We have tested a single instruction skip on all three instructions in the code.
- **Stuck-at fault:** In this fault model, the value of the destination register changes to a certain value, usually to all zeroes. Therefore, we have tested this value in our simulator.

After the output is produced under a faulty condition, it is analyzed by the output checker, which decides on its classification. Outputs can be of four types (*Corrected, Valid, Invalid, and Null*), and these types are described in detail in Sect. 7.2.2.

7.4.4 Simulated Results

Figure 7.2 shows plots for $C_{8,4,min4}$ and $C_{8,4,eq4}$, with and without the error correction. Instruction skip faults and stuck-at faults show zero success when

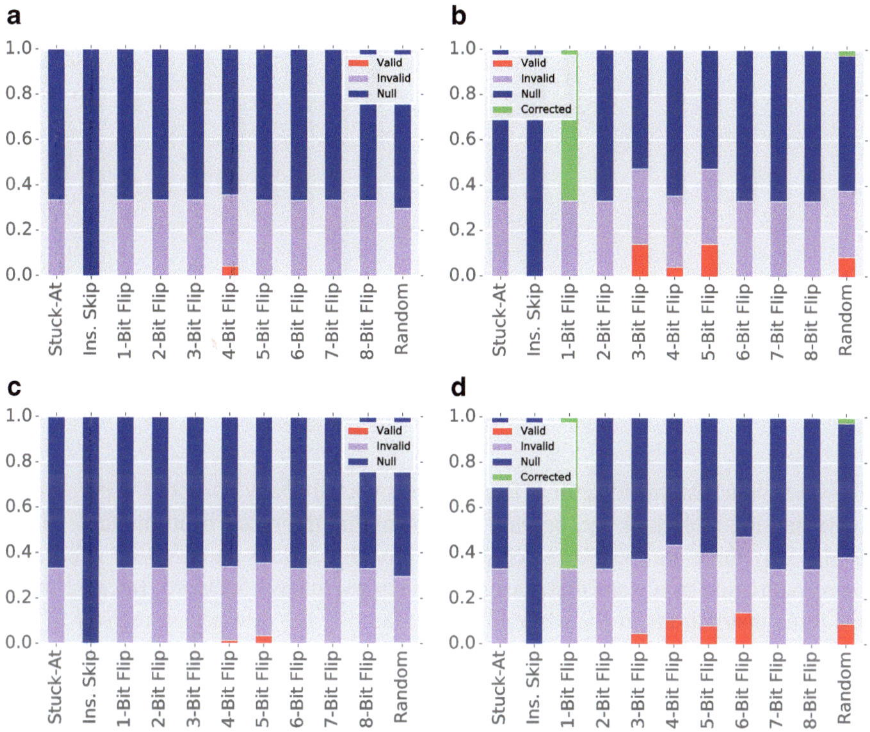

Fig. 7.2 Simulation results for $C_{8,4,eq4}$ with equidistant detection scheme in (**a**) and with equidistant correction scheme in (**b**); $C_{8,4,min4}$ with detection scheme in (**c**) and with correction scheme in (**d**)

attacking any of the generated codes. When it comes to bit flips, we can see that for better fault tolerance, one should not use the error-correction capabilities, since the properties of such codes allow changing the faulty codeword into another codeword, depending on the number of bit flips and minimum distance of the code. When deciding whether to choose an equidistant code or not, situation is the same as in Table 7.3—equidistant codes have slightly better fault detection properties, but worse side-channel leakage protection. Therefore, it depends on the implementer to choose a compromise between those two.

7.5 Discussion

First, we would like to explain the difference between the calculated results in Table 7.3 and the simulated results in Fig. 7.2 in equidistant code $C_{8,4,min4}$. Table 7.3 shows theoretical results assuming that error happens before using the look-up table.

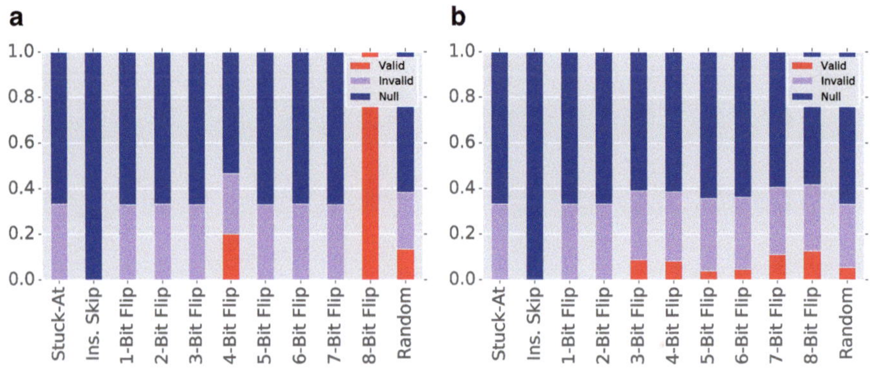

Fig. 7.3 Simulation results for the codes: (**a**) $C_{8,16,min4}$ and (**b**) $C_{8,16,min3}$

However, in a real-world setting, fault can be injected at any point of the execution, including the table look-up, or even after obtaining the result from the table. That is also why there are *Invalid* faults, despite the table always outputs *Null* in case of being addressed by a word that does not correspond to any codeword. Because there are three instructions in the assembly code, faulting the destination register of the last one after returning the value from the table results into 1/3 of *Invalid* faults in all the cases except instruction skips.

To explain the condition on lines 7–8 of Algorithm 1, we can take the code with $n = 8$, $M = 16$, and $d = 4$ as an example. The simulation result for this code is stated in Fig. 7.3a. Full results for this code are then in Table 7.5 in the appendix. There are no codes with these parameters that could satisfy the abovementioned condition—all 480 codes that can be constructed have the property that if any codeword is faulted by n bit flip, it will change to other codeword. Therefore, such codes are not suitable for protecting implementations against fault attacks. For this reason, it is more suitable to use the $C_{8,16,min3}$ code, stated in Fig. 7.3b, that does not suffer from such property.

To summarize the evaluation results, we point out the following findings:

- Correction scheme is not suitable for fault tolerant implementations—while it can be helpful in non-adversary environments, where it can be statistically verified, how many bits are usually faulted, and therefore, a proper error-correction function can be specified, in adversary-based settings, one cannot estimate the attacker capabilities. In case of correcting 1-bit error, for example, attacker who can flip multiple bits will have a higher probability of producing *Valid* faults, compared to using detection scheme with the same code.

- We can find an optimal code either from the fault tolerance perspective or from side-channel tolerance perspective—if we consider both, a compromise has to be made, depending on which attack is more likely to happen or how powerful an attacker can be in either setting. If we sacrifice the fault tolerance, we will normally get a code with distance 2 (e.g., side-channel resistant codes in [13] all have distance 2 and they are not equidistant codes); therefore, such codes will be vulnerable to 2-bit faults. On the other hand, by relaxing the power consumption variance condition, we will be able to choose codes with bigger distance, being able to resist higher number of bit faults.
- Both types of resistances can be improved if we sacrifice the memory and choose codes with greater lengths.
- Equidistant detection schemes is a good option in case the implementation can be protected against certain number of bit flips—because all the *Valid* faults are achieved only if the attacker flips the same number of bits as is the distance. However, this condition does not hold in case of equidistant correction schemes.

7.6 Chapter Summary

In this chapter, we provided a necessary background for constructing side-channel and fault attack resistant software encoding schemes. Current encoding schemes only cover side-channel resistance, and either do not discuss fault resistance or only state it as a side product of the construction, such as [19]. Our work defines theoretical bounds for fault detection and correction and provides an automated way to construct efficient codes that are capable of protecting the underlying computation against both physical attack classes.

To support our result with a practical case study, we designed an automated simulator to evaluate the table look-up operation under faulty conditions, by using a microcontroller assembly code. As expected, the codes constructed using the stated algorithm provide robust fault resistance, while keeping the side-channel leakage at the minimum.

Appendix 1: Generated Codes

In this section, we state the remaining codes generated by Algorithm 1, for $M = 16$ and $n = 8, 9, 10$ (Tables 7.4 and 7.5).

Table 7.4 Codes generated by Algorithm 1

Code	Length	Distance	Denoted by
0x0E, 0x4D, 0xF1, 0xEC, 0x2D, 0x26, 0x86, 0x8D, 0xA5, 0x46, 0xD9, 0x13, 0xD2, 0x79, 0x72, 0x5A	8	>= 2	$C_{8,16,min2}$
0x4D, 0x8B, 0x96, 0x43, 0xE9, 0xE2, 0xBA, 0xD5, 0x33, 0x2E, 0x3D, 0xFC, 0xA5, 0x5A, 0x76, 0xCE	8	>= 3	$C_{8,16,min3}$
0xBA, 0xD9, 0xEF, 0x73, 0x1F, 0xD6, 0x83, 0xB5, 0x26, 0x4A, 0x7C, 0x45, 0x29, 0x8C, 0xE0, 0x10	8	>= 4	$C_{8,16,min4}$
0x145, 0x15A, 0x1CA, 0x95, 0xCC, 0xDA, 0xC5, 0x18C, 0x0E, 0xD3, 0x19A, 0x185, 0x07, 0x193, 0x9C, 0x153	9	>= 2	$C_{9,16,min2}$
0x07, 0xF3, 0x146, 0xB5, 0xEC, 0x2E, 0x1BA, 0x165, 0x13C, 0x1D, 0x1D9, 0x5B, 0x1D4, 0x18B, 0x96, 0x185	9	>= 3	$C_{9,16,min3}$
0x3B, 0x75, 0x9D, 0x14B, 0x1D4, 0x1A5, 0xEC, 0x13C, 0x1F9, 0x193, 0x07, 0xDA, 0x166, 0xB6, 0x1AA, 0xE3	9	>= 4	$C_{9,16,min4}$
0x5D, 0xDC, 0x34B, 0x25C, 0x1CB, 0x359, 0xCE, 0x3CA, 0x3E6, 0x1F5, 0x1E7, 0x3F4, 0x375, 0x24E, 0x4F, 0x1D9	10	>= 2	$C_{10,16,min2}$
0xA7, 0x235, 0x3C8, 0x22A, 0x14C, 0x39, 0x298, 0x3C5, 0x3B1, 0x8B, 0x1B4, 0x1C, 0x326, 0x156, 0x169, 0x353	10	>= 3	$C_{10,16,min3}$
0x2D, 0x16A, 0x18C, 0x97, 0x136, 0x21A, 0x347, 0x3D4, 0x3A5, 0x159, 0x275, 0x2E6, 0xCB, 0xF8, 0x1F3, 0x24C	10	>= 4	$C_{10,16,min4}$

Appendix 2: Fault Resistance Probabilities

In this section, we show the detailed theoretical calculations of fault resistance probabilities and the overall resistance index (with error) for some specific examples.

Equidistant Detection Scheme
Using Lemma 7.1, we list the values of p_ms and p_{rand} in Table 7.6 for $(8, 4, 2)$ and $(8, 4, 4)$ equidistant binary codes.

Table 7.5 Side-channel and fault properties of the codes from Table 7.4

α	0.613331, 0.644584, 0.602531, 0.190986, 0.586268, 0.890951, 1.838814, 1.257943, 0.899922, 0.614699								
Code	$C_{8,16,min2}$	$C_{8,16,min3}$	$C_{8,16,min4}$	$C_{9,16,min2}$	$C_{9,16,min3}$	$C_{9,16,min4}$	$C_{10,16,min2}$	$C_{10,16,min3}$	$C_{10,16,min4}$
Codeword variance	0.0013	0.1190	1.8231	2.935×10^{-4}	0.0091	0.1043	3.920×10^{-5}	0.0017	0.0134
Highest leakage	3.2607	3.2607	0.1910	3.9510	3.9967	3.5960	4.7812	3.8545	3.9011
Bit variance	0.4657	0.3949	0.3367	0.2552	0.3571	0.3366	0.3170	0.3875	0.3929
p_1	1	1	1	1	1	1	1	1	1
p_2	0.9241	1	1	0.9167	1	1	0.9417	1	1
p_3	0.9286	0.9129	1	0.9732	0.9628	1	0.9854	0.9844	1
p_4	0.9714	0.9179	0.8	0.9752	0.9692	0.9147	0.9857	0.9857	0.9780
p_5	0.9308	0.9621	1	0.9683	0.9692	1	0.9866	0.9861	0.9767
p_6	0.9241	0.9554	1	0.9881	0.9643	0.9613	0.9911	0.9839	0.9911
p_7	1	0.8906	1	0.9549	0.9722	1	0.9854	0.9721	0.9865
p_8	0.1250	0.8750	0	1	0.9861	0.8889	0.9861	0.9861	0.9861
p_9	–	–	–	1	1	1	1	1	0.9625
p_{10}	–	–	–	–	–	–	1	1	1
p_{rand}	0.8505	0.9392	0.85	0.9771	0.9804	0.9739	0.9862	0.9904	0.9881
$p_{1,(e)}$	–	1	1	–	1	1	–	1	1
$p_{2,(e)}$	–	0.4778	1	–	0.7400	1	–	0.8750	1
$p_{3,(e)}$	–	0.5022	0	–	0.7782	0.4881	–	0.8844	0.8458
$p_{4,(e)}$	–	0.4179	0.8	–	0.6667	0.9147	–	0.8399	0.8381
$p_{5,(e)}$	–	0.4174	0	–	0.6726	0.4187	–	0.8343	0.8219
$p_{6,(e)}$	–	0.5089	1	–	0.6964	0.9613	–	0.8131	0.7970
$p_{7,(e)}$	–	0.4531	0	–	0.6944	0.5069	–	0.8240	0.8823
$p_{8,(e)}$	–	0	0	–	0.7639	0.8889	–	0.8111	0.8028
$p_{9,(e)}$	–	–	–	–	0.8750	0	–	0.8750	0.8375
$p_{10,(e)}$	–	–	–	–	–	–	–	1	0.6250
$p_{rand,(e)}$	–	0.4722	0.4750	–	0.7652	0.6865	–	0.8757	0.8450

Table 7.6 Theoretical values of p_m for (n, M, d)-equidistant binary code

(n, M, d)	p_1	p_2	p_3	p_4	p_5	p_6	p_7	p_8	p_{rand}
(8, 4, 2)	1	0.8929	1	1	1	1	1	1	0.9866
(8, 4, 4)	1	1	1	0.9571	1	1	1	1	0.9946

Table 7.7 Distance between each pair of codewords in the (8, 4, 4)-binary code $C_{8,4,min4}$

dis $(\cdot, \cdot)$	00011001	00100111	10001010	10110100
00011001	0	5	4	5
00100111	5	0	5	4
10001010	4	5	0	5
10110100	5	4	5	0

Detection Scheme

Since we require that dis $(C) \geq 2$ for Detection Scheme, for 1-bit fault, we expect the results to be *Null*, which means $p_1 = 1$. Now we give a theoretical calculation for the (8, 4, 4)-binary code $C_{8,4,min4} = \{00011001, 00100111, 10001010, 10110100\}$. We first list the distance between every pair of codewords in Table 7.7.

By Eq. (7.3), we can then calculate the m-bit fault resistance probabilities and the overall resistance index for C:

$$p_2 = p_3 = 1 - \frac{1}{4}(0+0+0+0) = 1,$$

$$p_4 = 1 - \frac{1}{4\binom{8}{4}}(2+0+1+1) = \frac{69}{70} \approx 0.9857,$$

$$p_5 = 1 - \frac{1}{4\binom{8}{5}}(2+2+2+2) = \frac{27}{28} \approx 0.9643,$$

$$p_6 = p_7 = p_8 = 1 - \frac{1}{4}(0+0+0+0) = 1, \quad p_{\text{rand}} = \sum_{m=1}^{8} \frac{1}{8} p_m = 0.9938.$$

Correction Scheme

m-bit fault resistance probabilities with error correction for the same (8, 4, 4)-binary code $C_{8,4,min4} = \{00011001, 00100111, 10001010, 10110100\}$. As dis $(C) = 4$, by Remark 7.1 it is an 1-error-correcting code. By Eq. (7.2), $p_{m,(e)} = 1$ for $m = 1$. To calculate $p_{m,(e)}$ for $m \geq 2$, we first list the table of cardinalities of $F_{\mathbf{c},m}$ for $\mathbf{c} \in C$ and $m = 2, 3, \ldots, 8$ in Table 7.8.

By Eq. (7.2), we can then calculate the m-bit fault resistance probabilities with error correction as well as the overall resistance index with error correction for C.

Table 7.8 Cardinality of $F_{c,m}$ for $m = 2, 3, \ldots, 8$ and $\boldsymbol{c} \in C_{8,4,min4}$

	$\lvert F_{c,2}\rvert$	$\lvert F_{c,3}\rvert$	$\lvert F_{c,4}\rvert$	$\lvert F_{c,5}\rvert$	$\lvert F_{c,6}\rvert$	$\lvert F_{c,7}\rvert$	$\lvert F_{c,8}\rvert$
00011001	0	4	11	6	6	0	0
00100111	0	4	11	6	6	0	0
10001010	0	4	11	6	6	0	0
10110100	0	4	11	6	6	0	0

$$p_{2,(e)} = 1 - \frac{1}{4\binom{8}{2}}(0+0+0+0) = 1,$$

$$p_{3,(e)} = 1 - \frac{1}{4\binom{8}{3}}(4+4+4+4) = \frac{13}{14} \approx 0.9286,$$

$$p_{4,(e)} = 1 - \frac{1}{4\binom{8}{4}}(11+11+11+11) = \frac{59}{70} \approx 0.8429,$$

$$p_{5,(e)} = 1 - \frac{1}{4\binom{8}{5}}(6+6+6+6) = \frac{25}{28} \approx 0.8929,$$

$$p_{6,(e)} = 1 - \frac{1}{4\binom{8}{6}}(6+6+6+6) = \frac{11}{14} \approx 0.7857,$$

$$p_{7,(e)} = p_{8,(e)} = 1 - \frac{1}{4}(0+0+0+0) = 1,$$

$$p_{\text{rand},(e)} = \sum_{m-1}^{8} \frac{1}{8} p_{m,(e)} = 0.9313.$$

References

1. J. Breier, On analyzing program behavior under fault injection attacks, in *2016 Eleventh International Conference on Availability, Reliability and Security (ARES)* (IEEE, Piscataway, 2016), pp. 1–5
2. J. Breier, X. Hou, Feeding two cats with one bowl: on designing a fault and side-channel resistant software encoding scheme, in *Cryptographers' Track at the RSA Conference* (Springer, Berlin, 2017), pp. 77–94
3. J. Breier, D. Jap, C.-N. Chen, Laser profiling for the back-side fault attacks: with a practical laser skip instruction attack on AES, in *Proceedings of the 1st ACM Workshop on Cyber-Physical System Security (CPSS '15)* (ACM, New York, 2015), pp. 99–103
4. J. Breier, D. Jap, S. Bhasin, The other side of the coin: analyzing software encoding schemes against fault injection attacks, in *2016 IEEE International Symposium on Hardware Oriented Security and Trust (HOST)* (IEEE, Piscataway, 2016), pp. 209–216
5. A.E. Brouwer, J.B. Shearer, N.J.A. Sloane, W.D. Smith, A new table of constant weight codes. IEEE Trans. Inf. Theory **36**(6), 1334–1380 (1990)

6. C. Chen, T. Eisenbarth, A. Shahverdi, X. Ye, Balanced encoding to mitigate power analysis: a case study, in *International Conference on Smart Card Research and Advanced Applications*. Lecture Notes in Computer Science (Springer, Berlin, 2014), pp. 49–63
7. J.H. Conway, N.J.A. Sloane, *Sphere Packings, Lattices and Groups*, vol. 290 (Springer, Berlin, 2013)
8. F.-W. Fu, T. Kløve, Y. Luo, V.K. Wei, On equidistant constant weight codes. Discret. Appl. Math. **128**(1), 157–164 (2003)
9. L. Goubin, J. Patarin, DES and differential power analysis. The "duplication" method, in *International Workshop on Cryptographic Hardware and Embedded Systems (CHES)*. Lecture Notes in Computer Science (Springer, Berlin, 1999), pp. 158–172
10. P. Hoogvorst, J.-L. Danger, G. Duc, Software implementation of dual-rail representation, in *Second International Workshop on Constructive Side-Channel Analysis and Secure Design (COSADE)*, Darmstadt (2011)
11. S. Ling, C. Xing, *Coding Theory: A First Course* (Cambridge University Press, Cambridge, 2004)
12. F.J. MacWilliams, N.J.A. Sloane, *The Theory of Error Correcting Codes* (Elsevier, Amsterdam, 1977)
13. H. Maghrebi, V. Servant, J. Bringer, There is wisdom in harnessing the strengths of your enemy: customized encoding to thwart side-channel attacks – extended version–. Cryptology ePrint Archive, Report 2016/183, 2016. http://eprint.iacr.org/
14. P. Rauzy, S. Guilley, Z. Najm, Formally proved security of assembly code against leakage. IACR Cryptology ePrint Arch. **2013**, 554 (2013)
15. F. Regazzoni, L. Breveglieri, P. Ienne, I. Koren, Interaction between fault attack countermeasures and the resistance against power analysis attacks, in *Fault Analysis in Cryptography* (Springer, Berlin, 2012), pp. 257–272
16. W. Schindler, K. Lemke, C. Paar, A stochastic model for differential side-channel cryptanalysis, in *International Workshop on Cryptographic Hardware and Embedded Systems* (Springer, Berlin, 2005), pp. 30–46
17. T. Schneider, A. Moradi, T. Güneysu, ParTI – towards combined hardware countermeasures against side-channel and fault-injection attacks, in *Annual Cryptology Conference* (Springer, Berlin, 2016), pp. 302–332
18. N. Selmane, S. Bhasin, S. Guilley, T. Graba, J.-L. Danger, WDDL is protected against setup time violation attacks, in *2009 Workshop on Fault Diagnosis and Tolerance in Cryptography (FDTC)* (IEEE, Piscataway, 2009), pp. 73–83
19. V. Servant, N. Debande, H. Maghrebi, J. Bringer, Study of a novel software constant weight implementation, in *International Conference on Smart Card Research and Advanced Applications* (Springer, Berlin, 2014), pp. 35–48
20. K. Tiri, I. Verbauwhede, A logic level design methodology for a secure DPA resistant ASIC or FPGA implementation, in *Proceedings Design, Automation and Test in Europe Conference and Exhibition*, vol. 1 (IEEE, Piscataway, 2004), pp. 246–251
21. E. Trichina, R. Korkikyan, Multi fault laser attacks on protected CRT-RSA, in *2010 Workshop on Fault Diagnosis and Tolerance in Cryptography (FDTC)* (IEEE, Piscataway, 2010), pp. 75–86

Chapter 8
Idempotent Instructions to Counter Fault Analysis Attacks

Sikhar Patranabis and Debdeep Mukhopadhyay

8.1 Introduction

Embedded systems are increasingly representing the largest segment of the consumer electronic market, and are now a critical part of daily life with the growth of smart cards, mobile devices, autonomous vehicles, sensor-based appliances and, in a broader sense, the Internet of Things (IoT). Products such as credit cards, smart phones, and SIM cards are heavily reliant on embedded systems, while also storing and manipulating large volumes of customer data. This makes security an essential component of all embedded systems, and physical attacks such as side-channel attacks [7, 8] and fault analysis attacks [4, 5] represent a threat to such security guarantees. Hence, embedded systems should be implemented in a manner that resists such physical attacks. This requirement is especially applicable in the context of embedded systems implementing cryptographic algorithms, where implementation attacks can cause potentially devastating leakages.

Fault analysis attacks on cryptographic implementations are typically characterized by three major features—the fault model, the fault injection mechanism, and the analysis technique. The fault model typically relates to the spatio-temporal nature of the injected fault. For example, if the target is a block cipher implementation (such as AES), the fault model could be a "single-bit" fault model, a "single-byte" fault model, or a "multiple-byte" fault model, depending on the number of state bites/bytes affected by the fault injection. Any fault injection is typically followed by an analysis in an attempt to recover the secret key, and depending on such techniques, fault attacks may be classified as "differential fault analysis" (DFA), "differential fault intensity analysis" (DFIA), and "safe-error analysis" (SEA).

S. Patranabis (✉) · D. Mukhopadhyay
Department of Computer Science and Engineering, Indian Institute of Technology, Kharagpur, India
e-mail: sikhar.patranabis@iitkgp.ac.in; debdeep@iitkgp.ac.in

J. Breier et al. (eds.), *Automated Methods in Cryptographic Fault Analysis*,
https://doi.org/10.1007/978-3-030-11333-9_8

Interestingly, in a practical attack scenario, the appropriate choice of fault model and analysis technique to attack a target device often depends on the fault injection mechanism. In hardware implementations built on ASICs/FPGAs, faults are often injected via voltage/clock glitches or EM/laser pulses, but the fundamental idea in most cases is to cause bit toggles in various registers via either setup time violations or directed pulse injections. In software-based implementations on platforms such as micro-controllers or embedded processors, a popular avenue of fault injection is "instruction skips." In this fault injection mechanism, the adversary skips one or more assembly instructions to induce a faulty computation. Recent studies have shown this model to be both practical and efficient on a variety of platforms [2, 10, 11] and using a variety of fault injection mechanisms [1, 6, 12]. Consequently, instruction skips are a realistic threat to cryptographic implementations targeting embedded processors and micro-controllers.

In this chapter, we present to the reader an instruction-level redundancy based countermeasure to thwart instruction skip attacks. The approach is based on the assumption that while it is easy to inject identical faults in independent executions of an algorithm, introducing faults in two instructions separated by a few clock cycles is significantly harder. The scheme involves rewriting each individual instruction by a sequence of instructions that are immune to single instruction skips. To make the analysis concrete, we illustrate the countermeasure approach using the Intel x86 processor as an example; however, in practice, this approach is applicable to a wide class of implementations across varying architectures (ISAs).

8.2 Classifying Assembly Instructions

The first step in the study of instruction-level redundancy is to classify the assembly instructions in any given ISA into certain sub-categories, depending on the ease with which they may be converted into redundant instruction sequences with the same functionality. We present here the classification proposed by Heydemann et al. in [9].

1. **Idempotent Instructions:** Idempotent instructions may be duplicated any number of times without affecting the final outcome of the program. Consider, for instance, the instruction `move r1,r2`, which is common across nearly all existing ISAs. The instruction operates on two registers—a source register `r1` and a destination register `r2`, and copies the content of $r1$ to $r2$ (the reverse may also be the case depending on the specific conventions of a given ISA). These instructions are thus the easiest to protect against instruction skip attacks—they may simply be duplicated as many times as necessary.
2. **Separable Instructions:** These instructions are by default non-idempotent, meaning that duplicating them does not preserve functionality. However, they may be alternatively expressed as a functionally equivalent sequence of one or

more idempotent instructions, which can then be made redundant via duplication. Consider, for example, the instructions `add r1,r2` and `add r1,r2,r3`, and assume hypothetically that both these instructions are present in the same ISA, with the following semantic meanings:

- The first instruction computes the sum of the contents in the registers `r1` and `r2`, and stores the result of this computation in `r2` itself.
- The second instruction `add r1,r2,r3` has a separate destination register `r3`, which is distinct from both source registers.

The first instruction is not idempotent, since it does not preserve the same output upon duplication. The second instruction, on the other hand, is idempotent. Now, one can re-write the non-idempotent instruction `add r1,r2` as the following sequence of idempotent instructions:

- `add r1,r2,r3`
- `mov r3, r2`

This sequence now may be made fault-tolerant by repeating each instruction as many times as necessary.

3. **Specific Instructions:** These are non-idempotent instructions that cannot be directly written as a generic sequence of idempotent instructions. However, one can construct functionally equivalent redundant instruction sequences corresponding to these instructions that are tolerant against single instruction skips. These instruction sequences are typically specific to the underlying ISA. In this chapter, we will show examples of such instructions in the Intel x86 ISA.
4. **Non-Replaceable and Partially Replaceable Instructions:** This is the final class of instructions that cannot be made sufficiently redundant by any functionally equivalent sequence of instructions, so as to resist instruction skip attacks with 100% probability. However, for such instructions, one could create a partially redundant instruction sequence, which makes it difficult but not impossible for the adversary to precisely inflict a skip attack that affects the eventual outcome of the program in an exploitable manner.

8.3 Examples from the x86 ISA

In this section, we discuss some examples of each class of instructions from the Intel x86 ISA, beginning with idempotent instructions. It turns out that most assembly instructions that are idempotent have the following common format: they have a disjoint set of source and destination operands, and the value of the destination operand after the execution of the instruction is determined solely by the value of the source operands. Finally, the control flow of the program does not typically depend on the outcome of such instructions.

The foremost instance of an idempotent instruction is the mov instruction, which either copies the content of one register to the other or loads/stores a certain value from/to a given memory location to/from a register. The following is a list of ways in which this instruction could be used as per the x86 ISA (note that the suffix "l" at the end of the instruction indicates that each operand is a 32-bit register):

- `movl %eax,%ebx` (copies the contents of the register `eax` to the register `ebx`)
- `movl %eax,-4(%ebp)` (stores the contents of the register `eax` at the address (`%ebp-4`))
- `movl -8(%ebp),%eax` (loads the value at the address (`%ebp-8`) to the register `eax`).

The reader may observe that the each of the aforementioned assembly instructions satisfies the properties of idempotent instructions and, as such, can be made tolerant to skip attacks via simple duplication, under the assumption that the adversary can only skip one of the two instructions.

A second instance of an idempotent instruction from the x86 ISA is the "load effective address" instruction, abbreviated as `lea`. This instruction is primarily meant for performing memory addressing calculations without actually accessing the memory content. The syntax of the instruction is as follows:

```
leal %esi, [%ebx + 8*%eax + 4]
```

which translates to operating on the addresses stored in the registers `eax` and `ebx` and storing the resultant address in the register `eax`. Interestingly, this instruction can be used not only for computations involving high level memory references, but also for simple addition operations. Finally, when the destination operand is different from the source operands, this instruction satisfies all the properties of an idempotent instruction. As explained later, this instruction is very useful in re-writing separable instructions as sequences of idempotent instructions.

Table 8.1 summarizes the idempotent instructions from the x86 ISA as described above, along with their corresponding fault-tolerant replacement sequences.

Table 8.1 Fault-tolerant replacement sequences for idempotent instructions

Instruction	Fault-tolerant replacement sequence
`movl %eax,%ebx` (copies `%eax` to `%ebx`)	`movl %eax,%ebx` `movl %eax, %ebx`
`movl %eax,-4(%ebp)` (stores `%eax` at the address `%ebp-4`)	`movl %eax,-4(%ebp)` `movl %eax,-4(%ebp)`
`movl -8(%ebp),%eax` (loads the value at the address `%ebp-8` to `%eax`)	`movl -8(%ebp),%eax` `movl -8(%ebp),%eax`
`leal %esi, [ebx + 8*eax + 4]` (stores `(%ebx + 8*%eax + 4)` in `%esi`)	`leal %esi, [%ebx + 8*%eax + 4]` `leal %esi, [%ebx + 8*%eax + 4]`

8.3.1 *Separable Instructions in the x86 ISA*

We now present some examples of separable instructions in the x86 ISA. Recall that separable instructions are not by themselves idempotent; however, they can be written as a functionally equivalent sequence of idempotent instructions. The first example we consider of such an instruction is the `add` instruction, with the following syntax:

```
addl %eax, %ebx
```

which essentially translates to adding the content of the registers `eax` and `ebx`, and storing the outcome back in the register `eax`. Note that since the destination register is also a source register, the `add` instruction is not idempotent. However, one can work around this issue via a combination of idempotent instructions described above. In particular, consider the following sequence of instructions:

```
movl %eax, %ecx
leal %eax, [%ebx + %ecx]
```

The reader may observe that this sequence of idempotent instructions has exactly the same functional effect as the `add` instruction. One can now make this sequence fault-tolerant via simple duplication.

The next instruction that we consider is the `push` instruction, with the following syntax:

```
pushl %eax
```

which translates to pushing the content of the register `eax` to a new memory location at the top of the stack, and updating the stack pointer register `esp` to point to this new memory location. Note again that this instruction is not idempotent since pushing some register content to the top of the stack once is not functionally equivalent to pushing the same twice. Our aim is, however, to write it as a sequence of idempotent instructions. To do this, we first consider the following sequence of instructions:

```
subl %esp, $4
movl %eax, (%esp)
```

Observe that this sequence is indeed functionally equivalent to the `push` instruction; however, the `sub` instruction is similar to the `add` instruction in the sense that it is not idempotent. We work around this by assuming that there exists a designated register `rx` that is pre-loaded with the binary equivalent of -4 in two's complement notation at the beginning of the program. Under this assumption, consider the following sequence of instructions:

```
movl %esp, %ebx
leal %esp, [%ebx + %rx]
movl %eax, (%esp)
```

The reader may observe that this sequence of idempotent instructions has exactly the same functional effect as the push instruction. One can now make this sequence fault-tolerant via simple duplication.

The final instruction we consider is the pop instruction, with the following syntax:

```
popl %eax
```

which translates to copying the content at the top of the stack to the register eax, and updating the stack pointer register esp to point to the next memory location, which is the new stack-top. Now, consider the following sequence of instructions:

```
movl (%esp), %eax
addl %esp, $4
```

Observe that this sequence is indeed functionally equivalent to the pop instruction. Now replace addl by its idempotent equivalent, such that the resulting sequence of instructions takes the form

```
movl (%esp), %eax
movl $4, %ecx
leal %esp, [%bax + %ecx]
```

Once again, the reader may observe that this sequence of idempotent instructions has exactly the same functional effect as the pop instruction, and once again, one can now make this sequence fault-tolerant via simple duplication.

Table 8.2 summarizes how the aforementioned separable instructions can be replaced by functionally equivalent fault-tolerant instruction sequences.

8.3.2 Special Instructions in the x86 ISA

In certain cases, it may not be possible to write a given instruction as a sequence of idempotent instructions. However, this does not imply that such an instruction cannot be made fault-tolerant. In particular, there exist certain "special instructions" that, while not replaceable by a sequence of idempotent instructions, can still be made fault-tolerant by choosing certain specific instruction sequences.

Table 8.2 Fault-tolerant replacement sequences: separable instructions

Instruction	Fault-tolerant replacement sequence
`addl %eax,%ebx`	`movl %eax, %ecx`
	`movl %eax, %ecx`
	`leal %eax, [%ebx + %ecx]`
	`leal %eax, [%ebx + %ecx]`
`pushl %eax`	`movl %esp, %ebx`
	`movl %esp, %ebx`
	`leal %esp, [%ebx + %rx]`
	`leal %esp, [%bax + %rx]`
	`movl %eax, (%esp)`
	`movl %eax, (%esp)`
`popl %eax`	`movl (%esp), %eax`
	`movl (%esp), %eax`
	`movl %esp, %ebx`
	`movl %esp, %ebx`
	`movl $4, %ecx`
	`movl $4, %ecx`
	`leal %esp, [%ebx + %ecx]`
	`leal %esp, [%bax + %ecx]`

In this subsection, we present such a special instruction from the x86 ISA, namely the `call <function>` instruction. This instruction works as follows: first, the return address is pushed onto the stack, followed by an unconditional jump of the control to the beginning of the target function. We simulate this functionality by the following alternative sequence of instructions:

```
movl <returnlabel>, %ebx
movl $1, %ecx
leal %eax, [%ebx + %ecx]
movl %esp, %ebx
leal %esp, [%ebx + %rx]
movl %eax, (%esp)
jmp <function>
returnlabel:
```

We provide some insight into why the aforementioned sequence simulates `call <function>` instruction, although it is probably immediately evident to the observant reader. The sequence computes the appropriate return address post-function-execution by adding 1 to the address represented by `returnlabel`. Next, the return address is pushed onto the stack (here, we use the idempotent

Table 8.3 Fault-tolerant replacement sequence: call <function>

movl <returnlabel>, %ebx
movl <returnlabel>, %ebx
movl $1, %ecx
movl $1, %ecx
leal %eax, [%ebx + %ecx]
leal %eax, [%ebx + %ecx]
movl %esp, %ebx
movl %esp, %ebx
leal %esp, [%ebx + %rx]
leal %esp, [%ebx + %rx]
movl %eax, (%esp)
movl %eax, (%esp)
jmp <function>
jmp <function>
returnlabel:

sequence corresponding to the push instruction under the assumption that the register $\%rx$ holds the constant -4 in two's complement notation). Note that the only instruction which is not idempotent in this sequence is the jmp (unconditional jump) instruction. The challenge is thus to make this sequence tolerant to skip even in the presence of such a non-idempotent instruction.

To achieve this, we use the fault-tolerant sequence of instructions presented in Table 8.3, which basically duplicates the jmp instruction, while making sure that only one of the original and duplicate occurrences of the instruction is executed within a single execution of the call <function> instruction. We leave it as an exercise to argue that this is indeed the case.

8.3.3 Non-replaceable and Partially Replaceable Instructions in the x86 Instruction Set

Despite all the techniques described above, there remain a few instructions in nearly every ISA that cannot be replaced by a fault-tolerant sequence of instructions. One such example of a non-replaceable instruction in the x86 ISA is the jne (jump if not equal) instruction. To protect such instructions against skip attacks, more elaborate fault detection schemes appear necessary. Some other instructions may be "partially replaceable," meaning that they can be re-written using a combination of several idempotent instructions and a few non-idempotent ones. This helps reduce the probability that the adversary can skip precisely the non-idempotent instructions. One such example in the x86 ISA is the sub instruction, as illustrated in Table 8.4.

Table 8.4 Partially fault-tolerant replacement sequence: `subl %eax,%ebx`

Instruction	Partially fault-tolerant replacement sequence
`subl %eax,%ebx`	`compl %eax, %rx`
	`movl $1, %ry`
	`movl $1, %ry`
	`leal %rz, [%rx + %ry]`
	`leal %rz, [%rx + %ry]`
	`movl %eax, %ecx`
	`movl %eax, %ecx`
	`leal %eax, [%ecx + %rz]`
	`leal %eax, [%ecx + %rz]`

8.4 Implementation and Automation

In the aforementioned discussion, we have presented a countermeasure strategy against fault injection via skipping of assembly instructions. Any countermeasure idea is useful in the context of actual industrial products if it is amenable to implementations that are both cost-efficient and resistant to errors. Manual implementations by experts incur huge costs to the industry while also being prone to error, and are hence less desirable. A more cost-efficient and less erroneous alternative is to automate the process of implementing such countermeasures.

In this section, we present to the reader an LLVM compiler-based approach introduced in [3] to automate the implementation of instruction redundancy-based countermeasures. The approach modifies certain passes of the original LLVM-compiler and also introduces some new ones. The compiler takes as input an unprotected implementation of the target code to be executed, and outputs a protected binary, equipped with the instruction redundancy countermeasure.

8.4.1 Overview of Approach

The LLVM-based compilation approach proposed in [3] roughly consists of the following steps, each of which is automated by modifying/replacing certain parts of the LLVM compiler.

- **Identifying Idempotent Instructions:** Given an assembly level representation of the target code, the automation first identifies the set of idempotent instructions. As already discussed, these instructions can be immediately duplicated for fault tolerance. The identification may be done in various ways—either by looking up a table of all idempotent instructions in the ISA or by checking if the destination operand for a given instruction different from the source operand(s).

- **Replacing Non-Idempotent Instructions:** The next step is to convert certain classes of non-idempotent instructions, namely the replaceable and special instructions, into equivalent instruction sequences composed almost entirely of idempotent instructions. This step requires a modification to the *register allocator pass*. Once the transformation is complete, each instruction in the sequence is then duplicated for fault tolerance.

8.4.2 When to Replace Instructions?

Compiling a program using an LLVM compiler involves several stages, where each stage is associated with a unique representation of the original code, culminating in the final target platform-dependent representation. While the intermediate representation (IR) stage appears to be the most likely choice for instruction replacement (since it allows the countermeasure to be generalized to any language and architecture supported by the compiler), this is often difficult. As pointed out by the authors of [3], the IR representation may contain *static single assignments* (SSAs), which prevents updating a virtual register more than once in specific delimited regions of the code (such a region may be referred to as a "basic block"). The following example source code illustrates this scenario:

```
int mult(int x, int y, int z)
{
    return x * y * z;
}
```

The corresponding byte-code in the LLVM IR looks like the following:

```
%temp0 = mul %x, %y
%temp1 = mul %temp0, %z
ret %temp1
```

As per the techniques described above, and keeping the SSA restriction in mind, a fault-tolerant sequence corresponding to these instructions would be as follows:

```
%temp0 = mul %x, %y
%temp01 = mul %x, %y
%temp1 = mul %temp0, %z
%temp11 = mul %temp0, %z
ret %temp11
```

Note that each register is updated exactly once inside the basic block. However, the renamed virtual registers `%temp01` and `%temp11` are not used anywhere else in the block, and hence with very high probability, will trivially be removed by the dead code elimination (DCE) pass. Hence, the instruction replacement phase must occur post-SSA, preferably after the actual physical registers have already been allocated.

8.4.3 Modifying the Instruction Selection Pass

The first compilation pass that needs modification is the instruction selection pass (ISP). In this pass, the program is transformed into a low-level representation that is typically close to the target language. This pass interfaces with the application binary interface (ABI) to select the appropriate sequence of instructions for each operation described by the high-level development language (HDL). For example, upon encountering an instruction of the form

```
x = x*y + z;
```

the instruction selection pass would, by default, select the multiply and accumulate (`mla`) instruction, which is typically not idempotent. This phase is thus modified to prioritize the selection of idempotent/replaceable instructions wherever possible, while also optimizing register usage. For example, for the aforementioned instruction, the modified instruction selection pass would choose a sequence of `add` and `mov` instructions, which are replaceable and idempotent, respectively.

8.4.4 Modifying the Register Allocator

The register allocator is assigned the post-SSA task of optimally mapping a large number of virtual registers to a small number of actual physical registers. This is done through a technique called "liveness analysis," wherein the register allocator associates with each virtual register a "liveness interval," denoting the period during which this virtual register is initialized, read, and updated. Two or more virtual registers with disjoint liveness intervals may be mapped on to the same physical register.

Consider in isolation a post-SSA instruction of the form

```
add %x3, %x1, %x2
```

The liveness analysis phase would ascertain that the liveness intervals of the virtual registers $\%x1$ and $\%x2$ intersect, since they are both used in the same instruction simultaneously. This in turn implies that they need to be assigned to different physical registers. On the other hand, the liveness interval of the virtual register $\%x3$ does not intersect with that of $\%x1$ and $\%x2$, meaning that it could be assigned to the same register as either of the two.

In order to implement the instruction redundancy-based countermeasure, it is essential to ensure that given a post-SSA virtual-register-based instruction of the form

```
opcode [dest], [src1], [src2]
```

the register allocator allocates different physical registers for the source and destination operands as far as possible, so that the resulting instruction sequence can be made fault-tolerant by simple duplication.

8.4.5 Transforming Instructions

At this stage, an additional *instruction transformation* phase is incorporated in the compiler to carry out the instruction transformations as described in Sect. 8.3. Note that the ISA for LLVM is ARM Thumb2; however, in principle, the instruction replacement methodologies outlined in [3] are conceptually very similar to the ones we described with respect to the x86 ISA. The passes involved in this stage may be summarized as follows:

- **Replaceable Instruction Elimination Pass:** In this pass, all replaceable non-idempotent instructions, particularly `push` and `pop`, are substituted with equivalent sequences of idempotent instructions.
- **Special Instruction Elimination Pass:** In this pass, all special instructions such as function calls are substituted with equivalent sequences of instructions that, while not necessarily fully idempotent, can be made fault-tolerant via duplication.
- **Partial Instruction Elimination Pass:** In this pass, partially replaceable instructions are substituted with equivalent sequences consisting of many idempotent and only few non-idempotent instructions.
- **Pre-Instruction Duplication Pass:** This is a collection of passes that check the overall sequence of instructions for the presence of unprotected non-idempotent instructions.
- **Instruction Duplication Pass:** Finally, in this pass, all idempotent instructions are duplicated for fault tolerance against instruction skip attacks.

8.4.6 Scheduling Instructions

The role of an instruction scheduler is to re-arrange the order in which the different instructions of a program are executed. The goal of this step is to reduce execution latency while preserving functionality, based on an economical use of the processor pipeline. The re-arrangements are often motivated by data dependencies, individual instruction latencies, and the amount of parallelism supported by the underlying architecture. As mentioned by the authors of [3], it is typically advantageous to perform instruction duplication before the scheduling, since this ensures that the duplicated instructions are scheduled together with the original ones, which in turn ensures a more optimal usage of the processor pipeline, and reduces overall latency.

Finally, it is important to note that instruction duplication naturally incurs latency overheads, the exact value of which depends on the fraction of idempotent, replaceable, and special instructions in a given program. Trade-offs between security and efficiency may be achieved by directing the compiler to perform instruction replacements and duplication only in the most sensitive sections of the overall instruction sequence for a given program.

8.5 Chapter Summary

In this chapter, we presented an instruction-level redundancy-based countermeasure strategy to prevent instruction skip attacks—a potent and popular fault injection technique for software implementations. We described the concept of "idempotent" instructions, which are instructions that may be duplicated without affecting the eventual outcome of the overall program. However, not all instructions may be simply duplicated; in this regard, we showed the reader some non-trivial conversions of simple non-idempotent instructions into sequences of idempotent instructions that can then be made redundant via duplication. Finally, we discussed how to automate the process of compiling any program into a fault-tolerant sequence of assembly instructions via a case study on a modified LLVM compiler.

References

1. J. Balasch, B. Gierlichs, I. Verbauwhede, An in-depth and black-box characterization of the effects of clock glitches on 8-bit MCUs, in *2011 Workshop on, Fault Diagnosis and Tolerance in Cryptography (FDTC)* (IEEE, Piscataway, 2011), pp. 105–114
2. A. Barenghi, L. Breveglieri, I. Koren, D. Naccache, Fault injection attacks on cryptographic devices: theory, practice, and countermeasures. Proc. IEEE **100**(11), 3056–3076 (2012)
3. T. Barry, D. Couroussé, B. Robisson, Compilation of a countermeasure against instruction-skip fault attacks, in *Proceedings of the Third Workshop on Cryptography and Security in Computing Systems* (ACM, New York, 2016), pp. 1–6
4. E. Biham, A fast new DES implementation in software, in *International Workshop on Fast Software Encryption* (Springer, Berlin, 1997), pp. 260–272
5. D. Boneh, R.A. DeMillo, R.J. Lipton, On the importance of checking cryptographic protocols for faults (extended abstract), in *Proceeding of the Advances in Cryptology - EUROCRYPT '97, International Conference on the Theory and Application of Cryptographic Techniques*, Konstanz, 11–15 May 1997 (Springer, Berlin, 1997), pp. 37–51

6. A. Dehbaoui, J.-M Dutertre, B. Robisson, A. Tria, Electromagnetic transient faults injection on a hardware and a software implementations of AES, in *2012 Workshop on Fault Diagnosis and Tolerance in Cryptography (FDTC)* (IEEE, Piscataway, 2012), pp. 7–15
7. P.C. Kocher, Timing attacks on implementations of Diffie-Hellman, RSA, DSS, and other systems, in *Proceedings of the 16th Annual International Cryptology Conference on Advances in Cryptology (CRYPTO '96)* (Springer, Berlin, 1996), pp. 104–113
8. P. Kocher, J. Jaffe, B. Jun, Differential power analysis, in *Proceedings of the 19th Annual International Cryptology Conference on Advances in Cryptology (CRYPTO '99)* (Springer, Berlin, 1999), pp. 388–397
9. N. Moro, K. Heydemann, E. Encrenaz, B. Robisson, Formal verification of a software countermeasure against instruction skip attacks. J. Cryptogr. Eng. **4**(3), 145–156 (2014)
10. J. Schmidt, C. Herbst, A practical fault attack on square and multiply, in *2008 5th Workshop on Fault Diagnosis and Tolerance in Cryptography* (IEEE, Piscataway, 2008), pp. 53–58
11. J. Schmidt, M. Medwed, A fault attack on ECDSA, in *2009 Workshop on Fault Diagnosis and Tolerance in Cryptography (FDTC)* (IEEE, Piscataway, 2009), pp. 93–99
12. E. Trichina, R. Korkikyan, Multi fault laser attacks on protected CRT-RSA, in *2010 Workshop on Fault Diagnosis and Tolerance in Cryptography (FDTC)* (IEEE, Piscataway, 2010), pp. 75–86

Chapter 9
Differential Fault Attack Resistant Hardware Design Automation

Mustafa Khairallah, Jakub Breier, Shivam Bhasin, and Anupam Chattopadhyay

9.1 Introduction

Cryptographic implementations, while mathematically secure on paper, have been shown to be vulnerable to various implementation attacks. The same holds for security protocols, such as PIN verification. These attacks can be non-invasive or invasive—either they depend on passively observing the device characteristics, such as time of the execution or power consumption, in which case they are called side-channel attacks, or they try to alter the processed values by various fault injection techniques, such as laser or electromagnetic pulse, namely, fault attacks. In the era of ubiquitous computing and lightweight cryptography, security against

This research was conducted when author "Jakub Breier" was with Temasek Laboratories, NTU.

M. Khairallah
School of Physical and Mathematical Sciences, Nanyang Technological University, Singapore, Singapore
e-mail: mustafam001@e.ntu.edu.sg

J. Breier (✉)
Underwriters Laboratories, Singapore, Singapore
e-mail: jbreier@jbreier.com

S. Bhasin
Temasek Laboratories, Nanyang Technological University, Singapore, Singapore
e-mail: sbhasin@ntu.edu.sg

A. Chattopadhyay
School of Computer Science and Engineering, Nanyang Technological University, Singapore, Singapore
e-mail: anupam@ntu.edu.sg

J. Breier et al. (eds.), *Automated Methods in Cryptographic Fault Analysis*,
https://doi.org/10.1007/978-3-030-11333-9_9

both types of attacks is required. This chapter focuses on protecting against the fault injection attacks (FIA), which pose a powerful threat to implementations on integrated circuits.

While there are several implementation-level techniques, normally involving redundancy to protect against fault injections, these always depend on the precision of the attacker and his possibility to inject multiple faults. In other words, if we have a triple redundancy, an attacker can defeat it with three precisely injected faults. Developers of countermeasures normally rely on the fact that reproducing precise faults is not trivial and requires a high amount of expertise together with expensive equipment. However, equipment is getting cheaper and more automated over time, boosting the attackers' possibilities. It is, therefore, better to have protection on different levels, raising the chance to detect a malicious behavior.

It was shown that physical sensors are effective against different fault injection techniques and provide high detection rates when deployed properly. For example, the detection of laser fault injections can reach 100% [10]. Such sensors can be deployed on the upper layers of the chip, covering the underlying computation modules and providing the means of protection without slowing down the processing of the data. When it comes to production, it is crucial to have automated methods of placing a sensor when it is chosen for the protection. Manual placement and routing techniques used in [10] take time and human errors can be involved and could prevent some parts of the vulnerable modules to be covered by the sensor. In this chapter, we discuss a countermeasure that can be automated using existing commercial digital design tools.

The rest of the chapter is structured as follows. Sections 9.2 and 9.3 provide overview of state of the art in FIA and countermeasures, respectively. Section 9.4 provides details on laser fault injection sensor that is used as a basis for the differential fault attack resistant physical design automation (DFARPA) method, that is introduced in Sect. 9.5. Experimental results are provided in Sect. 9.6 and the chapter summary is stated in Sect. 9.7.

9.2 Fault Attacks: State of the Art

The most basic non-invasive technique is a clock/voltage glitch. Variations in the clock signal normally lead to shortening the cycle length and forcing a premature toggling of the signal [2]. This can introduce a transient fault in the algorithm execution, by introducing the setup time-constraint violations. Clock glitches work in a similar way, disturbing the instructions that can be either misinterpreted or skipped completely [2]. Both techniques are low cost and relatively easy to apply.

Another popular non-invasive technique is electromagnetic (EM) fault injection. There are two main methods: harmonic wave injection (for analog blocks) and pulse injection (for digital blocks). The first one is effective against analog blocks that are vulnerable to powerful harmonic waves, such as clock subsystem or random number generators. These waves create a parasitic signal biasing the behavior of the block. On the other hand, digital blocks, such as memories, can be disturbed by injecting a sudden and sharp EM pulse into the IC, introducing transients altering the behavior of logic cells. Cost of EM fault injection devices and expertise required varies with the desired precision. One can disturb an algorithm execution even by using a spark gap generator in case fault model is relaxed. However, for precise bit-flips, the attacker has to use expensive precise equipment and often needs an expertise in injection probe design. Optical techniques are the natural choice when it comes to semi-invasive fault injection. It was shown that ICs can be disturbed even with a camera flash. However, the most popular injection device in this field, despite its price, is a diode pulse laser. Laser fault injection is among the most precise injection techniques, capable of producing single bit faults [11]. Normally, the backside of the die is irradiated with (near-)infrared laser (around 1064 nm), so that the beam can access the components directly, compared to the front-side, where metallic layers block the light. For this purpose, a (near-)infrared laser source has to be used (1064 nm), so that it can penetrate the silicon on the back of the chip [4]. The precision of this technique is dependent on the quality of the source and on the magnification lens that is attached to it. It requires a greater amount of expertise compared to previous techniques and, depending on the precision required, the equipment price can go up to several hundred thousand US Dollars. We leave invasive techniques, such as focused ion beam microsurgery, out of the scope of this chapter, since those alter the circuit behavior in a permanent way.

9.3 Fault Attack Countermeasures

The countermeasures against fault attack revolve around two principles: detection and prevention. Any fault detection countermeasure tries to detect anomaly online to raise an alarm. The initial fault countermeasures used detection principles from communication theory and concurrent error detection like linear parity [13] or nonlinear (n,k) codes, etc. Concurrent error detection adds redundancy to the sensitive data processed, which allows detecting data modification under the given fault model. Such countermeasures are easy to implement and incur acceptable overhead. However, the protection is limited to specific fault models and the countermeasure can be bypassed [11]. Operation level redundancy can also be used for fault detection, e.g. duplicated computation-and-compare (for detection) or triple modular redundancy for correction. It is a classical operation redundancy technique used in fault tolerance, where fault correction is done using majority voting from the triplicated operation, thus incurring minimum 200% hardware overhead. Use of randomization has also been proposed to boost the security of fault detection countermeasures [15]. Fault detection can also be done at the circuit level, by monitoring physical conditions that can be exploited for fault injection. This essentially involves the design of physical sensors which often work in a plug and play configuration staying algorithm independent [10].

Fault prevention is a rather less-researched topic. Infection is one way to protect against fault attacks. It causes deeper diffusion (or pollution) of faulty value upon detection such that the faulty value is no longer usable by the attacker. Some infection approaches simply replace the faulty value by a random number. Public key encryption schemes can profit from strong arithmetic structure to prevent faults [8]. Otherwise, on the protocol level key, refreshing techniques are proposed to prevent fault and side-channel attacks. The protocol updates session key every time with a fixed master key [6, 7, 16]. Since DFA requires several correct and faulty ciphertext pairs, the attack is prevented. Another approach is to include internal randomness in the cipher in order to prevent encrypting the same plaintext twice [1].

9.4 LFI Sensors

As mentioned earlier, one way to detect laser fault injection (LFI) is to include a laser detection sensor. In [12], He et al. described two requirements for the countermeasure to be effective:

1. Spatial requirement: The sensor must cover an area larger than the area covered by the circuit that needs to be protected.
2. Sensitivity requirement: The sensor must be able to detect laser beams of strength lower than the strength needed to affect the circuit that needs to be protected.

The authors show that the frequency/phase of a watch-dog ring oscillator (WRO) can be easily affected by the laser beam used to inject faults into the circuit. Hence, using a phase-locked loop (PLL) to detect the frequency/phase of the WRO can be used to raise an alarm signal when laser beams are detected. They implemented the sensor on FPGA using a PLL IP and they used manual routing in order to implement the WRO to cover the protected circuit. Their experiments showed that more than 92% of the fault injections could be detected.

9.4.1 Low-Cost Digital LFI Sensor

While the PLL-based sensor is effective in detecting laser beams, it poses two challenges:

1. It is an analog component that requires special expertise in order to integrate into a digital circuit, which is not a straightforward process.
2. The PLL IP can be expensive and scarce. Besides, it can incur a big area overhead.

To address these two challenges, He et al. [10] proposed a new sensor design. The new sensor uses an all-digital phase detector (PD) instead of the PLL. The phase detector consists of two flip-flops (FFs) and one AND gate, as shown in Fig. 9.1. On FPGA, it requires 2 Look Up Tables (LUTs) and 2 FFs, instead of using the PLL IP. The experiments show that the sensor implementation considered in [10] can

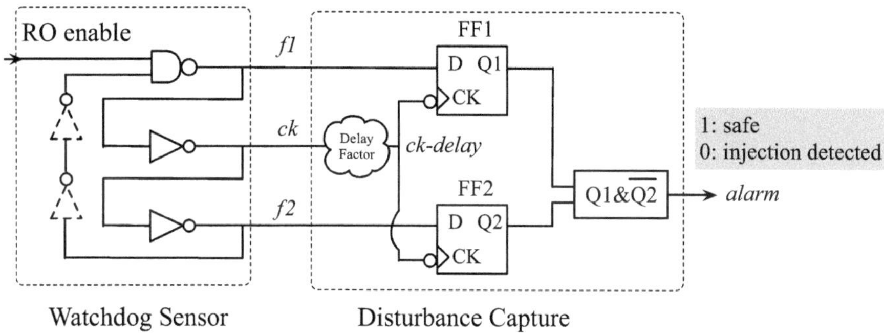

Fig. 9.1 Topology of the schemed fault injection sensor system [10]

achieve 100% detection rate when protecting only the critical FFs of the PRESENT cipher, and 94% when protecting the whole cipher circuit. However, the authors did not address implementing the sensor for ASIC circuits or how to automate the design process.

The idea of the sensor is to connect the outputs of three consecutive inverters in WRO as the inputs for the detection part, named as f_1, ck, f_2. Normally, the signals will have the same frequency and a fixed phase shift, and an opposite polarity to signal ck, w.r.t. f_1 and f_2. The FFs are both triggered by the falling edge of ck. In absence of signal delay from RO to flip-flops, the sampled values for FF1 and FF2 are, respectively, 1 and 0. Noticeably, the ripples in this RO will identically affect three frequencies, leading to no impact on the disturbance capture and thus giving false negatives. Hence, a delay factor is added to ck, such that at the fault injection moment, the clock used as input to the FFs is glitch free, since the glitch in ck will be delayed.

Using the Xilinx FPGA Editor, He et al. [10] implemented the sensor using manual routing in order to surround the registers of a PRESENT-80 [3] round based implementation, as shown in Fig. 9.2.

9.5 DFARPA Routing Flow

The DFARPA routing flow was first introduced in [14]. It targets protection against sophisticated fault injection methods, such as laser and EM pulse. These mechanisms can be used to inject single bit faults. This scenario requires reactive countermeasures that behave as physical sensors to detect high-energy injections like laser beam and EM pulse [10]. Once an injection is detected, the sensitive computation is halted. The sensor is composed of a watchdog ring oscillator (WRO) and a phase detection (PD) circuit. High energy injections impact signal propagation delay, which disturbs the phase of WRO. The change is detected by the PD circuit to raise an alarm and halt sensitive computation. While [12] uses phase-locked loop (PLL)

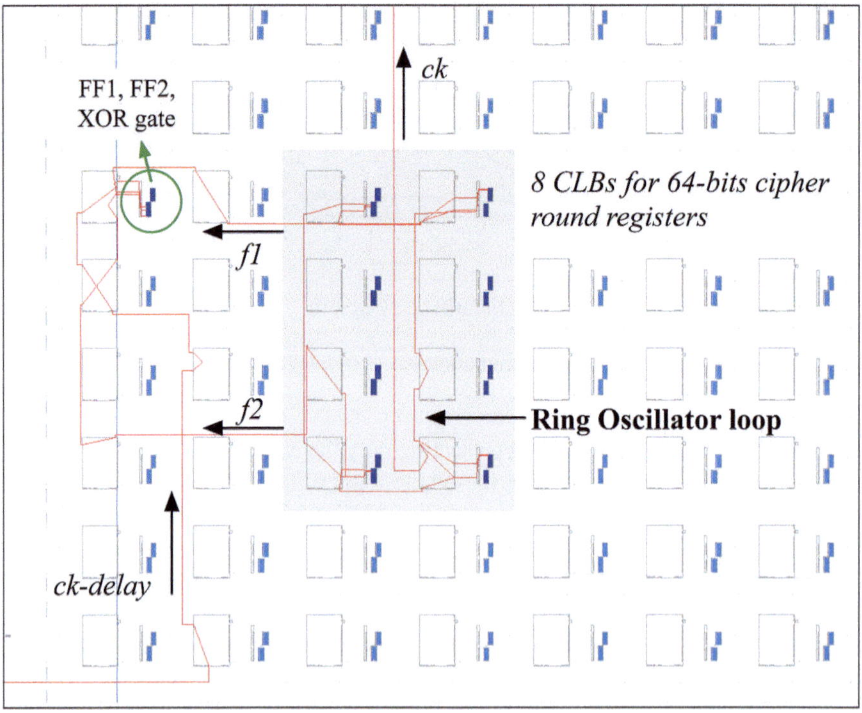

Fig. 9.2 Routing of the WRO sensor around the FFs in FPGA [10]

as a PD circuit, an all-digital PD is proposed in [10], allowing even higher detection rates. Its design is independent of the underlying circuit, so it can be implemented in a plug and play manner. This makes it also a good candidate for protecting non-cryptographic circuits, such as ALUs, multipliers, micro-controllers, etc.

A high-level design of the sensor is depicted in Fig. 9.3. It is assumed that the attack is performed from the front-side of the chip, backside attacks remain out-of-scope here, as we assume a ball-grid array package. Hence, routing the WRO on top-metal layer facilitates detection. The functioning and detection capability of the WRO-based sensor were tested on FPGA target in [10]. The sensitivity of top-metal layer to faults was validated on a prototype chip in [17] and the use of top-metal layer for deploying a shield against fault attack was previously proposed and validated on a prototype chip in [17].

9.5.1 Design Flow

The design of a custom WRO, which is the key component in the proposed countermeasure, in an automated digital design flow, is a tricky task, as the tools

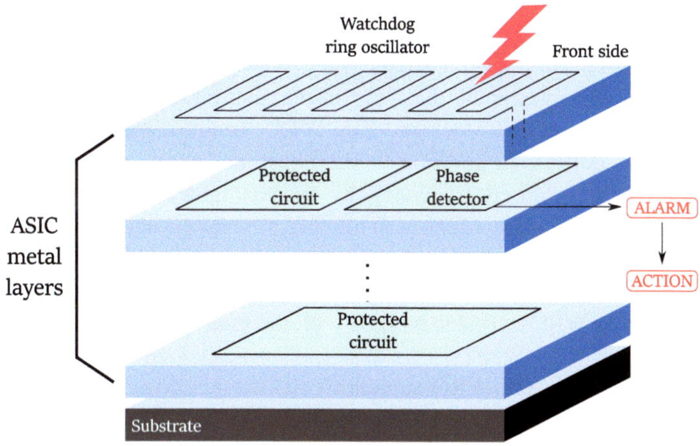

Fig. 9.3 Ring oscillator sensor deployed on top of the protected circuit

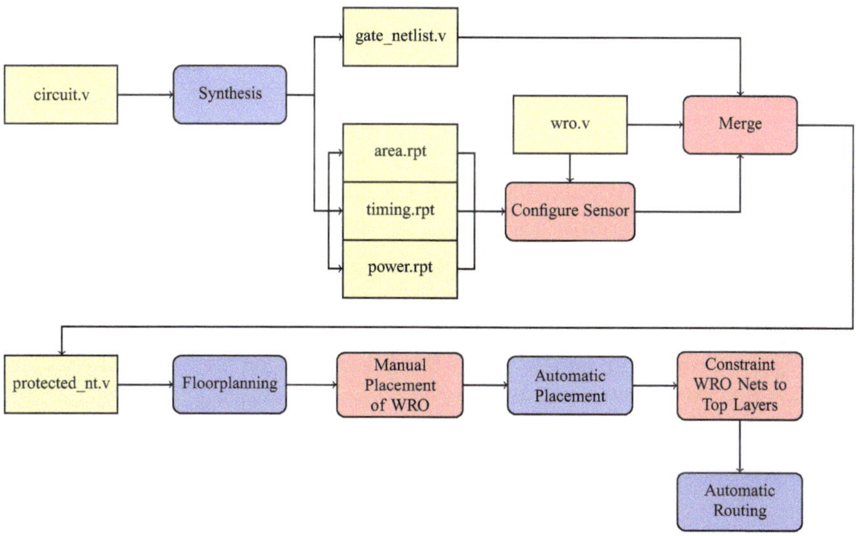

Fig. 9.4 DFARPA routing flow

would optimize WRO with multiple inverters. The reason is that the front-end synthesis tools are designed to remove logical redundancy. Specifically, these tools consider one inverter and an odd number of inverters to be exactly the same circuit. Consequently, a slightly modified flow is adopted, adding a new sub-flow to include the WRO. Similar to the conventional digital design flow, the proposed sub-flow consists of two parts, front-end and back-end. The design flow shown in Fig. 9.4 enables the implementation of the all-digital sensor, without the ring oscillator being trimmed due to logic redundancy and without using analog/mixed-signal design flows. The blue boxes represent the conventional steps and the red boxes indicate the newly adopted steps. These new steps are:

1. After the circuit is designed, the area, timing, and power information is used to configure a generic WRO design. This WRO design uses standard-cell-based inverters and buffers, in addition to the phase detection circuit. The designer can control the required frequency by increasing/decreasing the number of inverters/buffers. Besides, the overall number of inverters and buffers can be used to control the trade-off between power consumption and sensitivity of the sensor, which we will explain in this section.
2. The configured WRO netlist is merged with the gate-level netlist of the required circuit.
3. After floorplanning, the WRO standard cells are placed in order to create the sensor layout. This can be done either manually or by a user-defined script.
4. The WRO Nets are constrained to use only the top metal layers.

Figure 9.5 shows an example of the floorplanning/placement of a WRO with 38 buffers/inverters, where every square indicates one standard cell. The area covered by the sensor is $L * W$, and is divided into $2 * L/N$ steps, where N is the number of standard cells of the WRO.

9.6 Experimental Results of DFARPA Routing Flow

In order to show the overhead of implementing the sensor using the DFARPA routing flow, we have applied it to several circuits: an implementation of the plantlet stream cipher [14], a K-163 elliptic curve cryptography (ECC) multiplier [5], a 16-bit integer multiplier [9], and an 8-bit μProcessor based on the 8080 architecture.[1] The experiments have been done using Synopsys digital design flow and TSMC 65 nm technology Library with 9 metal layers. The implementation results are shown in Table 9.1. Similar to the observation in [14], the main overhead parameter is power consumption. However, it is shown that the overhead varies depending on the circuit. For the integer multiplier, the overhead is less than 6%, while for Plantlet it is more than 100%. Hence, the trade-off analysis has to be performed independently for every circuit. However, it is intuitive that smaller/cheaper circuits are harder to protect and the relative cost for protecting them is higher. An example of how the sensor layout looks like is shown in Fig. 9.6, which shows the WRO on top of the Plantlet circuit.

[1] https://opencores.org/project,sap.

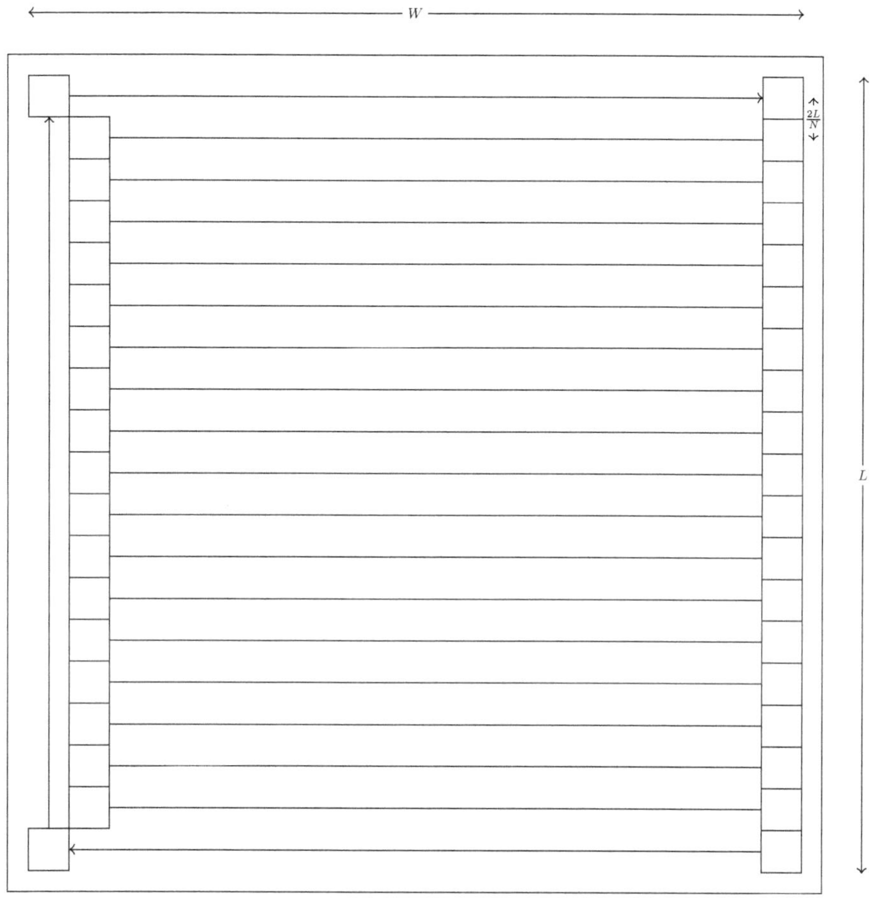

Fig. 9.5 Example of the floorplanning of a WRO with 38 inverters/buffers

Table 9.1 Experimental results of DFARPA routing

Circuit	Plantlet		K163 ECC	
Feature	Unprotected	Protected	Unprotected	Protected
Area (μm^2)	1293	1358 (5%)	44,055.4	44,401.3 (0.7%)
Max. path delay (ns)	0.61	0.62 (1%)	0.34	0.34 (0%)
Avg. dynamic power (μW)	259.26	551.5 (112%)	1380	1676 (21.7%)

Circuit	16-bit multiplier		8-bit μprocessor	
Feature	Unprotected	Protected	Unprotected	Protected
Area (μm^2)	2645.3	2675.5 (1.1%)	1194.5	1236.2 (3.5%)
Max. path delay (ns)	3.39	3.42 (1%)	0.15	0.23 (53.3%)
Avg. dynamic power (μW)	789	831 (5.3%)	44.8	80.2 (79%)

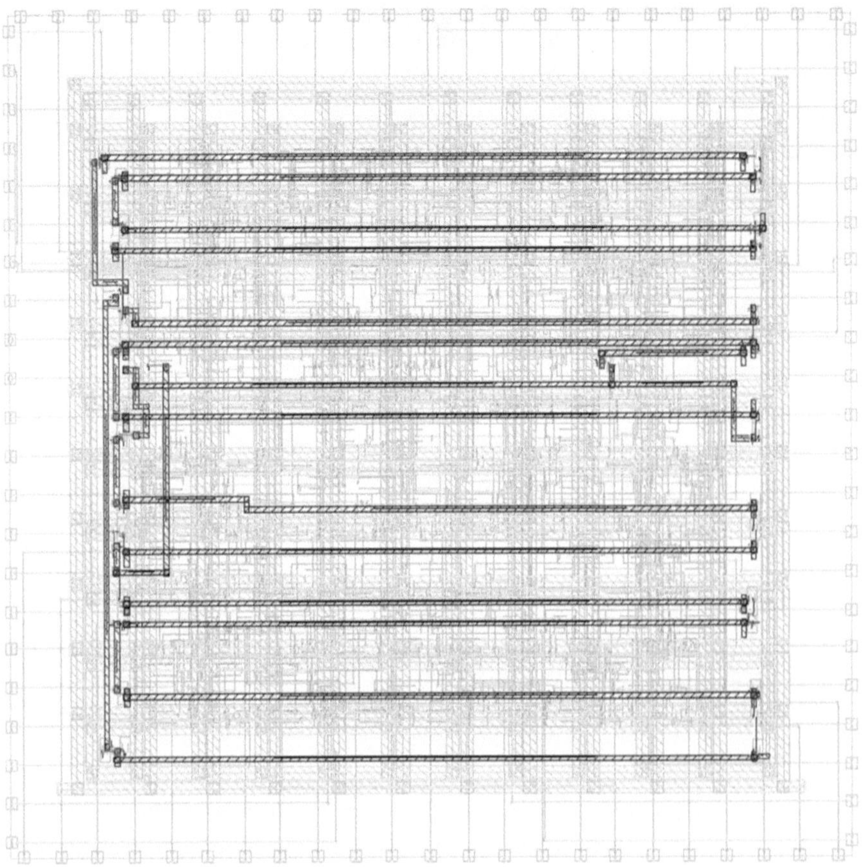

Fig. 9.6 Ring oscillator sensor deployed on top of the protected Plantlet circuit

9.7 Chapter Summary

In this chapter, we have provided an overview of automated protection of hardware circuits. DFARPA, detailed in previous sections, places a ring oscillator based sensor above the sensitive circuit automatically, providing reasonable overheads and good detection coverage.

References

1. A. Baksi, S. Bhasin, J. Breier, M. Khairallah, T. Peyrin, Protecting block ciphers against differential fault attacks without re-keying, in *2018 IEEE International Symposium on Hardware Oriented Security and Trust (HOST)* (IEEE, Piscataway, 2018), pp. 191–194
2. H. Bar-El, H. Choukri, D. Naccache, M. Tunstall, C. Whelan, The sorcerer's apprentice guide to fault attacks. Proc. IEEE **94**(2), 370–382 (2006)

3. A. Bogdanov, L.R. Knudsen, G. Leander, C. Paar, A. Poschmann, M.J. Robshaw, Y. Seurin, C. Vikkelsoe, PRESENT: an ultra-lightweight block cipher, in *Proceedings of the 9th International Workshop on Cryptographic Hardware and Embedded Systems, CHES '07* (Springer, Berlin, 2007), pp. 450–466
4. J. Breier, C.N. Chen, On determining optimal parameters for testing devices against laser fault attacks, in *2016 International Symposium on Integrated Circuits (ISIC)* (2016)
5. J.-P. Deschamps, J.L. Imaña, G.D. Sutter, *Hardware Implementation of Finite-Field Arithmetic* (McGraw-Hill, New York, 2009)
6. C. Dobraunig, M. Eichlseder, S. Mangard, F. Mendel, On the security of fresh re-keying to counteract side-channel and fault attacks, in *International Conference on Smart Card Research and Advanced Applications* (2014), pp. 233–244
7. C. Dobraunig, F. Koeune, S. Mangard, F. Mendel, F.-X. Standaert, Towards fresh and hybrid re-keying schemes with beyond birthday security, in *International Conference on Smart Card Research and Advanced Applications* (2015), pp. 225–241
8. P.-A. Fouque, R. Lercier, D. Réal, F. Valette, Fault attack on elliptic curve Montgomery ladder implementation, in *Proceedings of FDTC* (2008), pp. 92–98
9. M.C. Hansen, H. Yalcin, J.P. Hayes, Unveiling the ISCAS-85 benchmarks: a case study in reverse engineering. IEEE Des. Test Comput. **16**(3), 72–80 (1999)
10. W. He, J. Breier, S. Bhasin, Cheap and cheerful: a low-cost digital sensor for detecting laser fault injection attacks, in *International Conference on Security, Privacy, and Applied Cryptography Engineering* (2016), pp. 27–46
11. W. He, J. Breier, S. Bhasin, A. Chattopadhyay, Bypassing parity protected cryptography using laser fault injection in cyber-physical system, in *Proceedings of the 2nd ACM International Workshop on Cyber-Physical System Security* (2016), pp. 15–21
12. W. He, J. Breier, S. Bhasin, N. Miura, M. Nagata, Ring oscillator under laser: potential of PLL-based countermeasure against laser fault injection, in *Proceedings of FDTC* (IEEE, Piscataway, 2016), pp. 102–113
13. R. Karri, G. Kuznetsov, M. Goessel, Parity-based concurrent error detection of substitution-permutation network block ciphers, in *Proceedings of CHES* (2003), pp. 113–124
14. M. Khairallah, R. Sadhukhan, R. Samanta, J. Breier, S. Bhasin, R.S. Chakraborty, A. Chattopadhyay, D. Mukhopadhyay, DFARPA: differential fault attack resistant physical design automation, in *Design, Automation & Test in Europe Conference & Exhibition (DATE), 2018* (IEEE, Piscataway, 2018), pp. 1171–1174
15. V. Lomné, T. Roche, A. Thillard, On the need of randomness in fault attack countermeasures-application to AES, in *Proceedings of FDTC* (2012), pp. 85–94
16. M. Medwed, F.-X. Standaert, J. Großschädl, F. Regazzoni, Fresh re-keying: security against side-channel and fault attacks for low-cost devices, in *International Conference on Cryptology in Africa* (Springer, Berlin, 2010), pp. 279–296
17. X.T. Ngo, J.-L. Danger, S. Guilley, T. Graba, Y. Mathieu, Z. Najm, S. Bhasin, Cryptographically secure shield for security IPS protection. IEEE Trans. Comput. **66**, 354–360 (2017)

Part III
Automated Analysis of Fault Countermeasures

Chapter 10
Automated Evaluation of Software Encoding Schemes

Jakub Breier, Dirmanto Jap, and Shivam Bhasin

10.1 Introduction

In order to be able to mount a fault injection attack, the attacker needs to disturb the device during the computation. This can be done in various ways by different fault injection techniques. These range from very basic, inexpensive ones, such as varying the supply voltage, to advanced techniques, requiring device de-packaging and significant funds, such as laser fault injection (see Chap. 1 for more details).

Designing the experiment and getting plausible results is usually a long-term process, depending on the device, the equipment, and the fault model. Therefore, it is beneficial to model the fault behavior of the implemented algorithm to distinguish what fault models are possible and what is the probability of a particular fault occurrence.

The same holds for the other side—when designing a protection against these attacks, the security analyst needs to test the implementation for vulnerabilities. When it comes to countermeasures, there are two possible ways to protect the implementation. Either we try to protect the device itself by adding sensor or employing circuit-level countermeasures or by checking the software code for vulnerable points that can be exploited and fixing these points.

This research was conducted when author "Jakub Breier" was with Temasek Laboratories, NTU.

J. Breier (✉)
Underwriters Laboratories, Singapore, Singapore
e-mail: jbreier@jbreier.com

D. Jap · S. Bhasin
Temasek Laboratories, Nanyang Technological University, Singapore, Singapore
e-mail: djap@ntu.edu.sg; sbhasin@ntu.edu.sg

J. Breier et al. (eds.), *Automated Methods in Cryptographic Fault Analysis*,
https://doi.org/10.1007/978-3-030-11333-9_10

In this chapter, we focus on instruction set simulator (ISS), based on Java programming language, able to simulate a fault behavior of encoding based countermeasure-protected software implementations for microcontrollers. This simulator takes an assembly code as the input and checks all the possible fault models that may occur in the device. Based on this checking, it outputs an information about vulnerable instructions.

The case study in this chapter focuses on three software-based hiding countermeasures, originally proposed for side-channel protection, that can also be hardened against fault injection attacks. More specifically, it analyzes a bit-sliced software countermeasure following the dual-rail precharge logic (DPL) [9], a balanced encoding scheme providing constant side-channel leakage [3], and a customized encoding scheme built according to leakage model based on stochastic profiling [8].

Section 10.2.1 provides the necessary background for our work, outlining some works on instruction set simulation and detailing software encoding countermeasures proposed so far. Details on the automated code analyzer are stated in Sect. 10.3. Case studies are presented in Sect. 10.4. Experimental results are detailed in Sect. 10.5, followed by discussion provided in Sect. 10.6. Finally, Sect. 10.7 summarizes this chapter.

10.2 Background

This section first focuses on instruction set simulators, and later it details the encoding schemes used in the case study.

10.2.1 Instruction Set Simulators

In this part, we will provide an overview of works aiming at instruction set simulators. By inspecting these works, we will state the requirements for our simulator.

In [13], the authors adjust SID1 instruction set simulator to simulate effects of time-domain electromagnetic (EM) interference in a microcontroller. They used PIC C and PIC assembly to write their code. From the component point of view, simulator consisted of CPU with interrupt controller and external oscillator, bus component, and external memory. They estimated the EM emanation caused by particular instructions and fed the simulator with these values. As a result, they could predict different program behavior with respect to EM interference.

Authors of [6] created a high performance software framework based on multi-level hash table to enable development of more efficient ISS. They classified the instructions in the instruction set to construct a hash table and used a preprocessor to map relationships between instructions and hash table elements.

In [5], the authors use their own ISS in order to simulate source code of various cryptographic algorithms implementations on 8-bit microcontroller, allowing them an easy analysis and debugging. Based on these results, they could make the implementations more efficient by utilizing various extended instruction sets.

Simulating the fault behavior of instructions was previously used in [11], where authors implemented and analyzed 19 different strategies for fault attack countermeasures. They created their custom ISS for simulating ARM Cortex M-3 and performed benchmarking, allowing them to quantitatively compare the countermeasures. However, they did not provide any details about their simulator.

10.2.2 *Software Encoding Countermeasures*

The first proposal of *side-channel information hiding* in software was made by Hoogvorst et al. [7]. They suggested to adopt the dual-rail precharge logic (DPL) in the software implementation to reduce the dependence of the power consumption on the data. Their design uses a look-up table method—instead of computing the function value, the operands are concatenated and used as an address to the resulting value. The idea was explained on PRESENT implementation on AVR microcontroller.

Building on the idea of the seminal work, there were three notable works published in recent years. The rest of this section provides a short overview of each of them since they will be later used as a case study for the evaluation method detailed in this chapter.

10.2.2.1 Software DPL Countermeasure

In 2013, Rauzy et al. [9] published a work that follows DPL encoding by utilizing bit-sliced technique for assembly instructions. They developed a tool that converts various instructions to a balanced DPL, according to their design. In their implementation, each byte is used to carry only one bit of information, encoded either as "01" for "1" or "10" for "0." In the proposal, bits are chosen according to their leakage characteristics. In our work, we use the two least significant bits of the byte. This implementation uses look-up tables with balanced addressing instead of computing the operations directly. The assembly code we used in the code analysis is stated in Appendix 1. For the sake of simplicity, we refer to this implementation as to the *"Static-DPL XOR"* throughout the paper.

10.2.2.2 Balanced Encoding Countermeasure

Proposed in 2014 by Chen et al. [3], this work provides assembly-level protection against side-channel attacks by balancing the number of "1"s and "0"s in each instruction. The code proposed by the authors is aimed for 8-bit platforms and the constant leakage is achieved by adding complementary bit to every bit of information being processed. Therefore, in each instruction, there are four effective bits of information and four balancing complementary bits. Encoding structure looks as follows: $b_3\bar{b}_3b_2\bar{b}_2b_1\bar{b}_1b_0\bar{b}_0$. Order of bits may vary depending on the leakage model. For fault injection evaluation, it does not matter which format is chosen; therefore, all the data is transformed according to the structure above. The assembly code we used in the code analysis is stated in Appendix 2. In [3], two basic operations are used, i.e., XOR and look-up table (LUT). For the rest of this paper, we will refer to these operations as *"Static-Encoding XOR"* and *"Static-Encoding LUT."*

10.2.2.3 Device-Specific Encoding Countermeasure

In 2016, there was another encoding countermeasure proposal by Maghrebi et al. [8]. The proposed encoding aims to balance the side-channel leakage by minimizing the variance of the encoded intermediate values. Previous encoding proposals were based on the assumption of Hamming weight (HW) leakage model. However, the actual leakage model often deviates from HW, which leads to reduction in practical side-channel security of the encoding scheme. The proposal of [8] designs the encoding scheme by taking the actual leakage model into account.

The side-channel leakage is dependent on the device, and for the microcontroller case, each register leaks the information differently (though the paper argued that most of the registers have more or less similar leakage pattern). In general, the leakage normally depends on the processed intermediate value. The leakage can be formulated as follows:

$$T(x) = L(x) + \epsilon, \tag{10.1}$$

where L is the leakage function that maps the deterministic intermediate value (x) processed in the register to its side-channel leakage, and ϵ is the (assumed) mean-free Gaussian noise ($\epsilon \sim N(0, \sigma^2)$). The commonly used leakage function used is the n-bit representation. For example, in 8-bit microcontroller, the leakage could be represented as $L(x) = \beta_0 + \beta_1 x_1 + \cdots \beta_8 x_8$, where x_i is the ith bit of the intermediate value, and β_i is the ith bit weight leakage for specific register [10]. For HW model, β_1–β_8 are considered to be unity. In reality, due to several physical device parameters, β will deviate from unity in either polarity.

The deterministic part of the leakage can then be determined as $\tilde{L} = \mathbf{A} \cdot \beta$, where $\mathbf{A} = (x_{i,j})_{1 \le i \le N; 0 \le j \le n}$, with $\mathbf{x_i}$ as a row element of $\mathbf{A}$ and N denotes the number of measurements. We can then determine $\beta = (\beta_j)_{0 \le j \le 8}$ based on the set of traces $\mathbf{T}$, as follows:

$$\beta = (\mathbf{A}^T\mathbf{A})^{-1}\mathbf{A}^T\mathbf{T}. \tag{10.2}$$

After profiling of the device to obtain the weight leakages β, the encoding function can be calculated based on the method used in Algorithm 1.

Algorithm 1: Selection of the optimal encoding function [8]

Input : m: the codeword bit-length, n: the sensitive variable bit-length, β_i: the leakage bit weights of the register, where i in $[\![1, m]\!]$
Output: 2^n codewords of m-bit length

for X *in* $[\![0, 2^m - 1]\!]$ **do**
  Compute the estimated power consumption for each codeword X and store the result in table D: $D[X] = \Sigma_{i=1}^{m} \beta_i X[i]$;
  Store the corresponding value of the codeword in the index table $I : I[X] = X$;
Sort the estimated power consumption stored in table D and the index table I accordingly
for j *in* $[\![0, 2^m - 2^n]\!]$ **do**
  Find the $argmin$ of $[\![D[j] - D[j + 2^n]]\!]$;
return 2^n codewords corresponding to $[\![I[argmin], I[argmin + 2^n]]\!]$

Thus, the main aim of the algorithm is to choose a set of encoding, represented as a look-up table, which minimizes the variance of the estimated power leakage. This is done by considering the leakage bit weights, which are tightly connected to the device, specifically to its registers. Hence, for different registers, different encoding setup has to be considered.

The assembly code with look-up tables is stated in Appendix 3. In this paper, we will refer to this implementation as to the *"Device-Specific Encoding XOR,"* as the dependence on leakage model makes the encoding specific to a device register.

10.3 The Automated Code Analyzer

Before implementing fault injection capabilities, the instruction set simulator for a general-purpose microcontroller has to be specified. This section provides overview on mapping of particular hardware components to object model in software. Design of the solution used in this chapter is depicted in Fig. 10.1. Black components on the left side are the basic components of the Harvard architecture microcontroller [12]. On the right side, we can see a high-level class diagram of the simulator. In the following, we will explain each entity in the diagram:

- **MuC:** Microcontroller class encapsulates all the other entities constituting the device. It acts as a microcontroller itself, containing the instruction set, registers, and data memory and allows performing operations on these.
- **Instruction:** It is an abstract class, defining *execute()* method that is further specified by its subclasses. *MuC* contains a list of *Instruction* classes, loaded from a text file—this file acts as a program memory.
- **Instruction subclasses (MOV, ADD, ...):** Adding new instructions can be achieved simply by adding new subclasses of the Instruction class. This allows to simulate different architectures by using the same ISS.

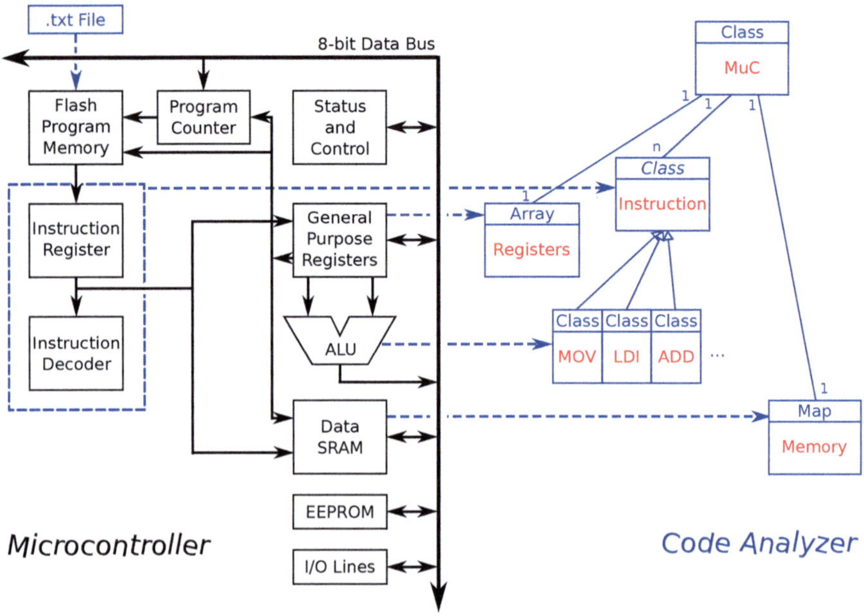

Fig. 10.1 Microcontroller architecture mapped to an object oriented computer program. Black parts indicate physical parts of the microcontroller, and red parts constitute a class diagram of the program

Table 10.1 Assembly text file example

Mnemonics	Operand 1	Operand 2
LDI	r16	*a*
LDI	r17	*b*
EOR	r16	r17
ST	X	r16

- **Registers:** Registers are simulated as an array of integers. Since the majority of IoT devices contains chips with constrained hardware, register sizes are either 8 or 16 bits; therefore, integer variables are enough for this purpose.
- **Memory:** Memory is simulated as a map, so that an instruction can define a variable name that serves as a key and links the value together with this key.

Text file contains the assembly code for the microcontroller and it is read and analyzed by the *MuC* class in order to assemble a program. The same class allows to run this code as well. An example of such file can be seen in Table 10.1. The first instruction is `LDI` (load immediate) and loads the value of *"a"* into register `r16`. The second instruction does the same with a different value and a different register. The third instruction computes an `xor` of the values in those two registers and stores the result in register `r16`. The last instruction stores the value in register `r16` in the memory, using the key *"X"* as a variable name.

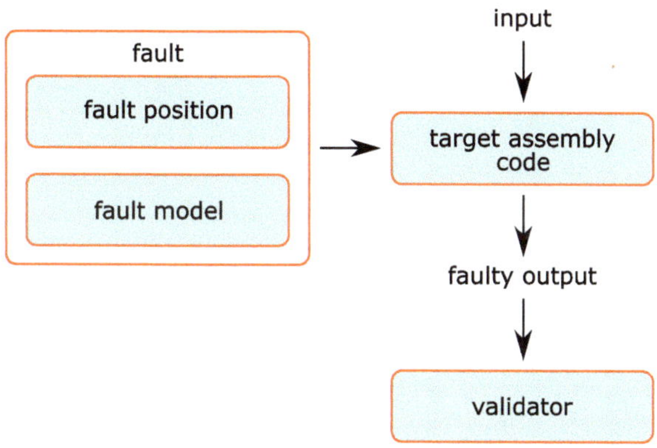

Fig. 10.2 Fault injection methodology for ISS

10.3.1 Fault Behavior Simulation

In order to add fault simulation capabilities to the ISS, first it has to be analyzed, what types of faults need to be simulated. Not all the faults are interesting for the fault analysis and also, we should only include fault models that are feasible to obtain by standard fault injection techniques.

Our fault injection methodology is depicted in Fig. 10.2. We will further explain each component of the picture in the following:

- **Input:** In assembly programs, variables are usually loaded either from the memory or as a constants (e.g., using `LDI` instruction as in Table 10.1). For testing the vulnerability against faults, we often have to try all the possible inputs. Therefore, the simulator allows to pre-load the inputs automatically in chosen registers without having to change the assembly code.
- **Faulty Output:** After every testing iteration, the faulty output is tested by the **Validator**. The tester can set up this component to check for certain types of faults, depending on he is aiming for. For example, in parity check countermeasure, it can be set up to check only the even number of bit flips in the output in order to keep the parity scheme working, however, with a faulty result.
- **Target Assembly Code:** The ultimate goal of the simulator is to test the assembly code. This code is fed to the program as a text file and can be first checked line by line if it works properly before it is tested.
- **Fault Position:** The simulator checks all the possible position for the fault to be injected. We check every instruction and every bit in the destination register—that ensures that all the bits used in the code will be tested.
- **Fault Model:** For every input and every position in the code, several fault models are tested. We have identified following fault models as the most commonly used in literature [1]:
  - *Bit flip*—This is, together with the random byte fault, the most commonly used fault model when it comes to attacking cryptographic algorithms. The

simulator tests every bit in the destination register of an instruction, using single and multiple bit flips, up to the number of bits used by the register.

- *Random byte fault*—This fault model expects flipping of random number of bits in the destination register. Because with the previous fault model we already test all the possible combinations, we do not have to use both tests at the same time. However, random byte fault is a weaker assumption for fault attacks (it is easier to achieve in a real device and it is harder to design an attack that recovers the secret key just from a random byte flip); therefore, if an attacker only needs this fault model, he can skip the bit flip testing in order to save time.
- *Instruction skip*—This fault model is not that popular in theoretical works, since they usually do not analyze concrete implementations of algorithms. However, as it was shown in [2], it is relatively easy to achieve this type of fault in microcontrollers and if used properly, this attack is very powerful. We test both single and multiple instruction skips, depending on the settings required by the tester.
- *Stuck-at fault*—This fault model changes the value in register to some specific value. Authors of [4] have shown that with different laser energy, it is possible to force certain memory bits either to *"0"* or to *"1,"* allowing precise stuck-at faults. Again, we test all the destination registers in the target code for stuck-at faults specified by the tester.

The **Validator** provides a human-readable output as a result of the code analysis. This output is in two forms—overview and a detailed view.

Overview shows the total number of faults and this number is then further divided by the tester's requirements. Usually, it is desired to have the output in some special form, e.g., specific encoding or some fault mask. The tester can then specify the output to be divided in two subsets—one that fulfills the requirement and the other that does not.

Detailed view provides insight on all the successful faults, i.e., only on those in the subset that fulfill the tester's requirements. Table 10.2 shows all the fields that are provided by the detailed view for each fault model. Please note that the output from faulty execution is always provided in any fault model that is selected.

Activity diagram for the whole process of code analysis is stated in Fig. 10.3. After writing the code it is necessary to check if the instruction set used is also

Table 10.2 Fields for different fault models provided in the detailed view

Fault model	Fields
Bit flip	Instruction number, Instruction mnemonics, Fault mask, Number of plaintexts affected
Random byte fault	Instruction number, Instruction mnemonics, Fault mask, Number of plaintexts affected
Instruction skip	Instruction number, Instruction mnemonics, Number of plaintexts affected
Stuck-at fault	Instruction number, Instruction mnemonics, Stuck-at mask, Number of plaintexts affected

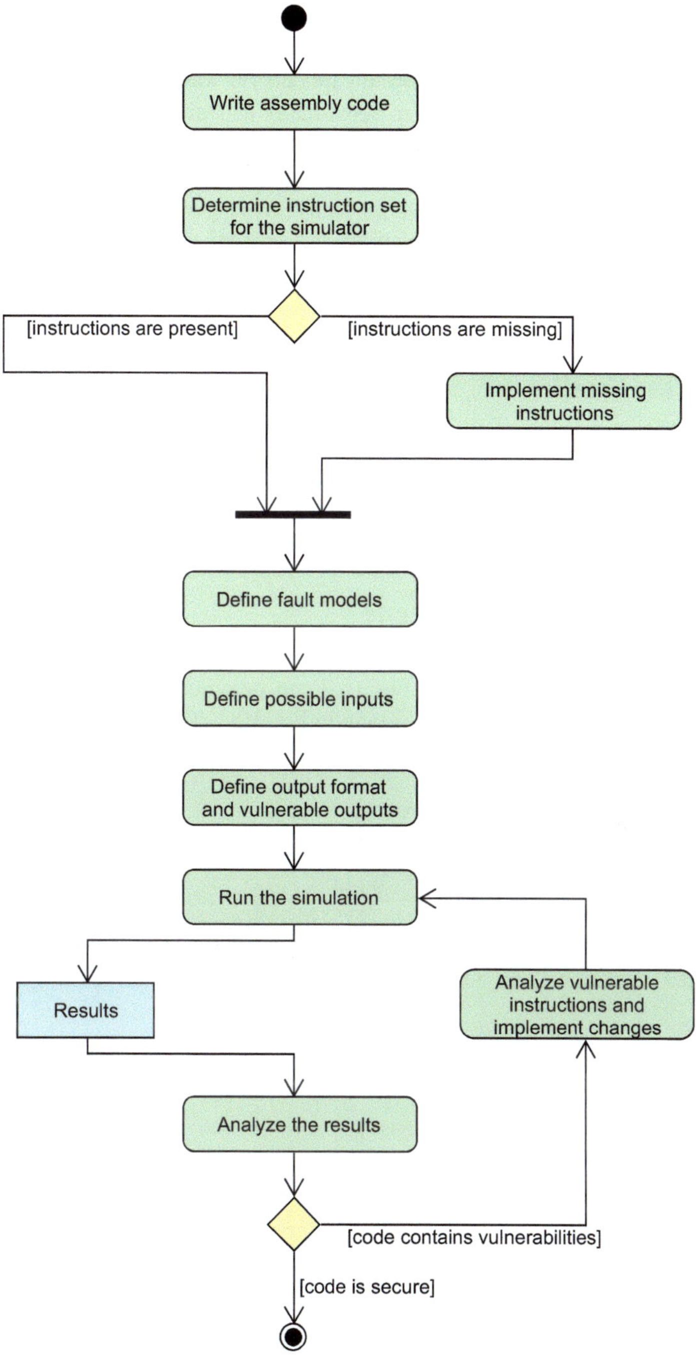

Fig. 10.3 Activity diagram for particular steps in the code design and analysis

implemented in the fault simulator. If not, the tester has to implement missing instructions. The modular architecture of the framework allows adding new instructions. Afterwards, he defines which fault models he want to test the code against and prepares a set of inputs, since in some cases it is not necessary to test all the possible inputs. Before running the simulator, he has to specify the output format and in some cases also a set of outputs that constitute a security risk so that the **Validator** could classify the outputs correctly. After getting the results, the tester should analyze the vulnerable instructions and propose changes in the code before re-running the simulations again.

10.4 Case Studies

In this section, the automated evaluation method from previous part is adjusted to evaluate the software encoding countermeasures.

10.4.1 Fault Injection Analysis

The high-level methodology from Fig. 10.2 can be adjusted to encoding countermeasures by tailoring each part, as shown in Fig. 10.4. The middle part serves as a

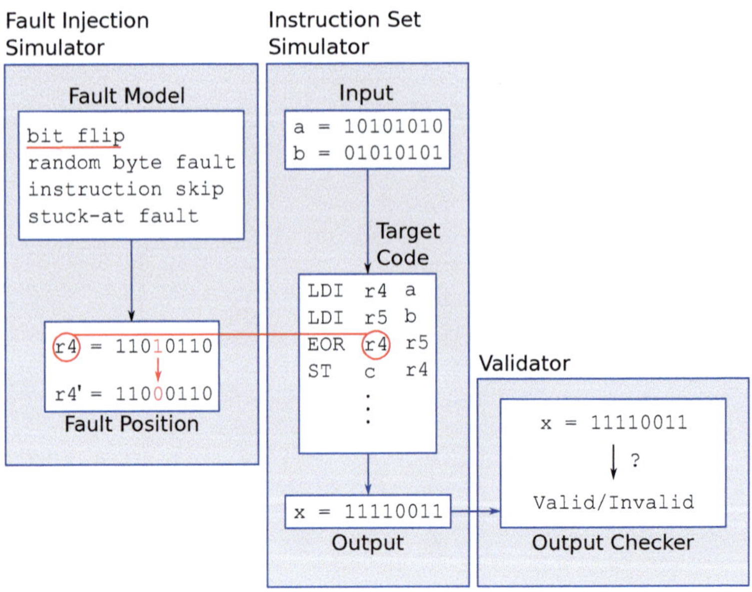

Fig. 10.4 Schematic of injecting a fault during the execution in the code analyzer

standard instruction set simulator, taking the given input, executing the code, and producing the output. This process is repeated for every possible combination of inputs for every instruction and every fault model. We are analyzing resistance against four basic fault models: bit flip, random byte fault, instruction skip, and stuck-at fault. After the output is produced, it is analyzed by the validator which decides whether the fault changed the resulting value and if this value is useful for fault attack. We consider inputs and outputs already encoded, and analyzing fault tolerance with respect to encoding/decoding is out of scope of this paper.

In the following, we will briefly describe parts of the code analyzer:

- **Instruction Set Simulator:** As stated previously, the assembly code is sent to the simulator as a text file. It accepts three different data encoding formats, according to what algorithm is currently being used. For the *Static-DPL XOR*, it accepts input in the bit-sliced complement form: $000000b_0\bar{b}_0$; therefore, there are 4 possible input combinations. The *Static-Encoding XOR* accepts four bit complement format: $b_3\bar{b}_3b_2\bar{b}_2b_1\bar{b}_1b_0\bar{b}_0$, resulting in 256 input combinations. The same number of combinations is analyzed for the *Device-Specific Encoding XOR*, where the number of codewords is 16 for 8-bit code.
- **Fault Injection Simulator:** In order to get the information about algorithm resistance against fault injection, we analyze four fault models. In the case of a fault being injected into the data, we change the content of the destination register of an instruction.

  In bit flip fault model, we inject single and double bit flips into the *Static* countermeasures. There is no need to test other multiple bit flips, since all of them are just a subset of those two, because of DPL properties. Therefore, e.g., if an algorithm is not vulnerable against single bit flips, it will not be vulnerable against other odd-number bit flips, and vice versa. In case of the *Device-Specific Encoding XOR*, we test all possible combinations of bit flips.

  Random byte fault model is a subset of bit flip fault model when it comes to code analysis; therefore, this model is already included in the previous testing.

  To analyze vulnerable parts against instruction skip attack, we skip either one or two instructions from the code, checking all the possible combinations. More complex instruction skip models are not considered because of the impracticability to implement them in the real environment.

  Finally, to analyze the resistance against the stuck-at-fault model, we change the value of the destination register either to `0x00` or to `0xFF`.
- **Validator:** The final part of the code analyzer checks the resulting output and assigns it to one of the following pre-defined groups:

  – *VALID:* This is the most useful type of output an attacker can get. Outputs in this category follow the proper encoding of analyzed algorithm, but the value deviates from the expected value with respect to given inputs. A *VALID* fault can be directly exploited with fault injection attack.
  – *INVALID:* This type of output does not follow the encoding. Therefore, it can be easily recognized by an output checker which can then decide to discard the value instead of further propagation.

– *NULL:* This type of fault has one of the following values: `0x00` or `0xFF`. These outputs are mostly produced by look-up table implementations and can be easily recognized as well.

10.4.2 Evaluation Methodology

To analyze different software encoding countermeasures against fault injection attacks, and show the utility of the automated code analyzer, we have implemented the basic operations of each previously discussed encoding scheme, i.e., *Static-DPL XOR*, *Static-Encoding XOR*, *Static-Encoding LUT*, and *Device-Specific Encoding XOR* implementation. The corresponding code is provided in the appendices. As the first step, we performed a comprehensive fault analysis by putting the code under custom designed code analyzer. Such analysis cannot be done in a practical setting due to limited control over the injected fault model for a given equipment setting. Albeit it is possible to inject all the discussed fault models, it is not easy to control the fault model at will.

10.4.3 Code Analysis Results

To analyze vulnerabilities in different schemes, the basic operations were fed to the code analyzer. The analyzer considers three different fault models, i.e., stuck-at, bit flips, and instruction skip. Both single and multiple bit flips are possible. The first three analyzed operations, i.e., *Static-DPL XOR*, *Static-Encoding XOR*, and *Static-Encoding LUT* are a special case, where more than 2-bit flips are equivalent to 1-bit or 2-bit flips eventually. The analyzer reports the impact on the final output in the presence of discussed fault models. This is represented as a normalized distribution of faulty output for each considered fault models. Three outputs are expected: `VALID`, `INVALID`, and `NULL`. `VALID` implies that final faulty output stays within the encoding. Similarly, `INVALID` refers to the faulty output which is no longer in the applied encoding. `NULL` faults are `0x00` or `0xFF` values at the output. While `VALID` faults stay within the encoding and can lead to differential fault analysis (DFA), it is rather less likely with `INVALID` faults. On the other hand, `NULL` deletes any data dependent information, disabling any further exploitation by DFA. Therefore, `VALID` faults must be prevented at all costs, while keeping `INVALID` in check and maximizing `NULL` faults. The analysis results for Static-DPL XOR, Static-Encoding XOR, and Static-Encoding LUT are shown in Fig. 10.5. We discuss each of the results in the following.

The fault distribution of *Static-Encoding XOR* is shown in Fig. 10.5a. This encoded operation does not produce any `VALID` faults for 1-instruction skip and 1-bit flip. The percentages of `VALID` faults for other fault models stay between 4 and 6%. Majority of the faults (92–100%) result in `INVALID` faults, while only

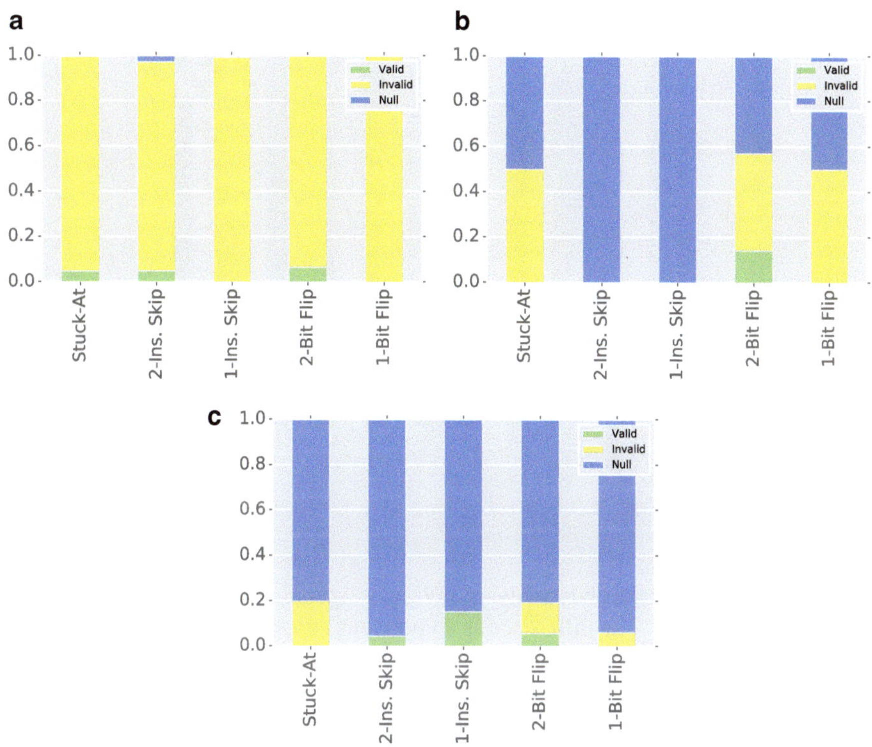

Fig. 10.5 Fault distributions of **(a)** *Static-Encoding XOR*, **(b)** *Static-Encoding LUT*, **(c)** *Static-DPL XOR* code analysis

double instruction skip results in a non-negligible `NULL` faults (2.7%). Although `INVALID` faults are more desirable than `VALID` faults, later we will show that some `INVALID` can be exploitable in particular for *Static-Encoding XOR*.

The *Static-Encoding LUT* shows an altogether different fault resistance (Fig. 10.5b). This encoded operation produces much more `NULL` faults than the previous case, which is a desirable property. Instruction skips result in 100% `NULL` faults, while stuck-at and 1-bit flips produce 50% `INVALID` and 50% `NULL` faults. The only way to produce `VALID` faults in this operation is to inject 2-bit flips which result in 14.2% `VALID` faults. Rest of the faults would result in `INVALID` or `NULL` faults with equal probability.

The analysis results of *Static-DPL XOR* are shown in Fig. 10.5c. While no `VALID` faults are possible for stuck-at and 1-bit flip, it stays below 6% for 1-instruction skips and 2-bit flips. The worst performance is under 2-instruction skip model, where the percentage of `VALID` faults is as high as 15.3%. The high vulnerability against 2-bit flips can be explained by the fact that 2-bit flips are the limit of the dual-rail encoding scheme. Apart from these, the other faults are more likely to be `NULL` rather than `INVALID`, which is desirable.

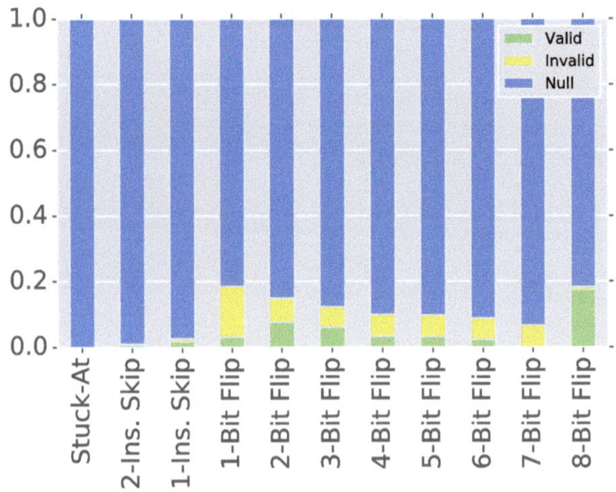

Fig. 10.6 Fault distributions of *Device-Specific Encoding XOR* code analysis

Finally we applied the code analysis on *Device-Specific Encoding XOR* operation as shown in Fig. 10.6. The difference from previous cases is that here multi-bit flips cannot be dealt as a subset of 1-bit and 2-bit flips. Therefore the analysis covers bit flips from 1 bit to 8 bits, i.e., the data width of the target processor. It can be easily observed that this encoding scheme is more likely to produce `NULL` faults which is highly desirable. For the `VALID` faults, stuck-at model produces none, while only <2% can be achieved by instruction skips. In case of bit flips, the percentage of `VALID` faults stays between 2 and 7% with the exception of 7 and 8 bit flips. However, for different β coefficients used in the leakage function, results on bit flips should be slightly different, but consistent with the expectations. The total value of *VALID* bit flips for all the possibilities ranges within 4.2–4.7%, but their distribution is different, depending on used coefficient.

10.5 Experimental Evaluations on Atmel Platform

Following the code analysis, the fault resistance was experimentally verified. In this step, the corresponding code was implemented on a real AVR microcontroller and tested under laser fault injection. The objective of practical validation was to check whether the simulated results correspond to what can be achieved in reality.

10.5.1 Chip Decapsulation

In order to conduct the experiments, the microcontroller had to be decapsulated. The microcontroller was de-packaged from the backside, using a precise milling equipment. While performing this process, there had to be several pauses in between the grinding sessions, to let the chip cool down. First, the epoxy layer had to be milled down. Then, the copper substrate was thinned, so that it could be peeled off. As the thinning process came to an end, there was still glue which is supposed to hold the chip on the copper substrate. This was removed by hard plastic tools. Finally, we obtained a sample ready for the experiments. One can still use more advanced milling/polishing tools in order to thinner the silicon substrate and make the surface even.

10.5.2 Laser Setup

The fault injection was done with a near-infrared diode pulse laser with a pulse power of 20 W (reduced to 8 W with 20× objective). The pulse repetition rate of the laser is 10 MHz and spot size is 30 × 12 μm (15 × 3.5 μm with 20× objective). Intentional nop instructions were inserted at the beginning of each code sequence to overcome the 100 ns delay between trigger and laser injection. The target platform was Atmel ATmega328P microcontroller, de-packaged and mounted on Arduino UNO development board. The surface area of the chip is 3 × 3 mm^2, manufactured in 350 nm CMOS technology. An X–Y positioning table with a step precision of 0.05 μm was used to scan the chip surface and perform laser injection. The timing of injection was synchronized with executed code using a code-generated trigger. The injection platform along with the target is shown in Fig. 10.7.

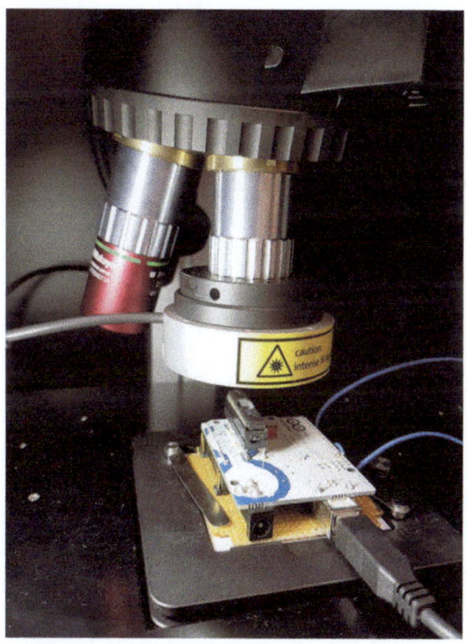

Fig. 10.7 ATmega328P device under a near-infrared diode laser injection setup

10.5.3 Laser Experiments Results

The prime difference from the previous analysis is that in the experimental validation, we cannot precisely control the fault model. Therefore, the fault models are not uniformly distributed. Before starting the real experiment, we performed some profiling on the target with basic assembly code and verified that all the fault models are possible to produce experimentally. Next, we flashed the assembly code of the four previously discussed software encoding operations. The percentage of `VALID`, `INVALID`, and `NULL` faults produced for each tested operation is stated in Fig. 10.8.

Static-Encoding XOR shows the best consistency with the simulated analysis previously (see Fig. 10.8a). While 93.56% of the faults are `INVALID`, only 5.88% `VALID` were produced. Moving towards *Static-Encoding LUT*, we observe a 32.42% `VALID` faults in Fig. 10.8b. Since a `VALID` fault in this implementation can

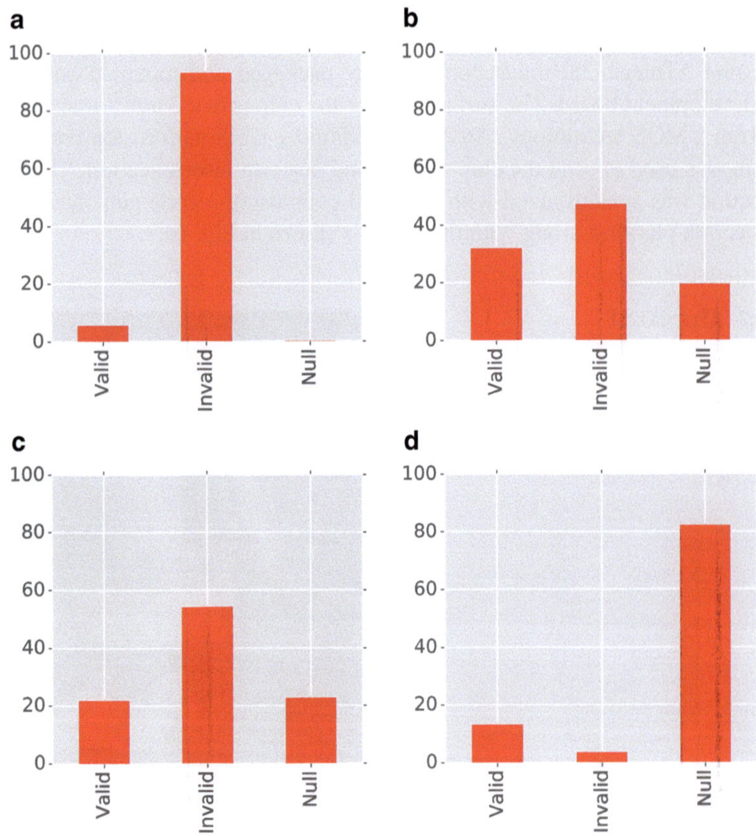

Fig. 10.8 Fault distributions of **(a)** *Static-Encoding XOR*, **(b)** *Static-Encoding LUT*, **(c)** *Static-DPL XOR*, and **(d)** *Device-Specific Encoding XOR* experiments

only result from even bit flips, this infers that the fault model distribution is biased towards multiple bit flips in our experiments. Similarly, we also observe a 22.2% `VALID` faults in *Static-DPL XOR* (Fig. 10.8c) owing to the prevalent multiple bit flip model.

When it comes to *Device-Specific Encoding XOR* (Fig. 10.8d), results show distribution very similar to the one obtained by the code analysis. Because it is more likely to produce bit flips when injecting faults in the microcontroller, at 13.5% an inflated number of *VALID* faults can be observed in this case, with a relatively small number of *INVALID* faults. As expected, *NULL* outputs are dominant, i.e., 82.5%, because of the look-up table properties.

10.6 Discussion

In this section, we will discuss some important parameters of particular encoding implementation with respect to fault injection attacks.

10.6.1 Selection of β Coefficients

We considered several parameters for the code analysis of *Device-Specific Encoding XOR*. We analyzed different β values scenarios. We considered the case where the variance of the β is relatively high (the βs might be cancelling each other), and the case where the variance of the β is low (almost Hamming weight).

The most significant difference can be observed in the result for implementation with β coefficients that do not follow Hamming weight leakage model (stated in Fig. 10.9a). From the figure, it can be observed that the number of 1-bit flips is

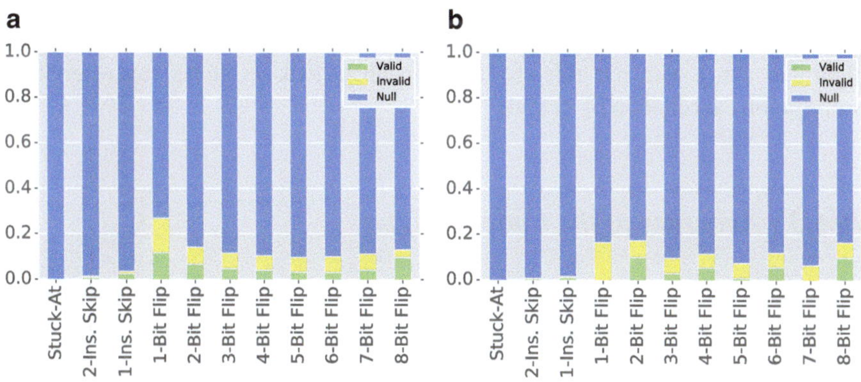

Fig. 10.9 Fault distributions of *Device-Specific Encoding XOR* code analysis with **(a)** high variance and **(b)** almost Hamming weight

inflated, compared to the almost Hamming weight case (stated in Fig. 10.9b). The behavior of the faults shows contrast between different beta values, which is not the case for other encoding schemes, and hence could be further investigated.

10.6.2 Fault Propagation

When considering security of different implementations, fault propagation is an important factor that can significantly affect the possibility to mount an attack. In case we want to prevent a successful fault attack, it is necessary to avoid the propagation of an *INVALID* output when it is fed as an input to a next iteration of the algorithm. Otherwise, this output could leak some information about the processed data and therefore allow an attacker to use the differential fault analysis.

From this point of view, look-up table implementations have an advantage, since every input that does not follow the encoding will be automatically converted to *NULL*. Analysis results of *Static-DPL XOR*, *Static-Encoding LUT*, and *Device-Specific Encoding XOR* show that if any of the inputs is either *INVALID* or *NULL*, it will always output *NULL*. Situation with the *Static-Encoding XOR* is different because of the algorithm design. There are several combinations of inputs that lead to *VALID* faults—more specifically, any combination of:

- Two *INVALID* inputs,
- Two *NULL* inputs, and
- *INVALID* and *NULL* inputs.

Moreover, a combination of *VALID* and *NULL* inputs leaks a complete information about the *VALID* input in the form $\bar{v_3}\bar{v_3}\bar{v_2}\bar{v_2}\bar{v_1}\bar{v_1}\bar{v_0}\bar{v_0}$, where $v_3v_2v_1v_0$ is the original input.

To summarize, table look-up implementations provide solid protection against fault attacks when it comes to fault propagation. Any other implementation that uses standard operations performed by using ALU can be vulnerable if it is not directly designed with such goal in mind. Therefore, when designing a fault resistant algorithm along with the side-channel resistance, look-up tables can offer fault propagation cancellation by default.

10.6.3 Accuracy of the Code Analyzer

After observing the experimental results, one can notice the (sometimes significant) difference, compared to simulated results. As we already mentioned in Sect. 10.5, when performing the experiment, the attacker often cannot precisely control the fault model. Moreover, different fault injection techniques lead to various outcomes on different devices under test.

The purpose of the code analyzer is to show possible vulnerabilities with respect to provided code. Once the probability of inducing a *VALID* fault into the code is greater than zero, it means the attacker will always be able to inject a fault if he chooses a proper fault injection equipment and uses correct settings, unless there is some other tamper protection implemented in the device.

On the other hand, if the probability of a *VALID* fault is zero in the simulated environment, it stays the same in the experimental settings, no matter what technique the attacker uses.

10.6.4 Simulating a Pipelined Architecture

When targeting a pipelined architecture, the fault can be injected in all the instructions in the pipeline at a current time. For example, Fig. 10.10 depicts a 4-stage pipeline, where a fault was injected at cycle 4. The fault can disturb up to four instructions, each at a different stage; however, in this case, only the instruction in the *Execute* stage was disturbed. As observed in [14], different pipeline stages can have a different fault sensitivity. An advanced attacker may be able to determine the sensitivity threshold by carefully profiling the device, making it possible to precisely target each stage. Another observation in [14] is that the nature of the fault is dependent on the stage. In the 7-stage RISC processor they used, disturbing the first two stages can cause instruction skips, while the later stages can cause data faults.

Crafting an analyzer with a pipeline fault simulation is possible, but it would require a deep analysis of the target platform and the code analysis result would be dependent on this platform. Therefore, it would not be possible to get universal result with respect to given fault models.

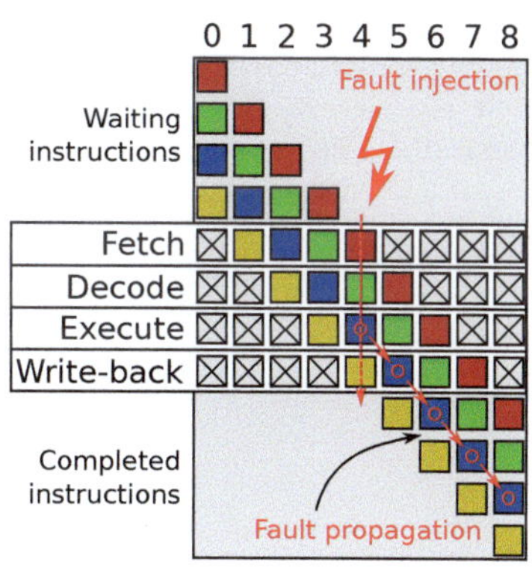

Fig. 10.10 Fault injection on a 4-stage pipelined architecture

10.7 Chapter Summary

This chapter presented an architecture of automated evaluation framework that can be used for testing protected cryptographic implementations.

A case study on evaluating fault attack resistance of three software-based encoding schemes that were introduced to prevent side-channel attacks was conducted to show the usefulness of the framework.

Appendix 1: Assembly Code For *Static-DPL XOR* Implementation

Table 10.3 in this section contains assembly code used for the code analysis. Note that there are several differences in comparison to the original paper. We precharge all the registers before the code execution; therefore, there is no need to use precharge instructions. The other change is in instructions 7 and 8, where we first load the operation code (can take values `01010101` for `and`, `10101010` for `or`, and 01100110 for `xor`) and then we execute `ldd` instruction using the destination register, operation code, and value. Look-up tables are stated in Table 10.4.

Table 10.3 Assembly code for *DPL XOR* in AVR

#	Instruction	#	Instruction
0	`ldi r1` *a*	5	`andi r2 00000011`
1	`ldi r2` *b*	6	`or r1 r2`
2	`andi r1 00000011`	7	`ldi r4` *operation*
3	`lsl r1 1`	8	`ldd r3 r4 r1`
4	`lsl r1 1`	9	`mov d r3`

Table 10.4 Look-up tables for `and`, `or`, and `xor`

index	and	or	xor
0000-0100	00	00	00
0101	01	01	10
0110	10	01	01
0111-1000	00	00	00
1001	10	01	01
1010	01	10	10
1011-1111	00	00	00

Appendix 2: Assembly Code for *Static-Encoding XOR* Implementation

The code stated in Table 10.5 follows the originally proposed algorithm for *Static-Encoding XOR*. This implementation uses several constants, either for clearing and precharging the registers before loading the data (e.g., `ldi r16 11110000`) or for changing the data to proper encoding format (e.g., `ldi r17 01011010`).

Appendix 3: Assembly Code for *Device-Specific Encoding XOR* Implementation

In this section, we describe the code used for *Device-Specific Encoding XOR*. After determining the bit leakage weights, and computing the encoding based on Algorithm 1, several look-up tables are constructed.

According to the original paper [8], it is reasonable to split an n-bit variable into two different halves in order to avoid holding large look-up tables in memory. Therefore, we use two registers for processing each value.

Table 10.5 Assembly code for *Encoding XOR* in AVR

#	Instruction	#	Instruction
0	ldi r1 *a*	19	and r20 r1
1	ldi r2 *b*	20	and r21 r1
2	ldi r16 11110000	21	swap r21
3	ldi r17 11110000	22	or r20 r21
4	and r16 r1	23	ldi r22 00001111
5	and r17 r1	24	ldi r23 00001111
6	swap r17	25	and r22 r2
7	or r16 r17	26	and r23 r2
8	ldi r18 11110000	27	swap r23
9	ldi r19 11110000	28	or r22 r23
10	and r18 r2	29	ldi r21 10100101
11	and r19 r2	30	eor r20 r21
12	swap r19	31	eor r20 r22
13	or r18 r19	32	ldi r24 11110000
14	ldi r17 01011010	33	ldi r25 11110000
15	eor r16 r17	34	and r24 r16
16	eor r16 r18	35	and r25 r20
17	ldi r20 00001111	36	or r24 r25
18	ldi r21 00001111		

Table 10.6 Assembly pseudocode for *Device-Specific Encoding XOR* in 8-bit AVR

#	Instruction	#	Instruction
1	`ldi r1` *a*	12	`eor r4 r4`
2	`ldi r2` *b*	13	`ldd r4` *lutlb* `r1`
3	`eor r3 r3`	14	`eor r5 r5`
4	`ldd r3` *luthb* `r1`	15	`ldd r5` *lutshift* `r4`
5	`eor r4 r4`	16	`eor r6 r6`
6	`ldd r4` *lutshift* `r3`	17	`ldd r6` *lutlb* `r2`
7	`eor r5 r5`	18	`or r5 r6`
8	`ldd r5` *luthb* `r2`	19	`eor r4 r4`
9	`or r5 r4`	20	`ldd r4` *lutop* `r5`
10	`eor r3 r3`		
11	`ldd r3` *lutop* `r5`		

In Table 10.6, the pseudocode for the encoding is presented. First, the upper nibble is retrieved for inputs a and b (a_h and b_h) under the encoding format ($f(a_h)$ and $f(b_h)$), using the *luthb* table, followed by the look-up table *lutop* used to perform xor operation ($LUT(f(a_h) << 4||f(b_h)) = f(a_h \oplus b_h)$). Similar procedure is done for the lower nibble, using the *lutlb*.

References

1. A. Barenghi, L. Breveglieri, I. Koren, D. Naccache, Fault injection attacks on cryptographic devices: theory, practice, and countermeasures. Proc. IEEE **100**(11), 3056–3076 (2012)
2. J. Breier, D. Jap, C.-N. Chen, Laser profiling for the back-side fault attacks: with a practical laser skip instruction attack on AES, in *Proceedings of the 1st ACM Workshop on Cyber-Physical System Security, CPSS '15* (ACM, New York, 2015), pp. 99–103
3. C. Chen, T. Eisenbarth, A. Shahverdi, X. Ye, Balanced encoding to mitigate power analysis: a case study, in *CARDIS*. Lecture Notes in Computer Science (Springer, Paris, 2014)
4. F. Courbon, P. Loubet-Moundi, J.J.A. Fournier, A. Tria, Adjusting laser injections for fully controlled faults, in *International Workshop on Constructive Side-Channel Analysis and Secure Design* (Springer, Cham, 2014), pp. 229–242
5. H. Groß, T. Plos, On using instruction-set extensions for minimizing the hardware-implementation costs of symmetric-key algorithms on a low-resource microcontroller, in *International Workshop on Radio Frequency Identification: Security and Privacy Issues* (Springer, Berlin, 2012), pp. 149–164
6. Z. Hao, P. Chu, T. Zhang, D. Wang, C. Hou, A high-performance framework for instruction-set simulator, in *Recent Advances in Computer Science and Information Engineering* (Springer, Berlin, 2012), pp. 9–14
7. P. Hoogvorst, J.-L. Danger, G. Duc, Software implementation of dual-rail representation, in *COSADE*, Darmstadt (2011)
8. H. Maghrebi, V. Servant, J. Bringer, There is wisdom in harnessing the strengths of your enemy: customized encoding to thwart side-channel attacks, in *International Conference on Fast Software Encryption* (Springer, Berlin, 2016), pp. 223–243
9. P. Rauzy, S. Guilley, Z. Najm, Formally proved security of assembly code against leakage. IACR Cryptol. ePrint Arch. **2013**, 554 (2013)

10. W. Schindler, K. Lemke, C. Paar, A stochastic model for differential side channel cryptanalysis, in *International Workshop on Cryptographic Hardware and Embedded Systems* (Springer, Berlin, 2005), pp. 30–46
11. N. Theissing, D. Merli, M. Smola, F. Stumpf, G. Sigl, Comprehensive analysis of software countermeasures against fault attacks, in *Design, Automation Test in Europe Conference Exhibition (DATE), 2013* (2013), pp. 404–409
12. I. Yasui, Y. Shimazu, Microprocessor with Harvard architecture. US Patent 5034887, 23 July 1991
13. S.Y. Yuan, H.E. Chung, S.S. Liao, A microcontroller instruction set simulator for EMI prediction. IEEE Trans. Electromagn. Compat. **51**(3), 692–699 (2009)
14. B. Yuce, N.F. Ghalaty, P. Schaumont, Improving fault attacks on embedded software using RISC pipeline characterization, in *2015 Workshop on Fault Diagnosis and Tolerance in Cryptography (FDTC)* (2015), pp. 97–108

Chapter 11
Automated Evaluation of Concurrent Error Detection Code Protected Hardware Implementations

Dirmanto Jap, Jakub Breier, Shivam Bhasin, and Anupam Chattopadhyay

11.1 Introduction

Security of critical devices is typically realized by deploying modern cryptography in a form of theoretically proven algorithms and protocols. However, bad implementation of these algorithms and protocols can eventually lead to serious exploits. Conventional security researches have mostly neglected the hardware layer vulnerabilities [23–25, 31]. Physical attacks which target poor implementations of cryptography could compromise system security. Embedded devices are particularly vulnerable against attacks directly on lower level of hardware abstractions, such as power or EM based side-channel attack (SCA) [1, 19], hardware Trojan horses (HTHs) [26], and fault injection attacks (FIA) [5].

This research was conducted when the author "Jakub Breier" was with Temasek Laboratories, NTU.

D. Jap · S. Bhasin
Temasek Laboratories, Nanyang Technological University, Singapore, Singapore
e-mail: djap@ntu.edu.sg; sbhasin@ntu.edu.sg

J. Breier (✉)
Underwriters Laboratories, Singapore, Singapore
e-mail: jbreier@jbreier.com

A. Chattopadhyay
School of Computer Science and Engineering, Nanyang Technological University, Singapore, Singapore
e-mail: anupam@ntu.edu.sg

J. Breier et al. (eds.), *Automated Methods in Cryptographic Fault Analysis*,
https://doi.org/10.1007/978-3-030-11333-9_11

Many techniques have been proposed against the fault attacks [3, 12, 22]. Since the embedded devices are usually restricted by the limited power supply and insufficient computation power in the remote nodes, expensive countermeasures are often not applicable. To protect against faults in resource-constrained environments, the default choice for designers is concurrent error detection (CED) and correction which stem from information theory. CED is widely used by VLSI design and testing community for applications like memory testing [27], fault tolerance [7], etc.

The simplest form of CED is *parity* [8]. It has been previously proposed in the context of fault injection attacks [28]. One of the first works, dealing with block cipher AES, applied a single parity on the whole 128-bit block [29]. This was further improved by using a single parity bit per byte or word [17, 18]. Moreover, non-linear codes have also been applied in this context to achieve higher detection coverage [14].

Later, a system-on-chip architecture was proposed [30] to detect and prevent a hardware Trojan insertion. As Trojan can maliciously modify sensitive data, its impact can be considered similar to fault injection [6]. Thus, the proposed solution relies on randomized linear parity prediction to detect any data modification by a malicious Trojan in hardware. Although the tested implementation deployed linear CED techniques, it was claimed that randomized non-linear codes can achieve better detection capabilities.

In a basic single bit parity as countermeasure against FIA, the protection heavily depends on the capabilities of the adversary to trigger even number of faults in the target device. However, while most of the previous works only assess this scheme in theoretical setting, practical evaluation has only received little attention [10].

Thus, further evaluations of the capabilities of different CED schemes used as an FIA countermeasure need to be conducted before the actual deployment, which is the aim of this chapter. The provided automated method takes the fault injection characterization of the device into account, and evaluates the success rates for a given parity scheme. To show the practicality of such method, insights on different characteristics of linear and non-linear randomized parity codes under fault injection attacks are provided.

The rest of the chapter is structured as follows: Section 11.2 provides a necessary background on concurrent error detection. Section 11.3 details the automated evaluation method, followed by Sect. 11.4 which provided results on CED-protected PRESENT cipher. Finally, Sect. 11.5 summarizes this chapter.

11.2 Technical Background

In this section, the technical details of several common concurrent error detection schemes are provided.

11.2.1 Parity-Based Concurrent Error Detection

Generally, a fault detection capability is achieved at the expense of either time or space redundancy. In the case of time redundancy, errors are detected by repeating the encryption process. No extra logic is required, but the throughput is reduced to around 50%. Parity is one of the simplest solutions when it comes to error detection based on space redundancy. It is heavily used in communication systems. When implemented in block ciphers, it can detect abnormalities during en/decryption, while allowing efficient hardware implementation, which makes it a good candidate for resource-constrained systems with low computational power. Different parity schemes have been proposed in literature, aiming at covering different fault models, while keeping the low cost [4, 17, 18, 29].

The parity techniques can be implemented by two general approaches [9]:

- Parity-1: Only 1 parity bit is required for all the bits of datapath in each round of the cryptographic algorithm, e.g., one parity bit checks the errors for all the bits of the data vector, such as 1-bit parity for the 128 bits in AES-128 [29].
- Parity-n: n parity bits are employed, and each parity bit is responsible for the error checking of a single data block in the cryptographic algorithm, e.g., Parity-16 implements 1 parity bit per byte of AES-128 [17, 18].

Nevertheless, both of the above-mentioned schemes can only detect odd number of faults occurred in the 128 data bits in Parity-1, or in the data word of Parity-n. In other words, if even number of faults appears, the schemes will be compromised. Another scheme, proposed by Karpovsky et al. [14] provides wider fault coverage, relying on a prediction circuit comprised of linear predictor, linear compressor, and cubic function. Despite the uniform detection of both odd and even number of faults, the circuit overhead is too high to be applied in resource-constrained scenarios. Considering the implementation efficiency and the fact that majority of fault models aim at single bit-flips, we hereby focus on the Parity-1 scheme, and the conclusions directly apply to Parity-n as well.

11.2.2 Parity-1 Detection Scheme

There are several works proposing the usage of parity for fault detection [15, 16, 28], showing that despite not being able to detect more complex fault models, its simplicity still attracts attention. Figure 11.1 depicts the parity detection scheme proposed in [29], targeting AES. Actually it is universal to all secret key cryptography (SKC) constructed by substitution–permutation networks (SPNs). The round inputs are denoted as `X`. The round key `K` is added with `X` to produce `Y`. The non-linear substitution box (Sbox) substitutes `Y` by `Z`. The linear diffusion layer permutes the bits of `Z` to give `U`, that is, actually the input `X` for the next round (or the

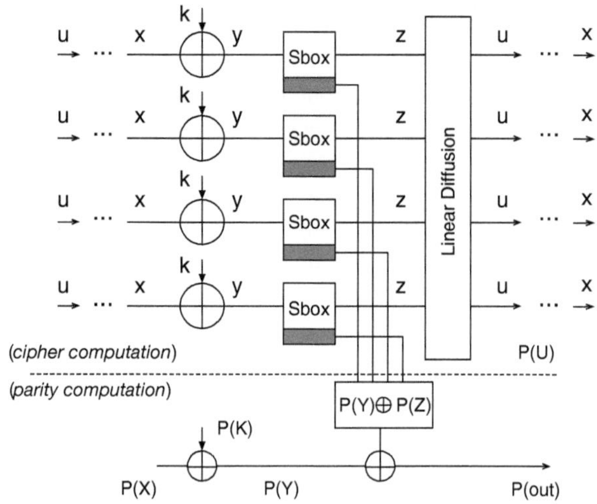

Fig. 11.1 Parity-based concurrent error detection in SPN block cipher

ciphertext in the case of the last round). This parity prediction mainly consists of three computations. For clarity, parity of a bit vector $\cdot$ is denoted as $P(\cdot)$. The computation then goes as follows:

1. In key addition, the round input parity $P(X)$ is XORed with round key parity $P(K)$ to get $P(Y)$.
2. In Sbox, $P(Y)$ is non-linearly changed. Since Sbox is normally fixed and public, the Sbox output parity $P(Z)$ can be pre-computed. To check the integrity of the processed data, the input and output parity can be combined by calculating $P(Y) \oplus P(Z)$ and pre-computed as an extension of a standard Sbox.
3. The linear diffusion layer simply permutes the bits, so the parity is not changed in this step.

This process is depicted in Fig. 11.1. Therefore, the computation goes as follows:

$$P(Y) \oplus (P(Y) \oplus P(Z)) = P(Z) = P(\text{out}). \tag{11.1}$$

If no odd number of errors occurs:

$$P(\text{out}) = P(U). \tag{11.2}$$

Otherwise,

$$P(Y) \oplus (P(Y^*) \oplus P(Z)) = P(\text{out}) \neq P(U), \tag{11.3}$$

where $P(Y^*)$ represents the error-infected bit vector Y. Similarly to key addition part,

$$P(Y) = P(X) \oplus P(K). \tag{11.4}$$

Since $P(X)$ is the $P(\text{out})$ of the previous round, we get

$$P(Y) = P(\text{out}) \oplus P(K). \tag{11.5}$$

By checking if $P(Y)$ is equal with $P(\text{out}) \oplus P(K)$, we can detect the encryption faults caused by the odd number of errors.

11.2.3 Advanced CED Techniques

To avoid the main shortcoming of the parity, which is limited fault coverage, advanced CED techniques need to be deployed. To recall, there are many different types of possible faults that could be injected in the hardware. To limit the scope of our investigation, we only consider fault models that are exploitable. The most commonly used fault model for fault analysis in the literature is the bit-flip, where one or several bit values are inverted. It could be formulated as $e = x \oplus x^*$, where $x \in \mathbb{F}_2^n$ is the original data, and $x^* \in \mathbb{F}_2^n$ is the faulty data. The number of bit-flips could be defined as a Hamming weight-$HW(e)$ (or number of "1"s) in e.

11.2.3.1 Linear Codes and Non-linear Codes

A linear code of length n, rank k is a linear subspace C of $\mathbb{F}_2^n$, and the vectors in C are called *codewords*. As a linear subspace over $\mathbb{F}_2^n$, the code C sometimes could be represented as the span of basis codewords from the rows of the generator matrix G, which has a standard form $G = [I_k|A]$. This sort of coding is called systematic encoding, and it has the original data present in the codeword. Here, I_k is a square identity matrix with dimension k, and A is $k \times (n-k)$ matrix. Hence, any codeword $y \in C$ can be written as $y = xG$, where $x \in \mathbb{F}_2^k$ is the message with dimension k.

A parity check matrix H for (n, k)-linear encoding is a matrix where

$$Hy^T = 0 \Leftrightarrow \exists x \in \mathbb{F}_2^k, \tag{11.6}$$

such that $xG = y$. If G is a generating matrix in standard form, then H can be written as $[P|I_r]$ (called linear systematic code), where $P = A^T$ is a $(n-k) \times k$ matrix and I_r is a square identity matrix with dimension r.

For non-linear codes, we consider the cubic codes proposed in [13] as well as the inverse code proposed in [20]. As mentioned earlier, for a linear code, the codeword y could be rewritten as

$$y = xG = [xI|xA] = [x|(Px^T)^T]. \tag{11.7}$$

For simplicity, since it is a vector, we could write it as $(x, (Px))$. In [13], it is shown that for binary code, the construction of cubic code is

$$C_V = \{(x, w) | x \in \mathbb{F}_2^k, w = (Px)^3 \in \mathbb{F}_2^r\}. \tag{11.8}$$

In [20], the construction for the inverse code is

$$C_V = \{(x, w) | x \in \mathbb{F}_2^k, w = (Px)^{-1} \in \mathbb{F}_2^r\} \tag{11.9}$$

instead, with $0^{-1} = 0$.

11.2.3.2 Randomized Parity Code

One example of a code is the parity code, defined as parity of a subset of the vector of length k, i.e., $y = a_1x_1 \oplus \cdots \oplus a_kx_k$, where $a_i \in \{0, 1\}$. Here, we have $n = k+1$. The (n, k) parity (linear) encoding is then defined as $g : x \rightarrow y$ $(x \in \mathbb{F}_2^k, y \in \mathbb{F}_2^n)$, where $n = k + r$, and r is the number of parity bits.

The construction of randomized parity codes, as defined in [30], is as follows:

- From the set of all (n, k)-linear systematic parity codes $R^{n\times k}$, with all rows and columns in the parity check matrix being non-zero, we sample uniformly at random.
- The code that has been sampled is called *randomized parity code*.
- The randomization is done in order to prevent the adversary from knowing the parity check matrix, while the non-zero constraint is to prevent the zero function as well as to ensure that each bit is included in the parity function.

The method used to obtain the randomized parity code can be described as follows: given the dimension k and parity bit r, output the randomized parity check matrix by choosing uniformly at random, a parity matrix H. If it satisfies the construction described earlier, output H, otherwise repeat and choose another parity matrix.

Figure 11.2 shows a system, proposed recently by Wu et al. [30], which is claimed to be robust against hardware Trojans. For every internal component in the system, the logic, memory, and the communication architecture, the corresponding CED techniques are elaborated. In this chapter, we particularly focus on the randomized parity encoding technique that was proposed to be a part of the memory and communication architecture (bus) protection. In Fig. 11.3, it is shown how to perform the selection of the bits for randomized parity [30].

It was argued that the detection rate could be higher if non-linear code is used at the expense of additional area. For non-linear randomized code, the parity matrix H is chosen similarly, following the construction of randomized parity code described earlier. However, the parity matrices are based on non-linear construction instead.

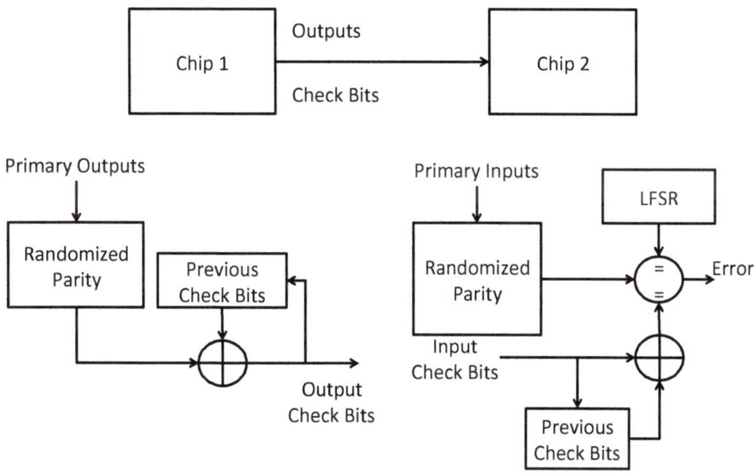

Fig. 11.2 Hardware Trojan detection based on randomized parity codes

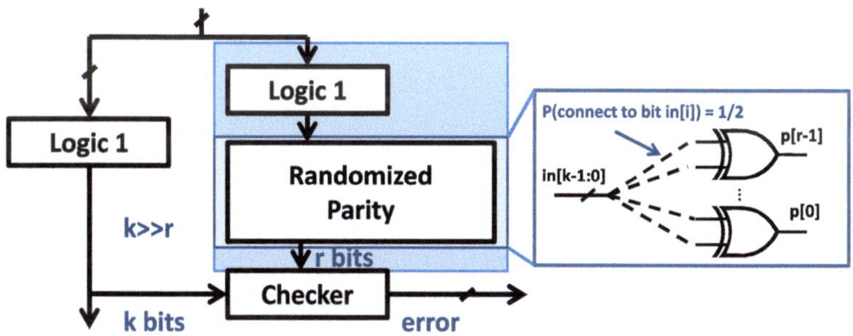

Fig. 11.3 Bit selection for randomized parity

From the algorithmic point of view, the parity scheme can be implemented across different bits. In this case, the parity can be calculated on either 128, 64, 32, 16, 8, or 4 bits data. As shown in the experiments later, we implemented the parity computation over 32 bits (for AES, adjusted to the MixColumn) and 16 bits (for PRESENT, adjusted to the pLayer). One consideration is that, for smaller bit size, it might improve the detection rate in the case of a localized fault. However, in this case, the area required to store the parity bits increases as well. For example, depending on the architecture, it will require additional register(s) or memory to store the additional parity bits (since the parity scheme is covering fewer bits, more parity bits are required to handle the data). Moreover, for the randomized parity, it requires randomness to select each individual parity check matrix, and thus, adds additional cost in terms of random number generation.

11.3 Automated Evaluation of the CED Schemes

In this section we first describe how to characterize the device to obtain the fault models which are achievable with given fault injection equipment. Later, we propose the automated evaluation method and show some results on several different CED schemes.

For 1-bit parity scheme, if the adversary is able to perform fault injection faulting even number of bits, it could bypass the parity countermeasure. This was previously shown to be feasible by using laser fault injection attack [10]. Hence, the protection offered is not sufficient enough.

As a more permanent and dependable solution, complex codes are a fair alternative. The CED codes can be made robust by varying different parameters. For instance, one can choose a longer code which is still linear in construction, like parity, to have better detection rate but still low implementation overhead. On the other hand, non-linear codes are more complex in construction and thus have high implementation footprint, but they are believed to be more robust. Another solution is to use randomization with codes to limit the fault injection capability of the attacker. In the rest of this section we focus on advanced CED techniques.

11.3.1 Attack Model and Device Characterization

Fault models depend on the injection techniques, injection parameters, and underlying target. Fault injection with the given model can be practically injected or simulated. Practical fault injection would represent non-uniformly distributed small subset of the simulations, preventing comprehension of overall trend. In contrast, simulation covers different fault scenarios, which could highlight the general trend; however, in case of practical setting, this might overestimate the trend, since most of the faults tend to be biased towards a specific model. Thus, a combination of simulation and practical evaluation usually has to be considered for validation purposes.

In order to operate on real-world values in the simulations, it is necessary to perform a device characterization before the actual evaluation of the CED-protected implementation. We provide the automated profiling method in Chap. 14 that can provide the necessary inputs describing the device characteristics for the evaluation in this chapter. Here, we will briefly state the results of the characterization.

For profiling purposes, we have implemented block cipher PRESENT, where 4 of the total 64 round registers reside in the target FPGA slice. Out of 10,000 experiments, we received 3918 faulty encryptions that were caused by flipping one or more bits in the registers. Percentages representing each bit-flip fault model are stated in Table 11.1.

Table 11.1 Experimental results from the FPGA slice scan targeting four round registers of PRESENT

Fault model	% of faults
1-bit flip	57.25
2-bit flip	24.17
3-bit flip	15.19
4-bit flip	3.45

Algorithm 1: Automated simulation method for randomized parity according to fault distribution D

Input : H: random parity matrix; r: number of faulted bits; k: number of inputs; f: implemented function; D: fault distribution (optional).
Output : p: detection accuracy.

Generate k random inputs T;
Generate k random r-bit faults E according to distribution D. If D is not present, assume uniform distribution.;
Set $detected := 0$;
for *int i:= 1 to k* **do**
  Calculate the faulty value $v = T[i] \oplus E[i]$;
  Calculate the parity $par_T = H(f(T[i]))$;
  Calculate the parity $par_v = H(f(v))$;
  if $par_T \;!= par_v$ **then**
    $detected{+}{+}$;
Calculate the detection accuracy as $p := \frac{detected}{k}$;
return: p – the detection accuracy value.

11.3.2 Automated Simulation Methodology

Algorithm 1 shows the automated method for computing the detection accuracy of evaluated random parity. In case the evaluator possesses the fault distribution of the target device obtained by a prior characterization, he can use these values as an input to the algorithm in a form of variable D. Otherwise, a random distribution will be assumed. In the rest of this chapter, we will use the values from Table 11.1. The algorithm iterates through random inputs and random faults and checks according to implemented function f (which can be any part of cryptographic algorithm) whether the parity value is equal or not. At the end, the detection accuracy is calculated.

As an alternative to Algorithm 1, in order to reduce the search space complexity to find faults that can bypass the parity check, some evolutionary algorithms, such as genetic algorithm, might be also employed instead of a random fault model. However, this might come with a trade-off, requiring higher time complexity.

Algorithm Implementation and Evaluation

Implementation of the algorithm was done in MATLAB. We varied the number of bit-flips from 1 to n, where n is the data bit-width. A multiple bit-flip fault was injected by performing a modulo-2 addition between the input x and a chosen fault mask m of the same bit-width. For small bit-width $n \leq 8$, combinations of x and m were chosen exhaustively. For bigger n, both x and m were chosen randomly from two independent uniformly distributed data for a representative number of scenarios.

We simulated the case for the randomized encoding, based on the construction described in [30]. The length of message was 120 bits and the parity bits were varied from 3 bits to 8 bits. We run repeated experiments (100,000×) with randomly selected message, and for each, we randomly selected the parity check matrix for the randomized encoding.

As shown in Fig. 11.4a, c, the linear $(x, p(x))$ and inverse $(x, p(x)^{-1})$ encodings perform similarly. For cubic $(x, p(x)^3)$ encoding (Fig. 11.4b), the detection probability for even parity ($r = 4, 6, 8$) is lower for small number of errors, which gradually improves when the number of bit-flips increases. For even parity ($r = 2q$), the cubic power mapping is not bijective. The number of elements which is cube can be calculated by

$$\frac{2^r - 1}{gcd(3, 2^r - 1)}. \tag{11.10}$$

For even r, $3|2^r - 1$, and thus, for lower number of error bits, the faulty predicted and calculated parity could have a collision, resulting to undetected faults. As the number of erroneous bits increases, the fault could affect the parity bit as well which allows the detection.

In general, it can be concluded that if the designer has the liberty to use multiple parity bits ($r \geq 5$), the performance of randomized encoding with linear or non-linear code is equivalent. For lower parity bits ($r \leq 4$), inverse stays similar to linear, while cubic can only outperform linear at a high number of bit-flips. In general, the detection is similar across different encoding schemes. This might be attributed to collisions due to the length of parity mapping. With longer parity, the effect of the collision could be reduced. However, considering the implementation perspective, it is common knowledge that non-linear encoding (cube or inverse) can be much more resource-consuming compared to basic linear encoding [27].

To improve the detection rate, authors in [30] suggest to introduce a *memory effect*. Memory effect means that instead of dealing with codewords individually, the current or ith codeword is computed as a combination of all $(i-1)$th codewords. With the memory effect, it becomes difficult for the attacker to manipulate several stages, thus improving the detection rate in practice. The simulation results for this scenario are shown in Fig. 11.4d–f. The detection probability was improved even for smaller codewords ($r = 3, 4$). Moreover, all the parity encoding performs similarly, supporting the linear encoding under implementation cost consideration.

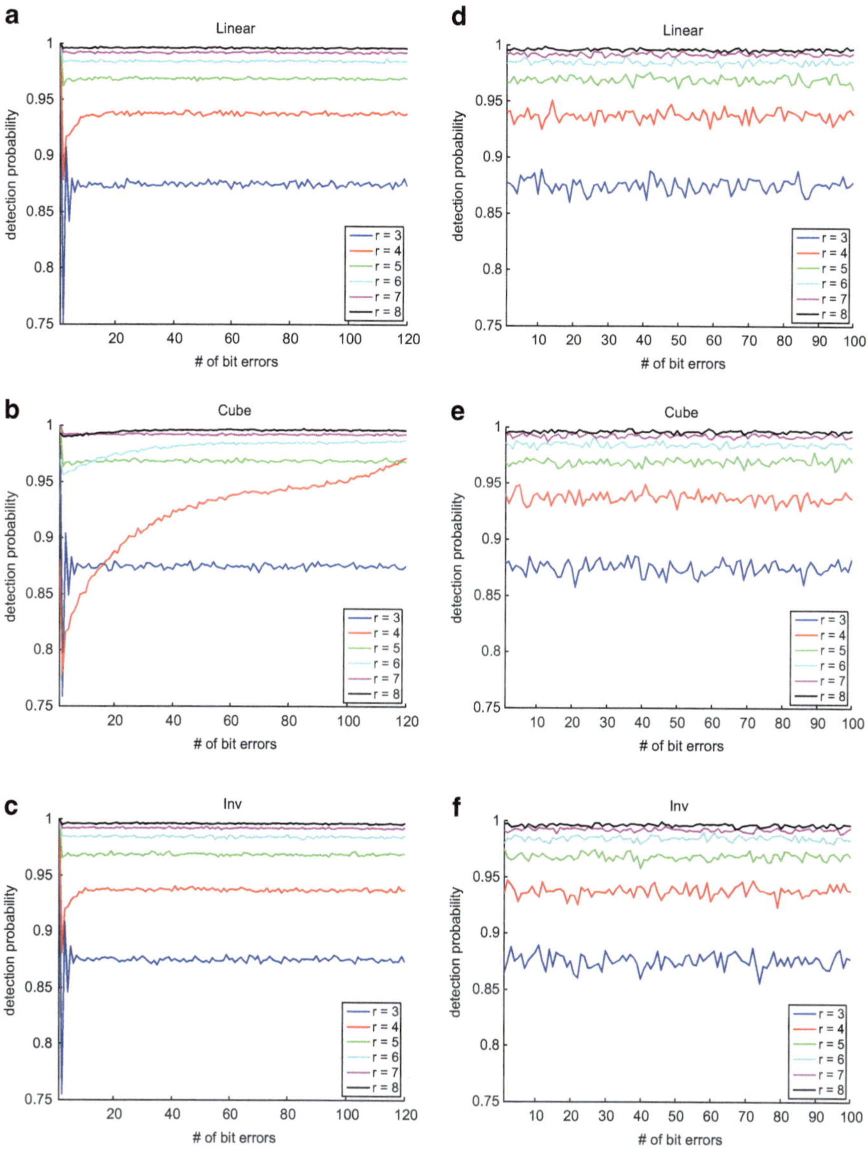

Fig. 11.4 Evaluation of randomized encodings: linear vs non-Linear. (**a**) linear, (**b**) cubic, (**c**) inverse without memory effect, while (**d–f**) with implemented memory effect

11.4 Case Study on PRESENT Cipher

The method from previous section was used to evaluate the fault coverage of CED-protected PRESENT cipher. Since the block cipher rounds use the same set of operations throughout the encryption, we simulated the fault propagation and coverage during one round of PRESENT. Since a PRESENT round could be divided into 4 groups of 4 nibbles (based on the properties of the SBox and permutation layer operations), we considered 16 bits input and output vectors.

11.4.1 Uniform Fault Distribution

First, we analyzed the PRESENT implementation considering a uniform fault distribution.

The r parity bits were varied from 2 to 8, under the memory effect, while $r = 1$ was ignored as it would be just a standard parity bit, for linear and non-linear encoding. The fault was then injected during the round computation. The parity computed over the faulty output was compared against the parity of the expected output. A match results in detection failure.

Based on the simulation results, as shown in Fig. 11.5, we can see that the fault detection rate for different encoding schemes is similar in all cases. Note that in our experiments, for longer message bit length, it was not possible to simulate all potential fault masks in a reasonable time. Hence, rather than exhaustive simulations, we carried the simulations until the error became negligible.

11.4.2 Profiled Fault Distribution

Further investigation were done based on the previously characterized fault model.

As previously shown, a fine grain scan of the DUT allowed us to achieve fault models shown in Table 11.1. The experiments were conducted using different parity lengths (4, 8, and 16 bits parity). The fault caused 1–4 bit-flips on a single nibble, randomly chosen from the 16 nibbles. The experiments were repeated 100,000×. In Table 11.2, we show the detection accuracy of random bit-flip faults in a nibble for different parity schemes. It can be observed that the accuracy for inverse function is similar to the linear encoding, and the cube function is performing worse than the others. In general, for 8-bit parity or longer, the accuracy of the achieved detection is greater than 98%. Given that the detection capability of randomized non-linear codes was no better than the linear counterpart (Table 11.2), we did not proceed with a real implementation. It is known that implementation overhead of non-linear codes is significantly higher than linear codes, with similar detection capabilities. Therefore, linear codes stand as an obvious choice.

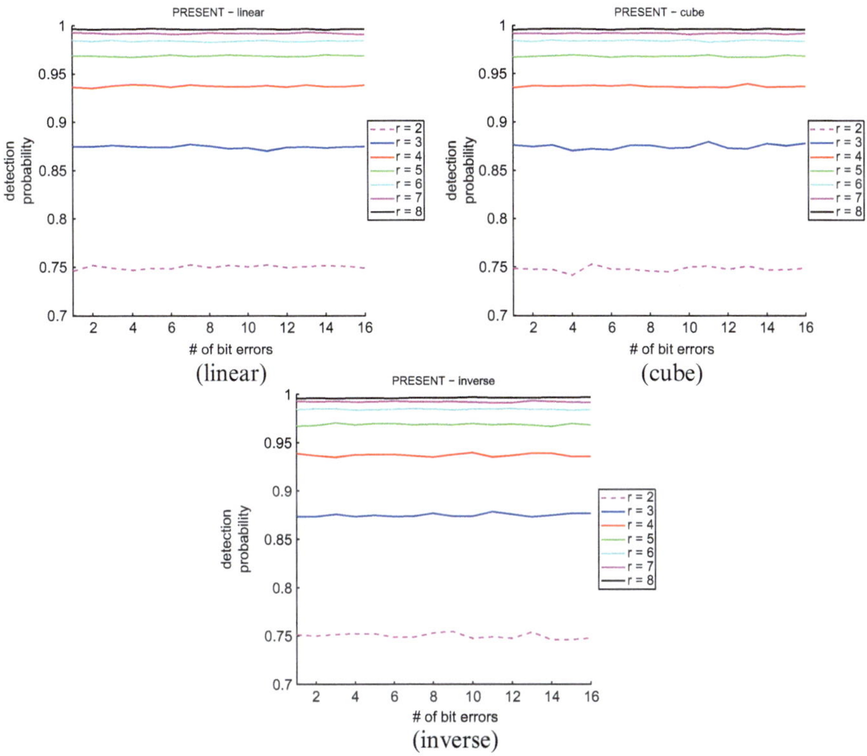

Fig. 11.5 Evaluation of different CED schemes during the operations of PRESENT cipher

Table 11.2 Detection accuracy of different randomized parity schemes (in %)

Parity/No. error	1	2	3	4
Linear randomized parity				
$r = 4$	93.79	95.76	96.45	97.04
$r = 8$	99.62	99.75	99.76	99.82
$r = 16$	100.00	100.00	100.00	100.00
Cube randomized parity				
$r = 4$	82.40	83.76	84.54	84.46
$r = 8$	98.81	99.07	99.09	98.95
$r = 16$	100.00	100.00	100.00	99.98
Inverse randomized parity				
$r = 4$	93.57	95.61	96.02	96.40
$r = 8$	99.61	99.72	99.82	99.84
$r = 16$	99.99	100.00	100.00	100.00

11.4.3 Discussion on Mitigation Solutions

Considering good design practices, even parity is an obvious solution. However, it does not eradicate the problem, only shifts the issue to different parameters, such as choice of fault injection technique or injection strength, etc. In the parity implementation in FPGA, a single slice was targeted for the evaluation, where four registers were used as the round registers in the PRESENT datapath. This does not incur any loss of generality since commercial placement and routing tools deploy the bits of the same bit vector close to each other, which is true for both ASIC and FPGA. This is owing to the requirement of area and timing optimizations in the late design phases. In our case, we noticed the 4 bits of a bit vector to be always located in the same slice. As a matter of fact, multiple bit flipping in registers or similar logics by a single laser injection is practical for the commercial chips [11].

Experimental results show that flipping an even number of bits using a single laser injection can be done easily [10]. Fault models based on this result also exist, such as the nibble fault model described in [2]. To circumvent these security vulnerabilities, a special attention should be paid during the implementation. It is highly recommended to carefully investigate the multiple bit-fault models against a specific cipher, and swap the unrelated bits of nibbles into different slices or deploy them far from each other during the placement phase of the FPGA or ASIC implementation.

The simulation of advanced (linear, non-linear, and randomized parity) schemes also shows vulnerabilities based on the attack result. Since the observations may differ significantly under different attack scenarios, it is necessary to thoroughly investigate the implemented devices and the possible attack vectors for choosing the adequate parity scheme and the implementation tactics.

Moreover, the experimental results we used in this chapter were obtained by using a relatively budget-friendly laser station ($\approx$100k EUR). As some other works show (e.g., [21]), by using more advanced setups, such as Hamamatsu PHEMOS-1000, the precision of the faults can be further improved.

11.5 Chapter Summary

As a typical intrinsic countermeasure against fault injections, parity concurrent error detection (CED) is often utilized for detecting faults in hardware [30]. Parity bit(s) basically flag the alarm once the number of flipped bits is satisfied depending on the used parity scheme. Generally, basic and randomized (linear and non-linear) encodings can be employed for constructing varying parity solutions.

In this chapter, we provided an evaluation method to estimate the detection probability of randomized parity-protected cipher. The methods take the device characteristics into account to provide accurate results on detection coverage. To show the applicability of such method, we detailed a case study on CED-protected PRESENT cipher and discussed the fault coverage.

References

1. D. Agrawal, B. Archambeault, J.R. Rao, P. Rohatgi, The EM side-channel(s), in *4th International Workshop on Cryptographic Hardware and Embedded Systems–CHES 2002*, Redwood Shores, CA, August 13–15, 2002, Revised Papers (Springer, Berlin, 2002), pp. 29–45
2. N. Bagheri, R. Ebrahimpour, N. Ghaedi, New differential fault analysis on present. EURASIP J. Adv. Signal Process. **2013**(1), 145 (2013)
3. A. Barenghi, L. Breveglieri, I. Koren, D. Naccache, Fault injection attacks on cryptographic devices: theory, practice, and countermeasures. Proc. IEEE **100**(11), 3056–3076 (2012)
4. G. Bertoni, L. Breveglieri, I. Koren, P. Maistri, V. Piuri, Error analysis and detection procedures for a hardware implementation of the advanced encryption standard. IEEE Trans. Comput. **52**(4), 492–505 (2003)
5. E. Biham, A. Shamir, Differential fault analysis of secret key cryptosystems, in *Advances in Cryptology–CRYPTO '97*. Lecture Notes in Computer Science, vol. 1294, ed. by B.S. Kaliski Jr. (Springer, Berlin, 1997), pp. 513–525
6. J. Breier, W. He, Multiple fault attack on present with a hardware Trojan implementation in FPGA, in *2015 International Workshop on Secure Internet of Things (SIoT)* (IEEE, Piscataway, 2015), pp. 58–64
7. J. Gaisler, Concurrent error-detection and modular fault-tolerance in a 32-bit processing core for embedded space flight applications, in *Digest of Papers: FTCS/24, The Twenty-Fourth Annual International Symposium on Fault-Tolerant Computing*, Austin, TX, June 15–17, 1994 (IEEE, Piscataway, 1994), pp. 128–130
8. R.G. Gallager, Low-density parity-check codes. IRE Trans. Inf. Theory **8**(1), 21–28 (1962)
9. X. Guo, D. Mukhopadhyay, R. Karri, Provably secure concurrent error detection against differential fault analysis. IACR Cryptol. ePrint Arch. **2012**, 552 (2012)
10. W. He, J. Breier, S. Bhasin, A. Chattopadhyay, Bypassing parity protected cryptography using laser fault injection in cyber-physical system, in *Proceedings of the 2nd ACM International Workshop on Cyber-Physical System Security* (ACM, New York, 2016), pp. 15–21
11. W. He, J. Breier, S. Bhasin, D. Jap, H.G. Ong, C.L. Gan, Comprehensive laser sensitivity profiling and data register bit-flips for cryptographic fault attacks in 65 nm FPGA, in *Proceedings of the 6th International Conference on Security, Privacy, and Applied Cryptography Engineering, SPACE 2016*, Hyderabad, December 14–18, 2016 (Springer, Cham, 2016), pp. 47–65
12. M. Joye, M. Tunstall, *Fault Analysis in Cryptography*, vol. 147 (Springer, Heidelberg, 2012)
13. M.G. Karpovsky, A. Taubin, New class of nonlinear systematic error detecting codes. IEEE Trans. Inf. Theory **50**(8), 1818–1820 (2004)
14. M.G. Karpovsky, K.J. Kulikowski, A. Taubin, Robust protection against fault-injection attacks on smart cards implementing the advanced encryption standard, in *Proceedings of the 2004 International Conference on Dependable Systems and Networks (DSN 2004)*, Florence, 28 June–1 July 2004 (IEEE, Piscataway, 2004), pp. 93–101
15. M.M. Kermani, A. Reyhani-Masoleh, Parity-based fault detection architecture of s-box for advanced encryption standard, in *21th IEEE International Symposium on Defect and Fault-Tolerance in VLSI Systems (DFT 2006)*, Arlington, VA, 4–6 October 2006 (IEEE, Piscataway, 2006), pp. 572–580
16. M.M. Kermani, A. Reyhani-Masoleh, A lightweight concurrent fault detection scheme for the AES s-boxes using normal basis, in *Proceedings of the 10th International Workshop on Cryptographic Hardware and Embedded Systems–CHES 2008*, Washington, DC, August 10–13, 2008 (Springer, Berlin, 2008), pp. 113–129
17. M.M. Kermani, A. Reyhani-Masoleh, Concurrent structure-independent fault detection schemes for the advanced encryption standard. IEEE Trans. Comput. **59**(5), 608–622 (2010)
18. M.M. Kermani, A. Reyhani-Masoleh, A lightweight high-performance fault detection scheme for the advanced encryption standard using composite fields. IEEE Trans. VLSI Syst. **19**(1), 85–91 (2011)

19. P.C. Kocher, J. Jaffe, B. Jun, P. Rohatgi, Introduction to differential power analysis. J. Cryptogr. Eng. **1**(1), 5–27 (2011)
20. K.J. Kulikowski, M.G. Karpovsky, A. Taubin, Fault attack resistant cryptographic hardware with uniform error detection, in *Proceedings of the Third International Workshop on Fault Diagnosis and Tolerance in Cryptography–FDTC 2006*, Yokohama, October 10, 2006, (Springer, Berlin, 2006), pp. 185–195
21. H. Lohrke, P. Scholz, C. Boit, S. Tajik, J.-P. Seifert, Automated detection of fault sensitive locations for reconfiguration attacks on programmable logic, in *Proceedings of the 42nd International Symposium for Testing and Failure Analysis* (ASM, New York, 2016), pp. 1–6
22. T. Malkin, F.-X. Standaert, M. Yung, A comparative cost/security analysis of fault attack countermeasures, in *Proceedings of the Third International Workshop Fault Diagnosis and Tolerance in Cryptography–FDTC 2006*, Yokohama, October 10, 2006 (Springer, Berlin, 2006), pp. 159–172
23. C. Perkins, G.A. Muller, Using discrete event simulation to model attacker interactions with cyber and physical security systems, in *Proceedings of the Conference on Complex Adaptive Systems 2015*, San Jose, CA, November 2–4, 2015 (Elsevier, Amsterdam, 2015), pp. 221–226
24. H. Sandberg, S. Amin, K.H. Johansson, Cyberphysical security in networked control systems: an introduction to the issue. IEEE Control Syst. **35**(1), 20–23 (2015)
25. C. Schmittner, Z. Ma, E. Schoitsch, T. Gruber, A case study of FMVEA and CHASSIS as safety and security co-analysis method for automotive cyber-physical systems, in *Proceedings of the 1st ACM Workshop on Cyber-Physical System Security–CPSS 2015*, Singapore, April 14–March 14, 2015 (ACM, New York, 2015), pp. 69–80
26. M. Tehranipoor, F. Koushanfar, A survey of hardware Trojan taxonomy and detection. IEEE Des. Test Comput. **27**(1), 10–25 (2010)
27. Z. Wang, M.G. Karpovsky, K.J. Kulikowski, Design of memories with concurrent error detection and correction by nonlinear SEC-DED codes. J. Electron. Test. **26**(5), 559–580 (2010)
28. L. Wen, W. Jiang, K. Jiang, X. Zhang, X. Pan, K. Zhou, Detecting fault injection attacks on embedded real-time applications: a system-level perspective, in *17th IEEE International Conference on High Performance Computing and Communications, HPCC 2015, 7th IEEE International Symposium on Cyberspace Safety and Security, CSS 2015, and 12th IEEE International Conference on Embedded Software and Systems, ICESS 2015*, New York, NY, August 24–26, 2015 (IEEE, Piscataway, 2015), pp. 700–705
29. K. Wu, R. Karri, G. Kuznetsov, M. Goessel, Low cost concurrent error detection for the advanced encryption standard, in *Proceedings of the International Test Conference, 2004–ITC 2004* (IEEE, Piscataway, 2004), pp.1242–1248
30. T.F. Wu, K. Ganesan, Y.A. Hu, H.-S.P. Wong, S.S. Wong, S. Mitra, TPAD: hardware Trojan prevention and detection for trusted integrated circuits. IEEE Trans. Comput. Aided Des. Integr. Circuits Syst. **35**(4), 521–534 (2016)
31. B. Zhu, A.D. Joseph, S. Sastry, A taxonomy of cyber attacks on SCADA systems, in *2011 IEEE International Conference on Internet of Things (iThings) & 4th IEEE International Conference on Cyber, Physical and Social Computing (CPSCom)*, Dalian, October 19–22, 2011 (IEEE, Piscataway, 2011), pp. 380–388

Chapter 12
Fault Analysis Assisted by Simulation

Kais Chibani, Adrien Facon, Sylvain Guilley, Damien Marion, Yves Mathieu, Laurent Sauvage, Youssef Souissi, and Sofiane Takarabt

12.1 Introduction

Embedded systems are based on hardware integrated circuits. Basically, any hardware design has its own conception life cycle that starts with the algorithm and architecture specification. In fact, the designer starts by describing sequentially the functional part of his design based on hardware description language (HDL). Then, we distinguish three abstraction levels in the design life cycle as follows:

- **Register transfer level** (RTL). It consists of specifying the logical operations and data-flow between registers. This level involves an explicit clock to synchronize the data-flow (clock/event accurate).

K. Chibani · D. Marion · Y. Souissi · S. Takarabt
Secure-IC S.A.S., Cesson-Sévigné, France
e-mail: kais.chibani@secure-ic.com; damien.marion@secure-ic.com; youssef.souissi@secure-ic.com; Sofiane.Takarabt@secure-ic.com

A. Facon · S. Guilley (✉)
Secure-IC S.A.S., Cesson-Sévigné, France

LTCI, CNRS, Télécom ParisTech, Université Paris-Saclay, Paris, France

École Normale Supérieure, département d'informatique, CNRS, PSL Research University, Paris, France
e-mail: adrien.facon@secure-ic.com; sylvain.guilley@secure-ic.com

Y. Mathieu
LTCI, CNRS, Télécom ParisTech, Université Paris-Saclay, Paris, France
e-mail: yves.mathieu@telecom-paristech.fr

L. Sauvage
Secure-IC S.A.S., Cesson-Sévigné, France

LTCI, CNRS, Télécom ParisTech, Université Paris-Saclay, Paris, France
e-mail: laurent.sauvage@telecom-paristech.fr

J. Breier et al. (eds.), *Automated Methods in Cryptographic Fault Analysis*,
https://doi.org/10.1007/978-3-030-11333-9_12

- **Post synthesis level** or gate level. It describes the timing properties of logical operations. In other words, it takes into consideration the delay propagation of signals within the circuit gates. Hence, this netlist level is technology dependent. Such netlist is generated by synthesis tools.
- **Place route level** or layout level. It comes with placed and routed cells and more timing information. In fact it takes into account the delay propagation into circuit's routing.

In the context of embedded security, the designer can conduct a security analysis hand-in-hand with the functional and timing verification. This is very useful as the designer will be able to think about the security testing at an early stage. He will not wait until the tapeout of a testing chip to start a security evaluation of his implementation. Moreover, he will be able to conduct such evaluation by himself without the need of high skills in physical security analysis. In fact, recently, the world of EDA arsenal of tools has come with a new tool, called VIRTUALYZR® [9, 11] that allows for such security analysis with seamless integration within the design life cycle as shown in Fig. 12.1. In the sequel, all the presented results are obtained with such pre-silicon assisted verification tool that we denote by its acronym PAVT.

12.2 Security Verification Assisted by Simulation: PAVT Workflow

PAVT deals with both SCA and FIA. The general workflow consists of two phases as illustrated in Fig. 12.2.

Simulation Phase During this phase the tool interacts with an HDL simulator to generate the so-called *virtual* activity of the design. The obtained virtual activity is an ideal image of the behavior of the cryptographic design during its execution. Usually, several queries are needed to conduct SCA or FIA analysis. PAVT manages the automatic configuration of input vectors (i.e., input parameters) based on the testbench of the design. This testbench needs not be written specifically for the purpose of using PAVT. It can either be the unitary testbench used for functional verification or a testbench generated automatically by EDA frameworks (such as Cadence Specman Elite). This phase allows for the generation of the fingerprint of the design activity. The fingerprint is just a dataset managed and organized by the database of the tool. At this point, the PAVT comes with two approaches to deal with obtained dataset. The first approach is called *real approach*. It consists of building one leakage trace from the overall activity of the design signals given one fixed input (e.g., one fixed key, one variable message for an AES implementation). This way, the tool generates a number of traces equal to the number of variable input messages. This approach is based on a high-level modeling of the electrical activity of the design. According to the literature [8], the basic model is composed

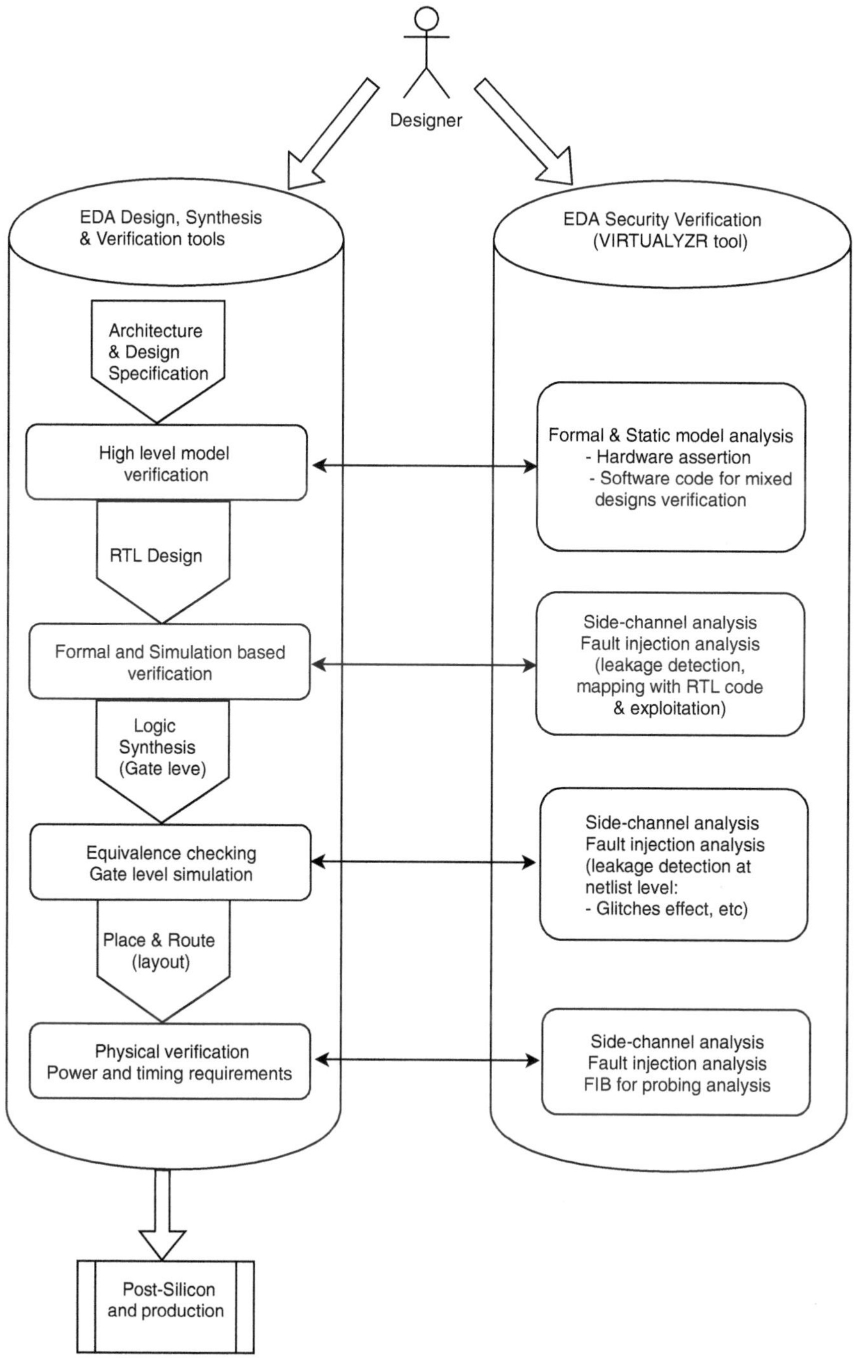

Fig. 12.1 Seamless security verification integration in the design life cycle workflow

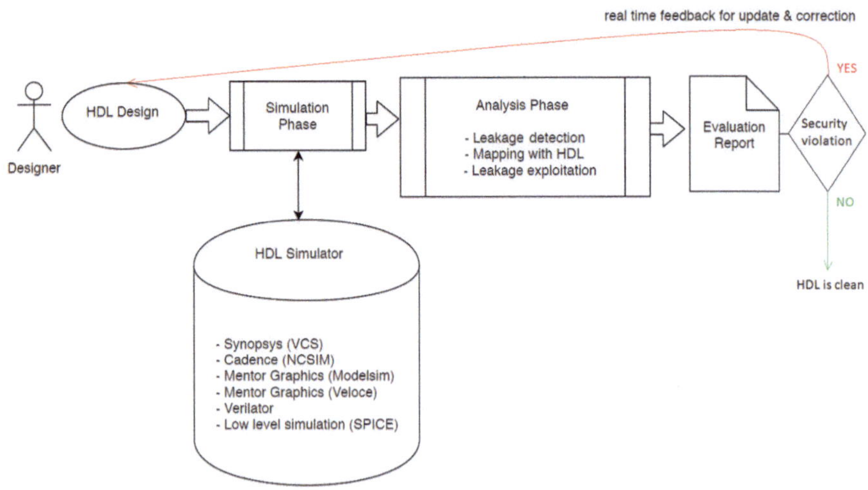

Fig. 12.2 PAVT: security analysis assisted by simulation general workflow (see Algorithm 1)

of a dynamic factor reflecting the transistors activity and a static factor reflecting the activity of the circuit while at rest. Digitally, the dynamic factor can be computed through a toggle count and the static factor is just the actual binary value of the signal at each simulated instant. In order to generate one trace, the sum is performed over all signals (so-called Hamming weight leakage model). Now, the second approach called *ideal approach* can be envisioned and is very fruitful in terms of security analysis in practice. It consists of dealing with the raw state of simulation without the need of any power consumption model. In fact, this wire-level approach allows detecting any security anomalies regarding each signal in the design. This approach is more complete and requires managing matrix-based traces. Besides, for both approaches, the analysis is always conducted in the best conditions as the designer here does not care about the impact of real factors like the measurement noise and configuration of real equipment such as oscilloscopes and pulse generators that require more skills and processing. Regarding fault(s) injection, erroneous values are forced from the simulation tool. The resulting traces are also termed "virtual."

Analysis Phase This phase consists of analyzing the obtained virtual trace for SCA or faulty log trace for FIA. It is noteworthy that the PAVT is not a tool designed for attackers, but rather a tool for designers. This tool allows a security analysis checkpoint at all the design levels.

Both phases are processed in an iterative manner until the HDL design is clean with no security violations (see Algorithm 1 and Fig. 12.2). Figure 12.3 shows deeper details about the internal workflow of the PAVT. First, it extracts the input design structure. Then, after having properly configured the user project, the tool runs a couple of simulations, which will produce a database of raw results. The

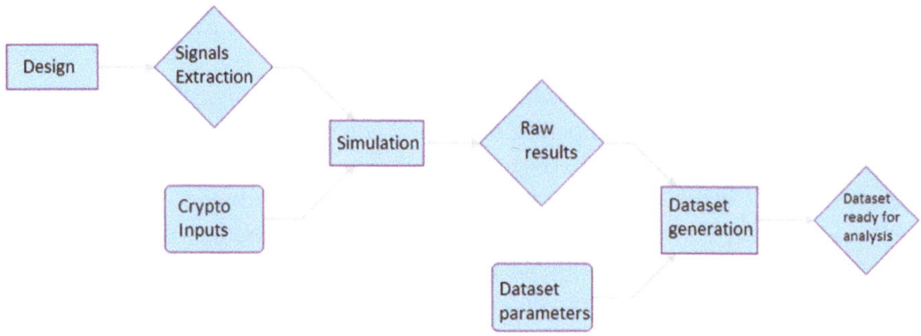

Fig. 12.3 PAVT internal workflow

Algorithm 1: Security verification and refinement of the HDL design by PAVT (see Fig. 12.2)

$HDL \leftarrow initial\ HDL\ design$;
while *HDL is not clean* **do**
    Simulation phase;
    Analysis phase;
    Evaluation report generation;
    if *Security violation* **then**
        $HDL \leftarrow fix\ the\ HDL\ design$
    else
        return: HDL

PAVT workflow distinguishes raw results generation and dataset generation as two separate operations. First, simulation results are computed and stored on a hard drive. Once simulation results are available, it is possible to use them to generate datasets with different properties. A dataset is associated with a given consumption model (for real approach) and a set of probed signals used for trace generation. To reflect this distinction, the tool uses the twin concepts of $target$ and $probe$:

- A project has only one $target$. It is the set of signals for which we store simulation results.
- A project can handle multiple $probes$. A probe is useful to study the activity of a sub-group of signals. Probing is useful to focus on sub-modules, to spot leaking signals, to simulate a cartography, etc.

The same notions naturally apply to fault injection simulation campaigns. In the following, we will focus on automated fault analysis of hardware designs, which is the main focus of this chapter.

12.3 Fault Analysis Assisted by Simulation

Fault attacks are active attacks, which need an adversary to induce errors into the target device, using some tampering means. This tampering can be accomplished in several ways, as extensively discussed in literature [6]. In general, tampering means (or fault injection techniques) are classified in two broad categories, i.e., *global* and *local*. Global fault injections [5] are, in general, low-cost techniques which create disturbances on global parameters like voltage, clock, temperature, etc. The resultant faults are more or less random in nature and the adversary might need several injections, to find required faults. On the other hand, local techniques (e.g., clock glitch, optical/electromagnetic injections, body bias injection [1]) are more precise in terms of fault location and model. However, this precision comes at the expense of costly and bespoke[1] equipment. The kind of injected fault can be defined as fault model. The fault model has two important parameters, namely *location* and *impact*. Location means the spatial and temporal location of fault injection during the execution of target algorithm. Depending on the type and precision of the technique, location can be at the level of bit, variable or random. Coming to the impact of fault, it is the effect on the target data. Commonly known fault injection impacts on target data can cause stuck-at, bit-flip, random-byte, or uniformly distributed random value.

12.3.1 Simulation-Based Fault Injection Evaluation

With the PAVT tool, it is possible to model the effects of several fault injection attacks, even those that require advanced technical skills and equipment as stated above. It is also possible to accurately analyze the intrinsic robustness of a digital circuit against such attacks early in the design flow. In the following, we detail two different use-cases related to local fault injection attacks on a hardware implementation of an unprotected AES 128 bit, which needs 10 clock cycles to perform an encryption.

12.3.1.1 Clock-Glitch Injection

The principle of the clock glitch injection consists of precisely modifying the period of one or more clock cycles of the target design during the AES execution. When the modified clock period is much shorter than what is expected in the normal clock, it shall create setup violation faults [10]. These faults can be exploited to retrieve the secret key. Since the modification of the clock frequency at RTL

[1]In Common Criteria parlance.

level is meaningless, we can perform clock glitch injections only with gate-level descriptions (i.e., post-synthesis level or place and route level). To this end, we synthesized the AES core to the gate level using ASIC 65 nm CMOS technology for 1.2 V supply voltage. After that, we configure the PAVT to perform clock glitch on a specific cycle during gate-level simulations to take into account the circuit delays (e.g., Standard Delay Format file). The configuration consists of defining some parameters needed to set the stimuli for simulations and the clock glitch parameters, in particular, the cycle target and the glitch duration. In our case, the main configuration was as follows:

- Target cycle: last round of the AES execution;
- Glitch duration: from 4 to 7 ns with steps of 100 ps;
- Number of simulations: 310.

Figure 12.4 shows a cartographic view of the effects of clock glitches in terms of erroneous bits observed in the final output (ciphertext). Based on such information, the evaluator can easily identify the maximal glitch duration that would lead to a final output error for a given cycle.

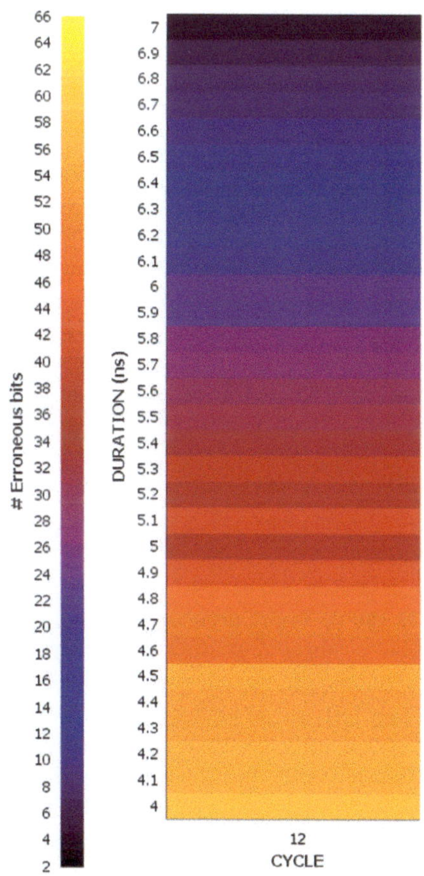

Fig. 12.4 Erroneous bits according to the glitch duration

Simulation results can be used to apply a set of differential fault analyses (DFA) which exploit differences between correct and faulty outputs to recover the key. The PAVT offers a set of DFA metrics which allow to analyze fault injection results. One example is the AES-128 DFA NUEVA (non-uniform error value analysis) metric [7] which measures the uniformity of error values injected before the last SubBytes operation in order to find the key. Another example is the AES-128 DFA using Giraud metric [4]: This fault analysis requires single-bit faults at the input of the last SubBytes operation. As shown in Fig. 12.5, the PAVT is able to recover the entire key with only 126 simulations using DFA of Giraud. A few more simulations are required to perform the full analysis with the NUEVA technique.

12.3.1.2 Laser Injection

Laser fault injection falls into optical fault injection methods which expose the device to an intense light for a brief period of time. The injection can be performed either through the front-side or the backside of the target chip. Laser attacks can be used to inject faults characterized by high locality and timing accuracy. In the PAVT, the laser injections can be modeled at the gate-level and functional level (i.e., register transfer level) by configuring parameters such as the fault type (e.g., permanent/transient), the fault model (e.g., bit-flip, bit-set, bit-reset, stuck-at-0/1), the fault location (e.g., wires, registers), and the fault time. When the time event occurs and the fault injection conditions are met, it becomes the fault time, and the fault model is injected into the fault location during simulation.

For this use-case, we have performed our analysis at RTL level with the following configuration: fault time (last round cycle of the AES execution), fault location (inputs of the SubBytes module), fault model (bit-flip model), number of simulations (100). The DFA results we obtained show that all key bytes are broken with only 10 simulations. Figure 12.6 illustrates the results of the analysis completed using the DFA metrics already presented in the previous section. We can see that all the key bytes are broken at the end of the simulation, in this case 10 with the DFA based on Giraud metric.

12.3.2 Case Study: Netlist Level Leakage Fault Detection

As shown in the previous part, the attacks based on malicious injection of faults can seriously degrade the security of a cryptosystem. Faults injected into the cryptographic modules during the encryption (or decryption) operation will very likely result in a number of errors in the encrypted/decrypted data. Such faults must be detected before their spread to avoid the transmission and use of incorrect data. Fault detection techniques represent therefore a possible countermeasure against fault injection attacks and a desirable property for preventing malicious attacks, aimed at extracting sensitive information from the device, like the secret key.

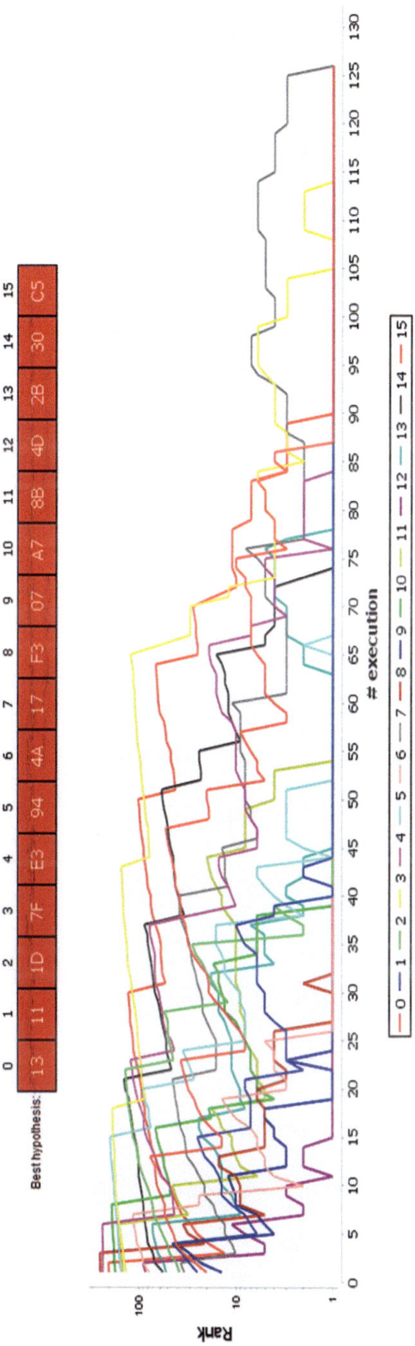

Fig. 12.5 Analysis of clock-glitch injection results using DFA AES-128 Giraud metric

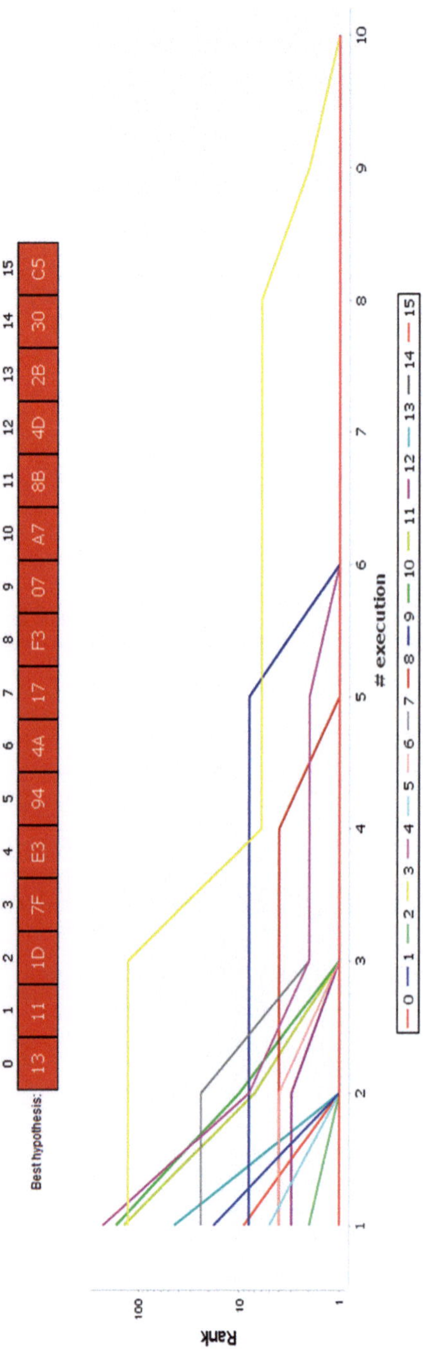

Fig. 12.6 Analysis of laser injection results using DFA AES-128 Giraud metric

For the AES block cipher, two main approaches have been proposed for achieving fault detection. The first one is based on temporal or spatial redundancy; in temporal redundancy, the same hardware is used to repeat the same process twice using the same input data. This technique uses minimum hardware overhead. Yet, it entails time overhead. In spatial redundancy, two copies of the hardware are used concurrently to perform the same computation on the same data. After each computation, the results are compared, and every difference is reported as a fault. The advantage of this technique is that it can detect all kinds of faults. However, it requires a significant hardware overhead. The second approach is concurrent error detection using error detecting codes (EDC). It employs circuit-level coding techniques, e.g., parity schemes, modular redundancy, etc., to produce and verify results after each computation.

From a security point of view, designers have to verify the effectiveness of a given implemented countermeasure and be sure that it prevents against fault analysis. Remark that all countermeasures detect faults only to some extent (e.g., up to a certain order, that is to say, up to a certain bit-wise multiplicity). For this purpose, we present our results based on the countermeasure presented by Bertoni et al. [3] which targets the datapath of the AES encryption module. This countermeasure uses a 4×4 parity matrix. Each bit corresponds to the byte state, and at each round the matrix is predicted and then compared with the computed one from the state. This countermeasure can detect all single errors and perhaps all odd errors and furthermore actually locate them. The hardware overhead is less than many other countermeasures (e.g., [2]) where a computation redundancy is required (2 times overhead). We designed an AES-128 encryption module implementing this countermeasure for the datapath. The control unit is also protected by computing the parity of the rounds counter. Then, we perform several simulation-based fault injection campaigns at the register transfer level (RTL) in order to evaluate the fault coverage of the proposed parity-based EDC scheme. One hundred thousand injections are performed using plaintexts and keys selected randomly. The fault model is a single bit-flip at the last round of the encryption operation. The obtained results show that the detection rate is equal to 100% as shown in [3]. Then, we launch the logic synthesis on a Virtex-V Xilinx FPGA as technology target in order to perform the same fault injection campaigns but at post-synthesis level (PS) (i.e., the post-map netlist is used during simulations). As expected, the detection rate is equal to 100%.

Thereafter, we re-synthesize the same RTL code but with different logic synthesis options to optimize the logic and to improve timing and design performances. As a matter of fact, the Xilinx synthesis technology (XST) synthesis tool allows designers to configure several options and properties that are taken into account during the synthesis process. These options target possible optimization for area, speed, or power consumption.

Figure 12.7 is an extract from the Xilinx synthesis settings dialog box. In our case, we activate some options to optimize the design such as the *−logic_opt* option which optimizes timing-critical connections through restructuring and re-synthesis, followed by incremental placement and incremental timing analysis.

Switch Name	Property Name	
-ol	Placer Effort Level	High
-xe	Placer Extra Effort	None
-t	Starting Placer Cost Table (1-100)	1
-logic_opt	Combinatorial Logic Optimization	✖
-register_duplication	Register Duplication	Off
-global_opt	Global Optimization	Power
-equivalent_register_removal	Equivalent Register Removal	✖
-x	Ignore User Timing Constraints	☐
-ntd	Timing Mode	Performance Evaluation
-u	Trim Unconnected Signals	✖
-ignore_keep_hierarchy	Allow Logic Optimization Across Hierarchy	☐
-cm	Optimization Strategy (Cover Mode)	Area
-detail	Generate Detailed MAP Report	☐
-ir	Use RLOC Constraints	Yes
-pr	Pack I/O Registers/Latches into IOBs	Off
-c	Maximum Compression	☐
-lc	LUT Combining	Auto
-bp	Map Slice Logic into Unused Block RAMs	☐
-power	Power Reduction	☒
-activityfile	Power Activity File	
-mt	Enable Multi-Threading	Off
	Other Map Command Line Options	

Fig. 12.7 Extract from the XST synthesis options for Xilinx FPGAs

Previous injection campaigns are performed based on the obtained netlist. However, results are not the same because the detection rate decreases from 100% to 18.75%. More precisely, only faults injected in the AES control unit are detected. All faults into the datapath are no longer detected due to the synthesis tool optimization, as shown in Fig. 12.8.

The countermeasure logic on the datapath was completely removed after the logical synthesis to optimize the design for area by reducing the total amount of logic used for design implementation. With obviously lower number of gates, an equivalent functionality is obtained, albeit with a lesser security. Indeed, the Sbox is left unprotected, simply because the synthesizer has been smart enough to eliminate some combinational schemes considered to be equivalent. Functionally speaking, there is no alteration. However, from a security standpoint, the complete SubBytes transformation is left unprotected.

For the optimization prevention of signal B in Fig. 12.8, we use the `DONT_TOUCH` attribute. This attribute prevents optimization where signals are either optimized or absorbed into logic blocks. It instructs the synthesis tool to keep the signal it was placed on, and that signal is placed in the netlist. Logic synthesis and fault injections are remade with the same options used during the previous experimentation. Results indicate that the detection rate increases from 18.75% to

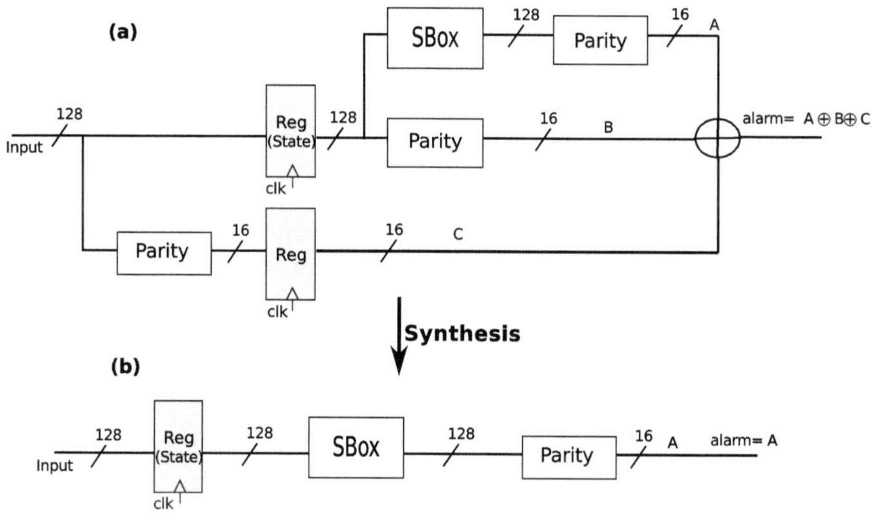

Fig. 12.8 Total simplification of fault detection logic upon synthesis. (**a**) RTL, (**b**) PS

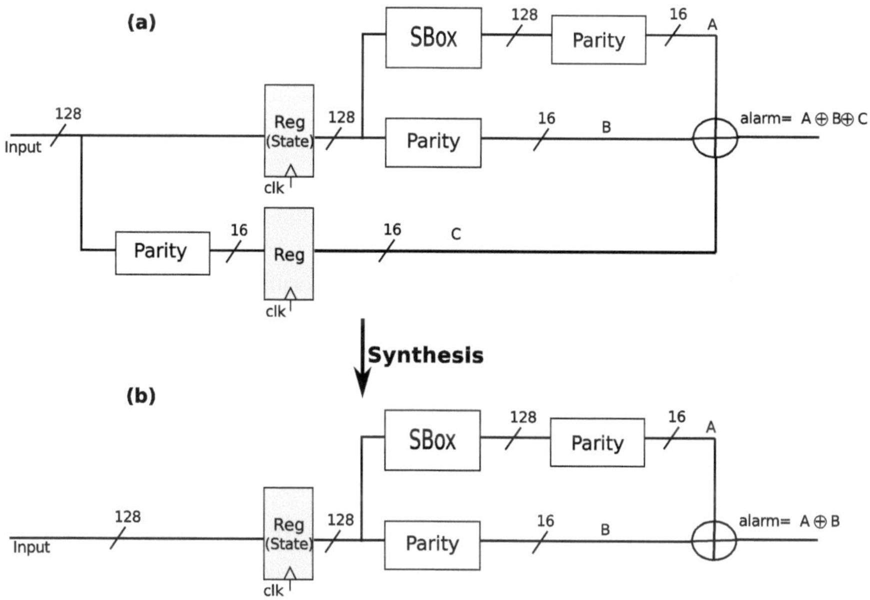

Fig. 12.9 Partial simplification of fault detection logic upon synthesis. (**a**) RTL, (**b**) PS

56.43%. Indeed, the synthesis tool has simplified partially the fault detection logic as shown in Fig. 12.9 by eliminating the combinational block producing C signal. Consequently, only faults injected in the state register are detected, which opens a large door for successful fault injection attacks within the combinational logic.

Table 12.1 Fault detection rate for RTL and post-synthesis levels

Level	RTL	PS (default options)	PS -logic_opt = true -xor_collapsing = true	PS -logic_opt = true DONT_TOUCH attribute
Detection rate	100%	100%	18.75%	56.43%

Table 12.1 summarizes the fault detection rate according to the analyzed level. From this, we conclude that the protection can be removed altogether during logical synthesis, thereby causing a security regression. This kind of mis-integration may happen in real case, where designers do not check the security evolution of their design at each stage of synthesis. Therefore, robustness of hardware cryptographic modules against fault injection attacks should be evaluated at each abstraction level in the design conception flow.

looseness-1Another reason for designers attention to be deflected from security is the requirements for testability. Clearly, in Fig. 12.9a, the alarm signal is not testable. Indeed, it is consistently equal to "0." Therefore, in a view to achieve DFT (Design For Test) requirements, some test logic to address independently the registers driving signals A, B, and C shall be added. But in the meantime, the designer might shift his focus so conscientiously that he might forget about the need for setting DONT_TOUCH attributes. Hence the need for an independent third-party verification tool.

12.4 Chapter Summary

SCA and FIA are serious threats to cryptographic algorithms [6]. Countermeasures have been developed against such attacks. Still, it is non-obvious how to implement such protections at source-code level. There are many options to configure the synthesis. Hence exploring their combinatorics is exponential. In practice, users select a few options. Some options can lead to total or partial simplification of the countermeasure. Using a simulation-based methodology, we manage to detect such alterations and we quantify the amount of degradation. In addition, we precisely pinpoint the residual leakage samples.

Acknowledgements This work has benefited from a funding via TeamPlay (https://teamplay-h2020.eu/), a project from European Union's Horizon2020 research and innovation programme, under grant agreement No. 779882. Besides, the French FUI program CSAFE+ also funded part of this work.

References

1. N. Beringuier-Boher, M. Lacruche, D. El-Baze, J.-M. Dutertre, J.-B. Rigaud, P. Maurine, Body biasing injection attacks in practice, in *Proceedings of the Third Workshop on Cryptography and Security in Computing Systems, CS2@HiPEAC*, Prague, January 20, 2016 (ACM, New York, 2016), pp. 49–54
2. G. Bertoni, L. Breveglieri, I. Koren, V. Piuri, Fault detection in the advanced encryption standard, in *Proceedings of the Conference on Massively Parallel Computing Systems (MPCS'02)* (2002), pp. 92–97
3. G. Bertoni, L. Breveglieri, I. Koren, P. Maistri, V. Piuri, Error analysis and detection procedures for a hardware implementation of the advanced encryption standard. IEEE Trans. Comput. **52**(4), 492–505 (2003)
4. C. Giraud, DFA on AES, in *4th International Conference on Advanced Encryption Standard–AES, AES 2004*, Bonn, May 10–12, 2004, Revised Selected and Invited Papers (Springer, Berlin, 2004), pp. 27–41
5. S. Guilley, J.-L. Danger, Global faults on cryptographic circuits, Chapter 17 in *Fault Analysis in Cryptography*, ed. by M. Joye, M. Tunstall, vol. 147 (Springer, Heidelberg, 2012)
6. M. Joye, M. Tunstall, *Fault Analysis in Cryptography*, vol. 147 (Springer, Heidelberg, 2012)
7. R. Lashermes, G. Reymond, J.-M. Dutertre, J. Fournier, B. Robisson, A. Tria, A DFA on AES based on the entropy of error distributions, in *2012 Workshop on Fault Diagnosis and Tolerance in Cryptography*, ed. by IEEE (IEEE, Piscataway, 2012), pp. 34–43
8. H. Li, A.T. Markettos, S.W. Moore, Security evaluation against electromagnetic analysis at design time, in *Proceedings of the 7th International Workshop on Cryptographic Hardware and Embedded Systems–CHES 2005*, Edinburgh, August 29–September 1, 2005 (Springer, Berlin, 2005), pp. 280–292
9. Secure-IC S.A.S. VIRTUALYZR® tool. http://www.secure-ic.com/solutions/virtualyzr/. Accessed 2 Nov 2018
10. N. Selmane, S. Guilley, J.-L. Danger, Setup time violation attacks on AES, in *EDCC, The seventh European Dependable Computing Conference* (IEEE, Piscataway, 2008), pp. 91–96, ISBN: 978-0-7695-3138-0. https://doi.org/10.1109/EDCC-7.2008.11
11. S. Takarabt, K. Chibani, A. Facon, S. Guilley, Y. Mathieu, L. Sauvage, Y. Souissi, Pre-silicon embedded system evaluation as new EDA tool for security verification, in *3rd IEEE International Verification and Security Workshop, IVSW 2018*, Costa Brava, July 2–4, 2018 (IEEE, Piscataway, 2018), pp. 74–79

Part IV
Automated Fault Attack Experiments

Chapter 13
Optimizing Electromagnetic Fault Injection with Genetic Algorithms

Antun Maldini, Niels Samwel, Stjepan Picek, and Lejla Batina

13.1 Introduction

Embedded security devices such as ID or bank cards, key immobilizers, and mobile phones are omnipresent in our lives and the threats to them directly affect the security and privacy of our data. The attackers often target the weaknesses of implementations rather than the algorithms running on those chips. Basically, those so-called implementation attacks do not focus on the algorithm itself but rather exploit some physical effects. Those effects, i.e. physical leakages become available due to the actual implementation of the algorithms on a platform that is typically constrained in terms of area, power, energy, etc. Two well-known types of implementation attacks are side-channel attacks (SCAs) and fault injection (FI) attacks. Side-channel attacks are passive, non-invasive attacks where the device under attack operates within specified conditions and the attacker simply observes the physical leakages produced. Fault injection attacks are, on the other hand, active, more invasive attacks where the attacker inserts faults (e.g. by glitching some parameters like voltage, power, clock, etc.) in order to disrupt the normal behavior of the algorithm.

A. Maldini (✉)
Faculty of Electrical Engineering and Computing, University of Zagreb, Zagreb, Croatia

N. Samwel · L. Batina
Digital Security Group, Radboud University, Nijmegen, The Netherlands
e-mail: nsamwel@cs.ru.nl; lejla@cs.ru.nl

S. Picek
Cyber Security Group, Delft University of Technology, Delft, The Netherlands
e-mail: s.picek@tudelft.nl

J. Breier et al. (eds.), *Automated Methods in Cryptographic Fault Analysis*,
https://doi.org/10.1007/978-3-030-11333-9_13

Side-channel attacks received a lot of attention in the last few decades where we saw successful exploitation of several side channels like timing [11], power consumption [12], and EM emanation [25]. To use that information and deliver powerful attacks that are even capable of recovering the secret keys, researchers devised various strategies. Not surprisingly, many of those strategies in the last few years are based on machine learning [14, 24] and deep learning [7].

When considering fault injection attacks, the situation is somewhat different in terms of attack methodology and analysis techniques. In certain scenarios, just one fault can lead to a total break of a cryptosystem as described in the first paper of Boneh et al. [6]. While it was clear that one fault can have devastating consequences on the security of a system, it was not immediately clear how complicated can be to insert the faults of interest. The first work that illustrates a practical fault injection attack on an actual RSA cryptosystem was done by Aumüller et al. [2]. They describe the engineering efforts behind a successful fault injection and possible countermeasures. Fault injection is often possible through glitching techniques. There are several sources of glitches possible such as laser pulses, electrical glitches, and electromagnetic radiation. A fault injection attack is successful if after exposing the device under attack to a specially crafted external interference, the device shows an unexpected behavior, i.e., a fault, which can be exploited by an attacker. Here, the challenge lies in selecting the appropriate parameters for a fault to succeed. If those parameters are not well chosen, the target will respond in a way that does not permit an actual fault analysis attack. When considering various sources of faults, we encounter different number of parameters and corresponding ranges. In general, the search space size of possible parameter values is large and relatively few points in the search space result in faults. Consequently, an interesting follow-up question is: how to find suitable parameter values or intervals, or more precisely, how to efficiently find the correct values in the search space? Surprisingly, the main options available so far are to use either random search or some sort of exhaustive search. This is mainly due to the fact that a complete exhaustive search is usually not feasible so the attacker should focus on specific regions with a certain precision, and perform a so-called grid search.

In our work, we start from this observation and turn to some well-known machine learning strategies to deal with what is basically an optimization problem with a large number of parameters. Another motivation comes from relevant experiences with side-channel analysis and machine learning. In principle, if it is possible to make SCA more powerful by using machine learning (and more generally, artificial intelligence) one would expect the same for fault injection.

In this chapter, we shed some light on the above-mentioned questions and observations. We discuss how one could use a special type of metaheuristics called genetic algorithms in order to find parameter values resulting in faults for electromagnetic fault injection. A somewhat similar research direction is followed in several previous works [8, 22, 23] but there the authors consider power glitching, which is only a subset of the search space we have to deal with for EMFI. In addition, they attack a PIN checking mechanism on a smartcard. In our research, we use the faults obtained via a novel technique to mount an algebraic fault attack on a SHA-3 implementation where we consider pulsed EMFI only resulting in a total of five parameters. We emphasize that our version of search algorithm differs significantly from previous works as detailed in the rest of this chapter. Finally, our code is available as an open source implementation.[1]

13.2 Related Work

Although a vast amount of work has been done on fault injection itself, see, e.g., [1, 6, 13, 19, 20, 26], only a small fraction of it concerns parameter optimization.

In [18], the authors develop an EMFI susceptibility criterion, which they use to rank the points of the chip surface depending on how sensitive they are to fault injection. The underlying assumption for the criterion is the sampling fault model, described in [21]. The criterion itself is a combination of power spectral density (measuring emitted power at the clock frequency) and magnitude squared incoherence (measuring how linked the emitted signal is to the data being processed). They use a grid scan (in two spatial dimensions) to measure all the points and rank them according to the criterion; a share α of the highest-ranking points are kept for further scanning; the rest is thrown away. They are able to reject over 50% of the chip surface (75% in their best case), while keeping 80% of the points causing faults. However, by *fault*, they mean any perturbation of the normal behavior of the algorithm.

In [8], the authors apply several different methods to the problem of parameter optimization for the supply voltage (VCC) glitching. They reduce the dimensionality of the problem by splitting the search into two stages. In the first stage, they look for the best (glitch voltage, glitch length) combination. In the second stage, ten most promising (voltage, length) combinations are tried at each point in the time range (which is discretized into 100 instants). All parameters not explicitly specified are set as random. The methods are compared at the first stage—random search, FastBoxing and Adaptive zoom&bound algorithms, and a genetic algorithm. This approach is a "smart" search in 2D with a grid search in 1D.

That work is extended in [23] where the authors use a combination of a genetic algorithm and local search (called memetic algorithm) in order to find faults even more efficiently. The authors consider power glitching with three parameters and are interested in a fast characterization of the search space.

[1]Github: https://github.com/geneticemfaults/geneticemfaults.

13.3 Preliminaries

13.3.1 Genetic Algorithms

Evolutionary algorithms represent population-based metaheuristic optimization techniques inspired by biological evolution and phenomena like mutation, recombination, and selection [3, 9, 10]. The solutions in the population compete and in that process improve their goodness as evaluated by a fitness function. Evolutionary algorithms often perform well in many types of problems because they ideally do not make assumptions about the underlying solutions' landscape. Today, there are many types of evolutionary algorithms, but probably the best known ones are genetic algorithms (GA). An instance of a genetic algorithm maps a real optimization problem to the natural concepts as follows:

1. The objective function (which we are optimizing) becomes the fitness function.
2. A solution (a point in the solution space) becomes an individual in the population.

The general pseudocode of an evolutionary algorithm is given in Algorithm 1. Note that this is general enough to cover any type of evolutionary algorithm, including genetic algorithms.

13.3.2 Keccak/SHA-3

In this work, we apply our attack on a cryptographic hash function, which is the new SHA-3 standard [5], and also known as Keccak. The Keccak main function is a sponge construction with a permutation as its core operation. Keccak is a cryptographic primitive that can be used in different modes (such as keyed and unkeyed) to compute hash values, MACs or to encrypt/decrypt data.

The core permutation named Keccak-$f[b]$ is defined by its width b and in our case, we use the full width where $b = 1600$. The permutation is described as a sequence of operations on a state a. The state is a three-dimensional array

Algorithm 1: Evolutionary algorithm pseudocode

Input : Parameters of the algorithm
Output : Optimal solution set

$t \leftarrow 0$;
$P(0) \leftarrow CreateInitialPopulation$;
while $TerminationCriterion$ **do**
    $t \leftarrow t + 1$;
    $P'(t) \leftarrow SelectMechanism\ (P(t-1))$;
    $P(t) \leftarrow VariationOperators(P'(t))$;
return: $OptimalSolutionSet(P)$

of elements in $GF(2)$. There are 5 rows and 5 columns, each of length 64, i.e. $a[5, 5, 64]$. Keccak-$f[b]$ is an iterated permutation over 24 rounds, the round function R is defined as follows:

$$R = \iota \circ \chi \circ \pi \circ \rho \circ \theta$$

In total there are five different steps that modify the state. An omitted index implies that the statement is valid for all values of that index.

$$\begin{aligned}
\theta: \quad a(x, y, z) &\leftarrow a(x, y, z) + \sum_{y'=0}^{4} a(x-1, y', z) + \sum_{y'=0}^{4} a(x+1, y', z-1),\\
\pi \text{ and } \rho: a(y, 2x + 3y) &\leftarrow \text{rot}(a(x, y), r(x, y)),\\
\chi: \quad a(x) &\leftarrow a(x) + (a(x + 1) + 1) \cdot a(x + 2),\\
\iota: \quad a(0, 0) &\leftarrow a(0, 0) + RC
\end{aligned}$$

All operations are carried out in $GF(2)$. The function $\text{rot}(W, i)$ is a bitwise cyclic shift operation, where the constants $r(x, y)$ are rotation offsets. The value RC is the round constant, there is a different value for each round. For more details, see [5].

13.4 Experimental Setup

For the target, we use a Cortex-M4 STM32F407IG (Riscure "Piñata") board running a C implementation of SHA-3. This implementation is taken from the WolfSSL library.[2] The board communicates with a PC by a serial interface and is powered by an external power supply. We use a Riscure EM probe and the VCGlitcher device that controls it. The whole setup is controlled by the code that is written in Python 2.x. For interfacing the Riscure equipment, we use Python bindings for the VCGlitcher C API (which is a 32-bit DLL). Figure 13.1 shows a photo of the setup.

The on-board code provides the trigger which signals that the cryptographic operation is in progress. This trigger is used as a reference point for injecting the fault. In case that the board gets stuck in an illegal state after a glitch, it needs to be reset. The only way to reliably reset this particular board is by cutting its power, which can take a significant fraction of a second, depending on the capacitors. We used 100 ms for this event.

[2]WolfSSL, an embedded SSL/TLS library. Available at: https://www.wolfssl.com/.

Fig. 13.1 The photo of the setup

The physical dimensions of the chip package are 24×24 mm. Repositioning error of our XYZ table is 0.05 mm, which gives us a spatial grid of at most 480×480. Note that the limiting factor in our case is likely the size of the probe tip, which is much larger.

13.4.1 Parameters

There are multiple parameters one can vary to affect fault probability as follows:

- the spatial position of the probe tip (there are three degrees of freedom; X, Y, Z);
- the moment when the EM pulse fires;
- the pulse intensity;
- the shape of EM probe and the angle w.r.t. the target;
- the shape of EM pulse w.r.t. time

In our experiments, we consider only a subset of those as listed below:

- We consider a position, as a pair of parameters (X, Y). We do not vary the distance from the board (Z), since this can be largely compensated for by a change in intensity. These parameters are real values in [0, 1] range.
- The glitch intensity regulates the voltage of the pulse. The SDK manual suggests it to be set to a percentage of power used. We use real values in the range [0, 1].
- We consider also time offset to be between 367 and 375 μs, because that is where the injection point must be, for the code we are running. We encode the offset at integer value (number of 2 ns ticks).
- A number of repetitions of the pulse is a primitive form of the pulse shape. We set this parameter to be in the range [1, 10].

We do not vary pulse duration, and we leave it to a fixed value of 40 ns. Similarly, we do not vary shape and angle of the probe tip, since changing those automatically is (for us) hard enough to make them unsuitable for automatic optimization.

13.4.2 *Search Space Size*

As mentioned above, it is not possible to conduct an exhaustive search when considering fault injection with realistic targets. Hence, the question that remains is the following: what is the search space size and how could we efficiently sample it? When considering X and Y position, there is a 0.05 mm repositioning error and 24×24 mm chip size, which gives max resolution of 480×480. For the time offset, with a 2 ns resolution and the range between 367 and 375 μs there are in total 4000 different values. We have no good rule for determining the smallest meaningful increment for the glitch intensity, but if we use a 1% increment, that gives us a range of 100 values. The repetitions are selected to be a random integer value in the range $[1, 10]$, which gives us ten values.

The total parameter set size is therefore $480 * 480 * 4000 * 100 * 10 \approx 10^{12}$. At ≈ 0.16 s per point, this results in 29,203 years to conduct an exhaustive search. Since trying the same parameters multiple times does not necessarily always yield the same response, we conduct five measurements for each point. Even if we would completely ignore everything except X, Y, and offset, we would still need 29.2 years to conduct an exhaustive search.

13.5 Search Algorithm

13.5.1 *Assumptions*

We consider the device to be a black-box. It assumes that the objective function is not a "golf-course-like" (i.e., to be too flat), in which case the lack of a significant gradient in the fitness landscape means that there is no driving bias toward fitness optimum. The reasoning behind is that a very weak EM pulse will not affect the target at all, and we will observe the normal behavior. Conversely, a very strong EM pulse will completely dishevel its operation and even potentially damage it. Consequently, we expect the faulty behavior to occur somewhere between those two extremes, i.e., along the class border. Additionally, offset ranges (min to max offset) are set by the user, based on a rough expectation of the duration of the cryptographic algorithm.

Usually, the objective function guides the optimization algorithm towards better solutions, and the algorithm ends when it finds the best one. Here, we do not want just a single "best" solution since not every fault we find will also be exploitable, and there are situations where more than one solution is required, so we aim to obtain multiple solutions.

13.5.2 GA Objectives

We require our algorithm to have the following two characteristics:

1. Good coverage of the parameter space—since we do not know where the exploitable faults are located, we need to explore the search space efficiently.
2. Speed—we require the algorithm to be fast in finding the faults, otherwise, there is no advantage of using it when compared to random search, for instance.

These requirements are somewhat conflicting with each other. Basically, as most of the parameter space is useless (i.e., has no faults), covering enough space to make sure we did not miss anything important implies potentially wasting a lot of measurements.

Next, we introduce the terminology we use when discussing the search and possible outputs of the algorithm. A point is a distinct set of parameters, i.e., a point in the parameter space. A measurement is the result of a single attempt at glitching the target with those parameters.

When counting the faulty measurements, we distinguish between:

1. the total number of faulty measurements,
2. the number of distinct faulty responses (i.e., "unique faulty measurements").

The difference is in the following: if a measurement results in a before-seen faulty output, the second one will not count again. As an example, assume we find a set of parameters resulting in a specific fault. Later, we find some other set of parameters that result in the same fault but we do not count that new faulty measurement into distinct faulty measurements. For the exploitability purposes, the second number is more interesting as detailed in the following sections.

We classify the board response in one of the following classes:

- NORMAL: for normal behavior, meaning the board performs as if we did not do anything.
- RESET: the board did not reply at all, requiring a reset to restore to normal operation.
- SUCCESS: the board produces an output/ciphertext/signature/hash different than the correct one.
- CHANGING: for each point we investigate, we perform five measurements. If all measurements are in the same class, the point is put into one of the first three classes; otherwise, it goes into the CHANGING class.

Fitness values are set according to the class: SUCCESS has the highest fitness (10), followed by the CHANGING, then RESET (5), and finally, NORMAL (2). CHANGING points' fitness depends on its underlying measurements: a mix of NORMAL and RESET measurements is somewhat better than an all-RESET or all-NORMAL point, but having individual measurements belonging to the class SUCCESS moves the fitness closer to an all-SUCCESS point. We calculate the fitness of a CHANGING point in the following way:

$$\text{fitness}_{\text{C}} = 4 + 1.2 * N_S + 0.2 * N_N + 0.5 * N_R. \tag{13.1}$$

Here, N_S represents the number of SUCCESS points, N_R the number of RESET points, and N_N the number of NORMAL points. Finally, factors 0.2 and 0.5 are chosen in analogy to the values for NORMAL and RESET (which are 2 and 5) while the rest of the factors are selected for the scaling reasons. For example, 4 NORMAL and 1 RESET measurements give fitness 5.3, which is higher than the fitness of a RESET point (with all 5 RESET measurements). Similarly, 4 SUCCESS and 1 RESET measurements give fitness 9.3, which is lower than the fitness of a SUCCESS point (with all 5 SUCCESS measurements).

13.5.3 Algorithm Definition

Despite the fact that we use genetic algorithms like similar to some previous works [8, 23], our custom-made algorithm is quite different. We discuss the specifics of our design in the following paragraphs. Our algorithm has several parameters to be determined. We selected those values on the basis of our tuning experiments and recommendations from [8, 23].

More in detail, our algorithm consists of two separate phases as follows:

1. The first phase is a genetic algorithm that we run for 20 generations with population size 50.
2. Only when the genetic algorithm is done, we start with the local search, which takes ten randomly chosen points in the neighborhood of each SUCCESS point. Note that this is a significant difference from related works where both GA and local search worked at the same time. We opted first to concentrate on exploration aspect—GA (to explore various regions of the target) and only after that on exploitability aspect—local search (to concentrate on more promising regions). Naturally, GA itself has also exploitability component that is especially manifested in the crossover operator but we also designed a custom-made crossover operator that promotes exploration perspective.

GA Phase

Our algorithm begins with a genetic algorithm that runs for N generations and has a population of M individuals. The initial population is selected uniformly at random within parameter ranges. We aim to maximize the fitness value, which corresponds to having as many as possible SUCCESS points.

The first phase of a GA is selection; we use roulette-wheel selection. In roulette-wheel selection, the fitness function assigns a fitness to possible solutions (in this case, points in parameter-space). This fitness level is used to associate a probability of being selected for each solution: the probability that an individual will be selected is directly proportional to its fitness. (More precisely, it is equal to its share in the overall fitness: for the i-th individual, $P_i = \frac{\text{fitness}_i}{\sum_k \text{fitness}_k}$.)

Note that, since the selection is done randomly, it is possible that low-quality individuals are picked. This is not troubling since a GA would not function otherwise. However, it allows for the possibility of an excellent, hard-found solution being accidentally lost. A common countermeasure is *elitism*: with elitism, a number of best-ranking individuals (called the *elite*) are always chosen. We use elitism, with the elite size equal to 1, so only the single best individual gets carried over.

We also experimented with a 3-tournament selection as used in [8, 23] but found it too restrictive, since it promotes a convergence too quickly, resulting in solutions being obtained from only a small part of the search space. This went directly against the objective of a good coverage of the search space.

After the selection phase finishes, the crossover can start. The purpose of crossover is to combine the existing solutions to produce better ones. There are many ways to do the crossover. When we imagine our individuals as points—or rather, vectors—in the 5-dimensional parameter space, then each of the parameters corresponds to one element of this vector. When combining two such individuals, a traditional crossover might take some elements from one parent, and the rest from the other parent. The version of crossover that we used instead picks a point in-between the respective parents' elements. This promotes explorability and enables GA to traverse large parts of the search space.[3] The pseudocode for this crossover is given in Algorithm 2, and the pseudocode for the mutation in Algorithm 3. The mutation rate *p_mutation* is set to 5%. In the case the parameter gets out of its ranges, it is clipped to the edges of its range.

[3]The parent points define an axis-aligned parallelepiped in parameter-space; the parents are placed on the diagonally opposite vertices. In a Hamming cube, these would be the all-zeros and all-ones vertices. The first crossover variant corresponds to picking one of its vertices, whereas the second crossover variant corresponds to picking a point within it.

Algorithm 2: Crossover operator

for *each parameter p* **do**
  $child.p = random\ value\ in\ range\ [parent1.p,\ parent2.p]$;

Algorithm 3: Mutation operator

$r_1, r_2 = uniformly\ random\ from\ [-0.5, 0.5]$;
for $param\ in\ [x, y, intensity]$ **do**
  with probability $p\_mutation$:
    $param = param + r_1$;
with probability $p\_mutation$:
  $offset = offset + r_2 * OFFSET\_RANGE$;
with probability $p\_mutation$:
  $repetitions = random\ integer\ from\ [1, 10]$;

Local Search Phase

After the GA is done, we use local search to focus on the promising parts of the explored search space as follows: the space around the intersection points (i.e., places where class values change), and the space around any faults that were already found. We define the neighborhood of a point as a cube centered in it, with edge length equal to 0.02. By length of 0.02 in parameter space, we mean 2% of the range of that parameter. Parameters x, y, and intensity are all within the range [0, 1]. For offset, it's 2% of OFFSET_RANGE (which is OFFSET_MAX - OFFSET_MIN). To determine the distance of values, we use the Euclidean distance.

13.5.4 Practical Considerations

Commonly, optimization algorithms (and nature-inspired metaheuristics in particular) rely on a large number of iterations. Another assumption usually made is that the evaluation of possible solution points is uniform. Here, we have expensive measurements, where the cost of evaluation depends not only on the property of the point itself, but also on the context of its evaluation. Although our algorithm consists of a genetic algorithm and local search, we denote it often as a genetic algorithm but we always consider it to have also local search phase. We do not call our technique a memetic algorithm since the GA and local search phases are separated.

When considering EMFI, there is the probe tip, which has to physically move to a different point. To do this with a sufficient precision requires a non-negligible amount of time—the exact amount varies depending on the setup, but it can be up to several seconds per measurement. In comparison, a reset requires just a fraction of a second (for our board, $\approx$100 ms to do it reliably). The measurement part itself is even faster—30 ms or less. Thus, the order in which points are evaluated matters. Even with an optimal routing for any batch of N points, more batches mean more time wasted. For population-based algorithms, this translates to small population sizes not being as efficient as large ones.

Additionally, we may want to get a glimpse of the results even before the scan is finished, especially for long-running scans. In case of a random or grid scan, this means splitting the scan into batches where each covers more or less the whole parameter space, since scanning points in the optimal order results in uneven coverage.

13.6 Results

In this section, we present our results when attacking SHA-3. First, we investigate how well is GA able to find faults (i.e., force the algorithm to output the wrong ciphertext) and then, whether such points can be used in order to obtain the state of the algorithm. For this purpose, we use algebraic fault analysis (AFA), as described in [17]. AFA eliminates the need for analysis of fault propagation as it is needed in differential fault analysis; instead, it uses a SAT solver to recover the state bits from a (clean output, faulty output) pair.

13.6.1 Finding Faults

The duration of GA is determined by the number of faults it finds. We conducted five independent runs with 2074, 2343, 3353, 3606, and 5132 points, respectively. Each individual run is different due to the stochastic nature of GA as well as the target response. To obtain statistically meaningful results, we report averaged values over all runs and report results in Table 13.1. For GA, on average, in each run, there are 3301.6 points, which means we conduct 16,508 individual measurements on average. Out of these, 9700.4 (58.8%) are faulty, and 3288.4 (19.9%) are unique/distinct. To compare, we use random search with 3302 points, which represents 16,510 individual measurements. Out of these, 228.2 (1.3%) are faulty, and 160.8 (1.0%) are unique/distinct.

Table 13.1 Statistical results for GA and random search

	GA	Random
NORMAL	662.8 (18.9%)	2995.8 (90.7%)
RESET	496.4 (15.0%)	65.0 (2.0%)
CHANGING	375.2 (11.4%)	232.4 (7.0%)
SUCCESS	1807.2 (54.7%)	8.8 (0.3%)

To conclude, when averaged over 5 runs, our GA algorithm gives 42.5 times more faulty measurements, and 20.5 times more distinct faulty ones. The somewhat lower share of distinct measurements for the GA algorithm can be explained by many SUCCESS points being close to each other due to the local search, thus being more likely to cause the same response.

In Table 13.2, we give results for Random search and genetic algorithms when considering 500, 1000, and 2000 points. Observe how the results for random search do not change significantly with more measurements. At the same time, we observe that GA is very successful already for the smallest case where we use only 500 points and as we add more points, the percentage of SUCCESS points increases.

We depict the search space after random search and GA in Fig. 13.2a–f. Figure 13.2a, b give results for X and Y parameters. Figure 13.2c–f also depict intensity as a parameter. We depict both cases with and without NORMAL points to improve the readability. The number of points for each figure is 3300 (figures not depicting NORMAL points have fewer points).

Finally, we show the results for GA as it progresses through the evolution process. More precisely, in Fig. 13.3a, we depict the search results for the first 500 points. We can observe that although there are several SUCCESS points, in this phase GA mainly finds NORMAL points spread across the search space. Figure 13.3b depicts results for 500–1000 points range. Here, we can see that GA does not find so many NORMAL points but actually manages to find a significant number of CHANGING and RESET points. We also see a good amount of SUCCESS points. In Fig. 13.3c, we show the results for points between 1000 and 2343 (end of search). Here, we see a large number of SUCCESS points where there are one large cluster and three smaller ones. CHANGING and RESET points occur in the same large cluster as the majority of SUCCESS points. NORMAL points occur in a number of small clusters surrounding the main cluster. To conclude, we see that GA is able to find SUCCESS points even with a small number of examined points but its true strength lies when there is a sufficient number of evaluations in order to guide the convergence.

13.6.2 SHA-3 Attack in Practice

To the best of our knowledge, SHA-3 implementation has not yet been attacked in practice. Attacks do exist, but only on simulated data such as [4]. The authors show that differential fault analysis (DFA) can be used to recover the complete state in around 80 faults on average if the attacker is able to inject single-bit faults in the input of the penultimate round (i.e., θ_i^{22}), though they rely on brute-forcing the last few bits. According to [16] (itself an extension of [15]), this is around 500 single-bit random faults for the whole state. The work in [16] generalizes the attack to a single-byte fault model, recovering the state in around 120 random faults.

Table 13.2 Random search and GA results for various search stages

Points	Algorithm	NORMAL	RESET	CHANGING	SUCCESS	#Faults	#Distinct
500	Random	452.6 (90.5%)	9.8 (2.0%)	36.0 (7.2%)	1.6 (0.3%)	33.4 (1.3%)	22.6 (0.9%)
	GA	315.2 (63.0%)	73.4 (14.7%)	79.0 (15.8%)	32.4 (6.5%)	260.8 (10.4%)	158.8 (6.3%)
1000	Random	910.4 (91.0%)	19.6 (2.0%)	67.2 (6.7%)	2.8 (0.3%)	58.8 (1.2%)	40.4 (0.8%)
	GA	381.8 (38.2%)	198.0 (19.8%)	169.2 (16.9%)	251.0 (25.1%)	1530.4 (30.6%)	956.6 (19.1%)
2000	Random	1814.6 (90.7%)	36.6 (1.8%)	144.2 (7.2%)	4.6 (0.2%)	130.6 (1.3%)	93.4 (0.9%)
	GA	541.6 (27.1%)	351.2 (17.6%)	285.0 (14.2%)	822.2 (41.1%)	4606.4 (46.0%)	2030.4 (20.3%)

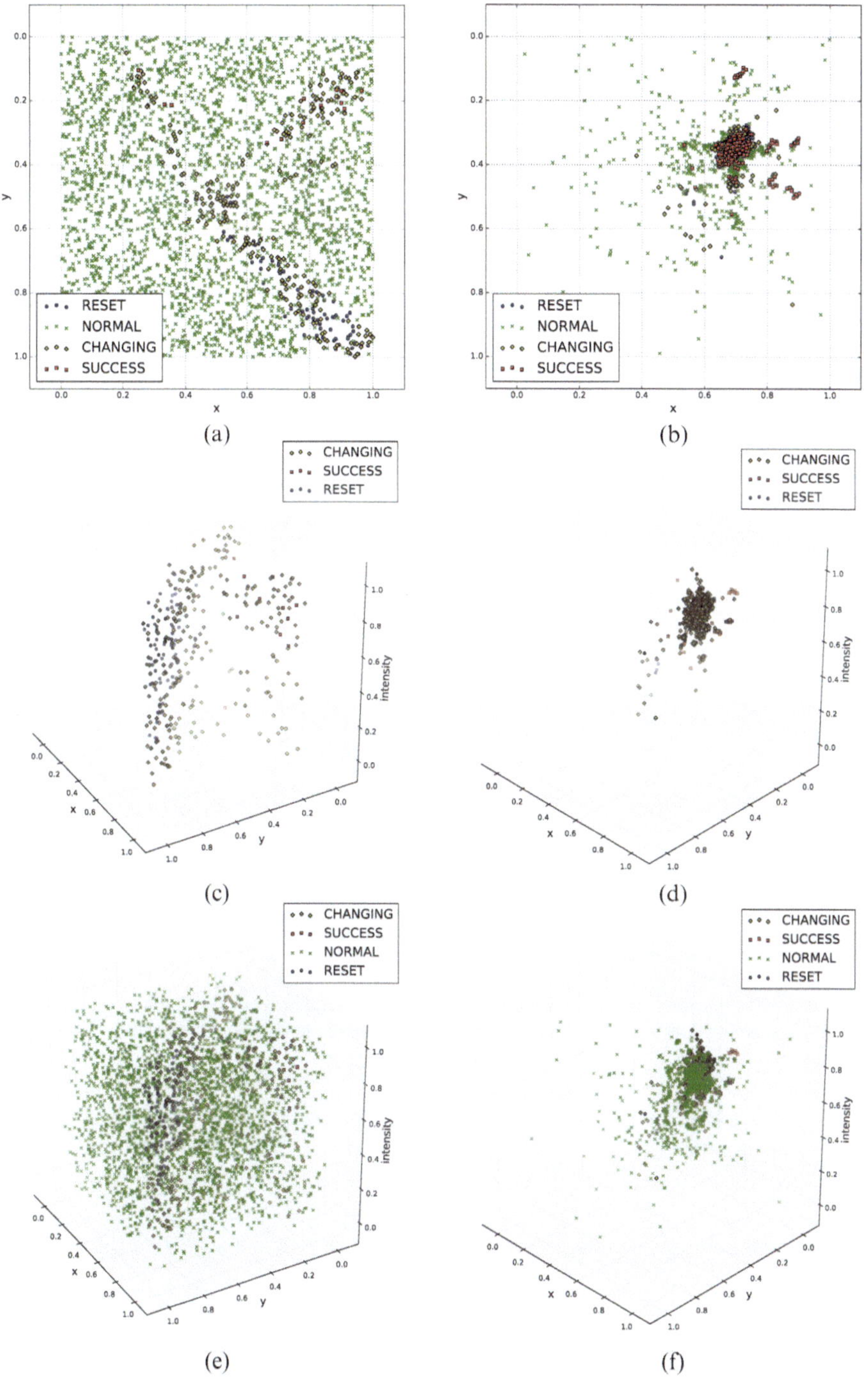

Fig. 13.2 Results for GA (with local search) and random search. (**a**) Random search in 2D. (**b**) GA and local search in 2D. (**c**) Random search without NORMAL points. (**d**) GA and local search without NORMAL points. (**e**) Random search. (**f**) GA and local search

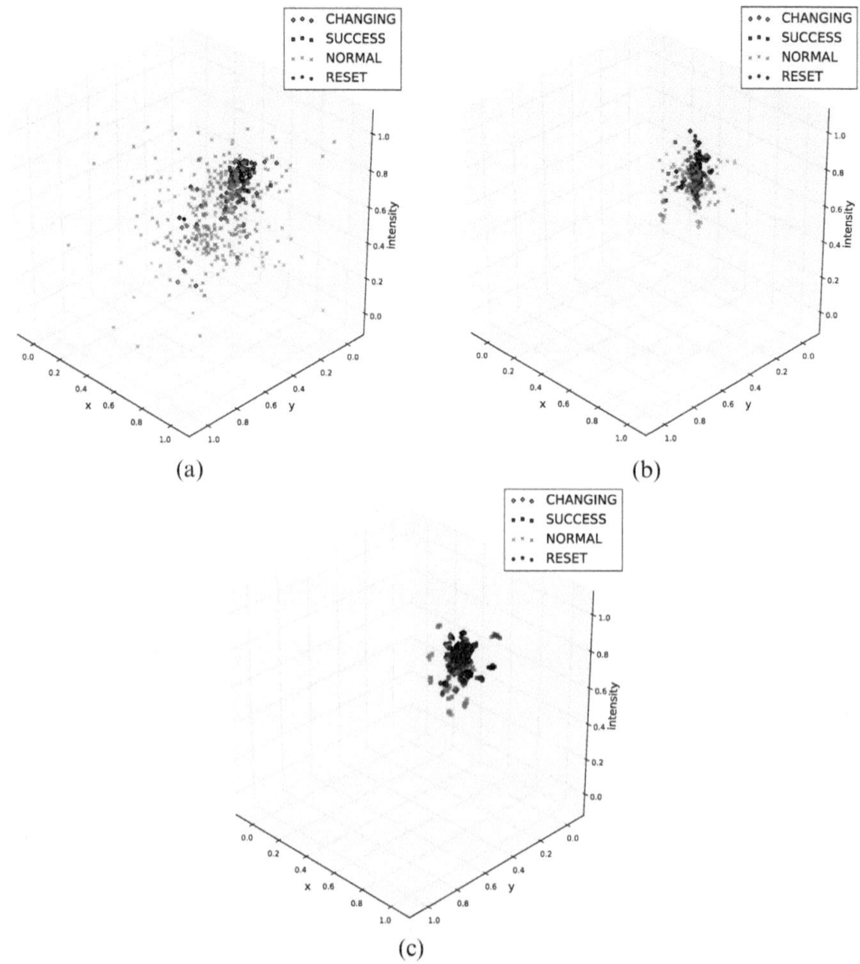

Fig. 13.3 Results for GA with local search depicting several stages of the search process. (**a**) Results for points 0–500. (**b**) Results for points 500–1000. (**c**) Results for points 1000–2343

Algebraic fault analysis (AFA) seems more promising. Luo et al. manage to bring down the number of faults needed to recover the internal state with SHA3-512 down to under 10 with AFA and the 32-bit fault model in the progress described in [15–17].

AFA has several advantages, besides being more efficient at recovering state, as follows:

- It does not require analysis of fault propagation through the algorithm, making it much easier to abstract the internal details.
- We can easily change the fault model, by just changing the appropriate constraints.
- Perhaps most importantly, it works for more relaxed fault models.

The attack in [17] allows the attacker to retrieve the state by injecting multiple faults in the input of the round 22 of Keccak. The faults are allowed to affect up to 1 unit of the state, where units are sized 8b, 16b, or 32b. As in the previous work, we use θ_i^{22} as the fault injection point, and χ_i^{22} as the target state to recover. We reused their C++ retrieval code for this purpose.

The general idea behind AFA on SHA-3/Keccak is simple: use a SAT solver to do the work for us: we just need to provide appropriate constraints for it. We start with 1600 Boolean variables representing the state (θ_i^{22}) and then provide constraints:

1. Fault Model—what kind of a fault do we cause? There's a separate set of (up to) 1600 Boolean variables ($\Delta\theta_i^{22}$) representing the induced fault. $\theta_i^{22} \oplus \Delta\theta_i^{22}$ is the faulted state, before propagating through the final two rounds of the algorithm. Depending on what the fault model is, we add constraints such as "exactly one bit of $\Delta\theta_i^{22}$ is non-zero," corresponding to a one-bit fault model, or slightly more verbose ones for specifying things such as "we faulted a word-aligned 32-bit word," which would correspond to a 32-bit fault model in [17].
2. Keccak—how the (faulted) internal state propagates? For Keccak, the internal transformations can be relatively simply encoded as Boolean expressions. This implicitly tells the solver everything it needs to know about fault propagation, regardless of the fault model constraints. There are two cases we consider:

$$H = \iota^{23} \circ \chi \circ \pi \circ \rho \circ \theta \circ \iota^{22} \circ \chi \circ \pi \circ \rho \circ \theta(\theta_i^{22})$$

   where H is the correct hash output, and

$$H' = \iota^{23} \circ \chi \circ \pi \circ \rho \circ \theta \circ \iota^{22} \circ \chi \circ \pi \circ \rho \circ \theta(\theta_i^{22} \oplus \Delta\theta_i^{22})$$

   where H' is the faulty hash output.
3. Outputs—which are the concrete outputs? We give the SAT solver the actual values of H and H'. After so constraining the SAT solver, we can let it find a solution—an internal state satisfying all the constraints. Once it finds the first such solution, we ban this newly found solution by adding it as an additional constraint and let the SAT solver find another one. This process is repeated until no new solutions can be found.

   The bits of the state which are the same in all solutions are the ones we can recover: as for those which take different values in different solutions, their values are not entailed by the combined constraints of the fault model, the algorithm, and the outputs (i.e., the "real" constraints).

   Depending on the fault model and the version of SHA-3 (SHA3-512, SHA3-224, etc.), these constraints may or may not be enough to recover part of the state.

In this case, additional constraints can be introduced, such as using two faulty hashes H_1' and H_2' at a time with a cost of extra Boolean variables and making it harder for the SAT solver (Method II in [17]), or first recovering part of the χ_i^{23} bits (Method III in [17]).

We applied the 32-bit fault model from [17] and Method III. The reason for this is a large number of potential faults to check while a short time for checking the exploitability of the induced faults is often an important factor. We tested the exploitability of all distinct faulty hashes obtained by our evolutionary algorithm, as well as of all those obtained by a random scan. While the share of distinct/unique faulty hashes depends on the size of the scan, the exploitability of a faulty hash does not. For this reason, we calculated the share of exploitable individual faults on all the samples we obtained (with the same hyperparameters).

The results are as follows: GA generated a total of 14,979 distinct faults (out of 82,540 individual measurements); 106 of these were exploitable 32-bit faults, for a share of 0.71%. Random search generated 947 distinct faults (out of 100,000 individual measurements); 110 of these were exploitable 32-bit faults, for a share of 11.61%. When translated into exploitable faults per individual measurement, this gives about 1.41×10^{-3} and 1.13×10^{-3} for GA and random search, respectively—an improvement of 24.6%.

Despite the fact that GA is still significantly more successful than the random search, we observe that actually most of the faults obtained with GA cannot be translated into exploitable faults. This results in a decrease between the performance difference of GA and random search. Still, such results are to be expected: since we never added the constraint of exploitability of faults into GA, it is hard to expect that GA will produce only such faults. Still, this could be addressed by having a fitness function that integrates an analysis of fault exploitability.

13.7 Conclusions and Future Work

In this chapter, we investigate how genetic algorithms can be used to facilitate faster and more powerful EMFI. When considering the search space size one needs to investigate, it is evident that both random search and exhaustive search should not be the methods of choice. Indeed, our custom-made algorithm is able to find more than 40 times more faults than random search. Those results enable us almost 25% more exploitable faults per individual measurements when considering SHA-3 and algebraic fault attack. To the best of our knowledge, our algorithm is the most powerful currently available technique for finding parameters for EMFI.

Since there are only a few works considering the parameter search that is leading to faults, this opens a number of potential research directions. We believe the following two would be the most interesting: (1) exploring laser fault injection, and adding the notion of exploitability to the fitness function. The latter means that, instead of running a local search on every SUCCESS point, we can first try to check whether it is exploitable, and only if it is, consider its neighborhood. Naturally, this also opens a question what is a good neighborhood to consider, or to state it differently: what is the best resolution for our search? Indeed, if all points within a certain neighborhood would result in no extra information we can use for exploitation, then there is no need for our algorithm to search within that region. (2) Next, in this paper, we concentrated on AFA Method III, but it would be interesting to investigate how our technique fares when used with Method II. Besides that, we also plan to investigate different targets and improve the performance of our algorithm.

Acknowledgements This work has been supported in part by Croatian Science Foundation under the project IP-2014-09-4882 and by the Technology Foundation TTW (Project 13499 TYPHOON), from the Dutch government.

References

1. A. Aghaie, A. Moradi, S. Rasoolzadeh, F. Schellenberg, T. Schneider, Impeccable circuits, Cryptology ePrint Archive, Report 2018/203, 2018. https://eprint.iacr.org/2018/203
2. C. Aumüller, P. Bier, W. Fischer, P. Hofreiter, J.-P. Seifert, Fault attacks on RSA with CRT: concrete results and practical countermeasures, in *CHES*, pp. 260–275 (2002)
3. T. Bäck, D.B. Fogel, Z. Michalewicz (eds.), *Evolutionary Computation 1: Basic Algorithms and Operators* (Institute of Physics Publishing, Bristol, 2000)
4. N. Bagheri, N. Ghaedi, S.K. Sanadhya, Differential fault analysis of SHA-3, in *Progress in Cryptology–INDOCRYPT 2015* (Springer, Cham, 2015), pp. 253–269
5. G. Bertoni, J. Daemen, M. Peeters, G. Van Assche, The Keccak reference, January 2011. http://keccak.noekeon.org/
6. D. Boneh, R.A. DeMillo, R.J. Lipton, On the importance of checking cryptographic protocols for faults (extended abstract), in *Advances in Cryptology - Proceeding of the EUROCRYPT '97, International Conference on the Theory and Application of Cryptographic Techniques*, Konstanz, May 11–15 (1997), pp. 37–51
7. E. Cagli, C. Dumas, E. Prouff, Convolutional neural networks with data augmentation against jitter-based countermeasures - profiling attacks without pre-processing, in *Cryptographic Hardware and Embedded Systems - CHES 2017 - Proceedings of the 19th International Conference, 2017*, Taipei, September 25–28 (2017), pp. 45–68
8. R.B. Carpi, S. Picek, L. Batina, F. Menarini, D. Jakobovic, M. Golub, Glitch it if you can: parameter search strategies for successful fault injection, in *Smart Card Research and Advanced Applications*, ed. by A. Francillon, P. Rohatgi (Springer, Cham, 2014), pp. 236–252
9. A.E. Eiben, J.E. Smith, *Introduction to Evolutionary Computing* (Springer, Berlin, 2003)
10. J.H. Holland, *Adaptation in Natural and Artificial Systems: An Introductory Analysis with Applications to Biology, Control, and Artificial Intelligence* (The MIT Press, Cambridge, 1992)
11. P.C. Kocher, Timing attacks on implementations of Diffie-Hellman, RSA, DSS, and other systems, in *CRYPTO '96: Proceedings of the 16th Annual International Cryptology Conference on Advances in Cryptology* (Springer, London, 1996), pp. 104–113
12. P. Kocher, J. Jaffe, B. Jun, Differential power analysis, in *Annual International Cryptology Conference* (Springer, Berlin, 1999), pp. 388–397
13. O. Kömmerling, M.G. Kuhn, Design principles for tamper-resistant smartcard processors, in *Proceedings of the USENIX Workshop on Smartcard Technology on USENIX Workshop on Smartcard Technology* (USENIX Association, Berkeley, 1999), pp. 2–2

14. L. Lerman, G. Bontempi, O. Markowitch, Side channel attack: an approach based on machine learning, in *Second International Workshop on Constructive SideChannel Analysis and Secure Design*, pp. 29–41 (Center for Advanced Security Research, Darmstadt, 2011)
15. P. Luo, Y. Fei, L. Zhang, A.A. Ding, Differential fault analysis of SHA3-224 and SHA3-256, in *2016 Workshop on Fault Diagnosis and Tolerance in Cryptography (FDTC)*, pp. 4–15 (2016)
16. P. Luo, Y. Fei, L. Zhang, A.A. Ding, Differential fault analysis of SHA-3 under relaxed fault models. J. Hardware Syst. Secur. **1**(2), 156–172 (2017)
17. P. Luo, K. Athanasiou, Y. Fei, T. Wahl, Algebraic fault analysis of SHA-3 under relaxed fault models. IEEE Trans. Inf. Forensics Secur. **13**, 1752–1761 (2018)
18. M. Madau, M. Agoyan, P. Maurine, An EM fault injection susceptibility criterion and its application to the localization of hotspots, in *International Conference on Smart Card Research and Advanced Applications* (Springer, Cham, 2017), pp. 180–195
19. H. Martín, T. Korak, E.S. Millán, M. Hutter, Fault attacks on STRNGs: impact of glitches, temperature, and underpowering on randomness. IEEE Trans. Inf. Forensics Secur. **10**(2), 266–277 (2015)
20. C. O'Flynn, Fault injection using crowbars on embedded systems, Cryptology ePrint Archive, Report 2016/810 (2016). https://eprint.iacr.org/2016/810
21. S. Ordas, L. Guillaume-Sage, P. Maurine, EM injection: fault model and locality, in *Workshop on Fault Diagnosis and Tolerance in Cryptography (FDTC), 2015* (IEEE, Piscataway, 2015), pp. 3–13
22. S. Picek, L. Batina, D. Jakobovic, R.B. Carpi, Evolving genetic algorithms for fault injection attacks, in *2014 37th International Convention on Information and Communication Technology, Electronics and Microelectronics (MIPRO)*, May (2014), pp. 1106–1111
23. S. Picek, L. Batina, P. Buzing, D. Jakobovic, Fault injection with a new flavor: memetic algorithms make a difference, in *Constructive Side-Channel Analysis and Secure Design*, ed. by S. Mangard, A.Y. Poschmann (Springer, Cham, 2015), pp. 159–173
24. S. Picek, A. Heuser, A. Jovic, S.A. Ludwig, S. Guilley, D. Jakobovic, N. Mentens, Side-channel analysis and machine learning: a practical perspective, in *2017 International Joint Conference on Neural Networks, IJCNN 2017*, Anchorage, AK, May 14–19 (2017), pp. 4095–4102
25. J.-J. Quisquater, D. Samyde, Electromagnetic analysis (EMA): measures and counter-measures for smart cards, in *Smart Card Programming and Security*, ed. by I. Attali, T. Jensen (Springer, Berlin, 2001), pp. 200–210
26. N. Samwel, L. Batina, Practical fault injection on deterministic signatures: the case of EdDSA, in *Progress in Cryptology – AFRICACRYPT 2018*, ed. by A. Joux, A. Nitaj, T. Rachidi (Springer, Cham, 2018), pp. 306–321

Chapter 14
Automated Profiling Method for Laser Fault Injection in FPGAs

Jakub Breier, Wei He, Shivam Bhasin, Dirmanto Jap, Samuel Chef, Hock Guan Ong, and Chee Lip Gan

14.1 Introduction

Before the actual fault injection attack, a thorough profiling of the device under test has to be performed, so that the achievable fault models are known as well as areas of interest on the chip. This process is repetitive in nature, but necessary when targeting a new chip. Also, it takes a non-negligible amount of time, since the whole area normally has to be scanned with varying parameters of the fault injection device. Therefore, it is an ideal candidate for automation.

Modern field programmable gate arrays (FPGAs) and programmable systems on chip (SoCs) come with interesting features, like rich logic resources, real-time reconfiguration, high-density memories, clock managers, environment sensors, etc. Owing to such features and low time-to-market, FPGAs are being deployed in variety of applications. FPGAs also find wide applications in security-critical

---

This research was conducted when the authors "J. Breier", "W. He" and "H. G. Ong" were with Temasek Laboratories, NTU.

J. Breier (✉)
Underwriters Laboratories, Singapore, Singapore
e-mail: jbreier@jbreier.com

W. He
Shield Laboratory, Huawei International Pte. Ltd., Singapore, Singapore
e-mail: hewei48@huawei.com

S. Bhasin · D. Jap · S. Chef · C. L. Gan
Temasek Laboratories, Nanyang Technological University, Singapore, Singapore
e-mail: sbhasin@ntu.edu.sg; djap@ntu.edu.sg; csamuel@ntu.edu.sg; clgan@ntu.edu.sg

H. G. Ong
Smart Memories Pte Ltd, Singapore, Singapore

J. Breier et al. (eds.), *Automated Methods in Cryptographic Fault Analysis*,
https://doi.org/10.1007/978-3-030-11333-9_14

domains due to constantly evolving protection requirements, such as aerospace, defense, etc. However, like other devices, FPGAs are also vulnerable to physical attacks, i.e., side-channel attacks [17], fault attacks [8], and probing [3].

Laser fault injection (LFI) falls into optical fault injection methods. It is a semi-invasive local perturbation technique, which requires decapsulation of the target device, followed by injection of a high intensity laser. The injection can be performed either through the frontside or the backside of the target chip. However, because of the dense metal wires covering the active logic layer, it is highly challenging to realize a successful fault perturbation from the frontside.

In this chapter, we provide an automated method to characterize the LFI on an FPGA. A fault injection-based laser sensitivity profiling of the exemplary 65-nm Virtex-5 FPGA is performed. The provided method is capable of finding areas to perform successful data register bit flips in logic arrays. We localize interesting logic within these blocks and sketch the laser sensitivity regions to demonstrate that the high-precision bit-flips in fundamental logic cells of the FPGA can be found automatically.

The entire process flow is depicted in Fig. 14.1, showing the steps required for the full chip profiling. First, it is necessary to open the chip package—this can be

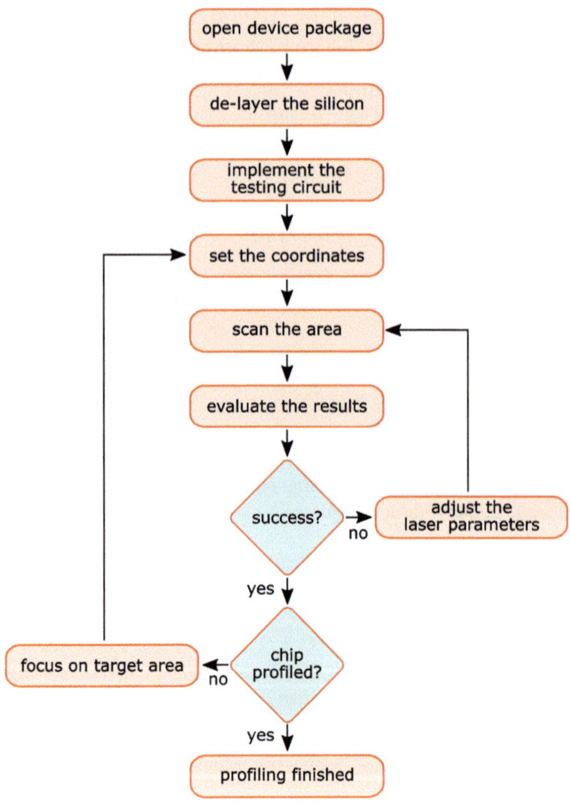

Fig. 14.1 Flow diagram of the automated profiling

either just a metal cover or an epoxy resin. This is followed by silicon delayering that helps to target the laser on the chip components. After these two mechanical steps, a testing circuit is implemented on a board, and the laser profiling scan is launched by following an automated procedure to adjust the parameters and evaluate the results. After the desired faults are found, the profiling finishes.

The rest of this chapter is organized as follows. Section 14.2 discusses previous work and outlines our contributions. In Sect. 14.3, the related work about optical properties on silicon, chip preparation, and configuration are presented. The profiling of laser sensitivity on chip and analysis methodologies are described in Sect. 14.4. Experimental results and further discussions are detailed in Sect. 14.5. Finally, summary is provided in Sect. 14.6.

14.2 Related Work

Many techniques have been proposed in the previous literature for disturbing values processed and stored in ICs [1, 9, 12, 14, 22, 23]. In general, results on microcontrollers show high degree of repeatability, mainly because of the stable clock and a possibility to predict the instruction order. Precision depends on the used CMOS technology and the size of the effective laser spot. Additionally to memory disturbances, it is also relatively easy to disturb the instruction execution on these devices, leading to instruction skip or alteration faults. Previous papers about fault injections on FPGAs mostly aim at memory disturbances both on configuration memory of SRAM FPGAs and data Block RAM [11, 21, 25]. Some of the works are detailed below.

Pouget et al. [21] proposed a laser platform for evaluating the sensitivity of SRAM-based FPGAs, where the test targets are the FPGA configuration memory bits, instead of the algorithmic data. They successfully injected single and multiple bit flips into configuration memory of a commercial FPGA manufactured on 1.5-μm technology.

Canivet et al. [11] conducted an attack on a protected AES implementation by using a laser with 20-μm spot size, targeting a 1.5-μm FPGA. Their results show that probability to flip a "1" is greater than the probability of flipping a "0." Also, they stated that the most vulnerable components within the CLB are the look-up-table (LUT) contents and the internal multiplexers.

Selmke et al. [25] presented a precise bit-level manipulations in BRAM for two different FPGAs, with 90- and 45-nm transistor sizes. The spot size of their laser was 4 μm, allowing comparatively higher precision faults on Spartan-3A and a bit lower precision on Spartan-6, where vulnerable areas for different bits were overlapping. Still, they could produce bit sets/resets in the latter case, only the success rate was lower.

The fault injection into the configuration memory of SRAM FPGAs intrinsically incurs the alterations on logic functions or routings, and hence leads to permanent circuit malfunction until the device is reconfigured with a new bitstream. The faults

Table 14.1 State of the art for laser fault injection

Work	Platform	Tech.	Target	Fault model	Position	Purpose
Dutertre et al. [1, 14, 23]	μC	350 nm	SRAM	Byte	Front	Attack
Courbon et al. [12]	ASIC	90 nm	Flip-Flops	Bit	Back	Attack
Breier and Jap [9]	μC	350 nm	Register	Bit	Back	Attack
Pouget et al. [21]	FPGA	150 nm	CLB/BRAM	Random	Back	Reliability
Canivet et al. [11]	FPGA	150 nm	Logic	Random	Back	Attack
Selmke et al. [25]	FPGA	90/45 nm	BRAM	Bit	Back	Attack
This work	FPGA	65 nm	Flip-flops	Bit	Back	Attack

are typically found and analyzed by a *readback* of the bitstream from the device after each fault injection to be compared with the unaffected *golden* sample [2, 16], in order to figure out the affected tiles on the logic array. The comparison efficiency is low and static, and furthermore, the method is becoming challenging to apply to newer FPGAs with more obscure bitstream formats.

Lohrke et al. [18] test CPLDs manufactured with 180-nm technology by using a high-end Hamamatsu PHEMOS-1000 laser scanning microscope. In their experiment, they show how to localize AND and XOR gates and apply this method in order map the location of a ring oscillator circuit. Later [27], they show how to attack physically unclonable functions by using this method.

Another direction in disturbing FPGAs is a bitstream fault injection. Swierczynski et al. [26] show malicious bitstream modifications of Xilinx Spartan-6 and Virtex-5, attacking AES. However, as authors have mentioned, in newer FPGAs, bitstream encryption has strengthened authentication, which can prevent such bitstream fault injection. Our method, on the other hand, does not have any assumptions on bitstream security, since it is applied directly on the logic components.

Some previous works are summarized in Table 14.1 and compared with this work. The comparison is drawn in terms of platform (μC, FPGA, and ASIC), technology node (Tech.), fault target (RAM, logic, and flip-flop), chip position (front-side, and back-side), fault precision (bit, and random), and the purpose of the fault injection.

14.3 Chip Preparation and Device Configuration

For modern FPGAs, two package styles are typically applied to encapsulate the naked dies. The first is the *bonded-wire* package in which the metal layer is placed up and the chip substrate is facing down to the PCB board. On the contrary, *flip-chip* package places the substrate up and metal layers down. Due to the metal layer placed above the active logic layer, laser injection can hardly affect the logic cells (active transistor layer) below. In this work, we target a 65-nm Virtex-5 FPGA (LX50T) with a flip-chip package on Digilent's *Genesys* board. The first step to allow an effective laser impact on the internal logic was to preprocess the FPGA chip by thinning down the substrate layer, by using a mechanical solution. This section explains the laser effects on silicon, sample preparation, and the description of the device under test.

14.3.1 Pulsed Laser Interaction with Silicon

The generation of carriers in semiconductor material by photoelectric effect has been used for decades in various fields such as failure analysis and defect localization [20], single event effect testing for space applications [10], and, as detailed in Sect. 14.2, security analysis.

When a pulsed laser irradiates silicon devices, two main mechanisms may occur:

- Single photon—Linear absorption (SPA). The photons have enough energy to induce a direct jump of the electrons from the valence band to the conduction band. The energy of the photons is bigger than the material bandgap in that case.
- Two photons—Nonlinear absorption (TPA). The free carriers generation results from the quasi-simultaneous absorption of two photons.

The dominant process will be qualified by the wavelength of light and the pulse duration. Generation of free carriers by SPA requires a wavelength shorter than the silicon bandgap ($\approx$1100 nm with undoped silicon). TPA has a quadratic relationship with the irradiance, meaning that a bigger number of carriers is generated compared to SPA. In addition, it happens in smaller volumes than SPA, providing resolution enhancement. One of the drawbacks is that triggering and detecting the effect can be more complex. Furthermore, TPA requires high peak power pulses achieved with a femtosecond laser which can be difficult to integrate to the test setup. More details about SPA and TPA can be found in [10]. In silicon, with pulses of duration within picosecond range or longer, and at the wavelengths shorter than 1100 nm, SPA will be the dominant mechanism.

Once carriers are generated, if no electric field exists, charges will recombine without further effect. On the other hand, when there is a high electric field, like in a reverse bias junction, carriers surviving prompt recombination will drift and establish a transient current. The latter can have important consequences on the device behavior such as upsets, latch-up, etc.

In a modern integrated circuit, the density and the number of metal layers forbids an irradiation from the frontside of the chip. When injecting photocurrent from the backside, it is mandatory to use a wavelength that can propagate further enough through the substrate and reach the sensitive volume. As a consequence, wavelengths close to the bandgap are commonly used: absorption is limited while still triggering photoelectric effect.

Spatial resolution is another factor to consider when choosing the laser wavelength. The spot size measured at $1/e^2$ of the maximum intensity is linked to the wavelength by the following equation:

$$2\omega_0 = \frac{4\lambda}{\pi NA}, \tag{14.1}$$

where ω_0 is the beam waist, λ is the wavelength, and NA is the numerical aperture of the objective. This equation shows that a smaller spot size is induced either by a higher numerical aperture or a shorter wavelength, so the shorter the better from the resolution point of view.

As a summary, laser wavelength needs to be shorter than bandgap wavelength to generate free carriers but not too short to limit absorption by the substrate. For this reason, a laser wavelength of 1064 nm is used in this work. While seeking for resolution enhancement, backside application of visible wavelength has been reported in other field of work [6], but it requires to thin the substrate down to few micrometers. Such thickness is even more complex to reach using mechanical tools when the device under test is soldered on a testboard.

14.3.2 Sample Preparation of Virtex-5

As detailed in the beginning of this section, the Virtex-5 device was mounted on a Genesys testboard. Removing the part from the board to prepare it for the backside analysis and then solder it again may result in damaging of the device. Thus, it was safer to prepare it while mounted on the testboard. As it is a flip-chip package, sample preparation from the top could be achieved. The compound was first removed using laser decapsulation until the metal heat-sink plate was revealed. The metal plate was then removed with tweezers to expose the silicon substrate. Before being diced and each sample individually packaged, silicon wafers are usually polished during the manufacturing process. The substrate surface quality is mirror-like, enabling IR inspection from the backside.

Therefore, in sample preparation, once the device is cleaned with chemicals to remove glue attaching the heat sink, the circuit can already be observed from the backside. However, if the doping is high, absorption can limit the image quality. This is also an issue for fault injection as part of the incident light is absorbed, resulting in higher energy requirements to induce upsets. In addition, die warpage leads to a nonuniformity of the substrate. Refraction of the light beam on nonplanar surface induces poorer image quality.

Thinning of the substrate aims to mitigate all these issues. For this experiment, thinning was achieved with the *Ultra Tec ASAP-1* mechanical processing system

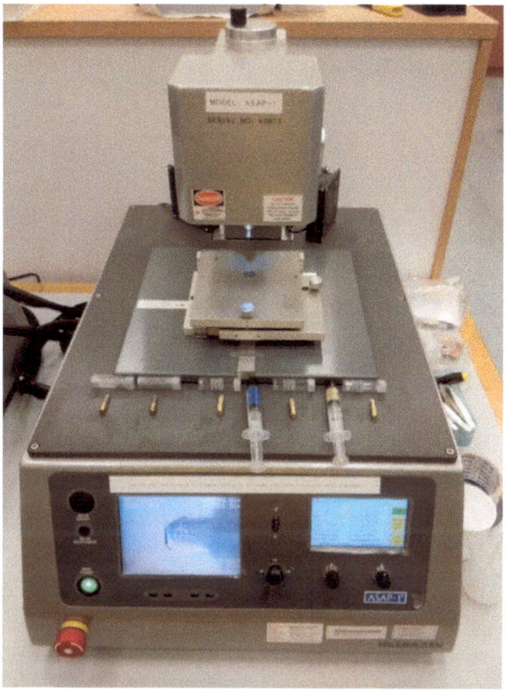

Fig. 14.2 Ultra Tec ASAP-1 polishing machine

(Fig. 14.2). The process involves two main steps: milling, to reach the desired thickness, and polishing, to achieve a mirror-like surface quality. The latter minimizes optical losses at the silicon/air interface, providing a better image quality. Depending on the step, tools of different material are used. For instance, the milling of the substrate is done with a diamond tool, while polishing involves Xylem and Xybove tools (Fig. 14.3).

Before machining, the substrate was estimated to be $\approx$300 μm thick. After processing, it was reduced to $\approx$130 μm. The estimation was performed using IR imaging and measuring the difference of focus level between the metal layers and the substrate surface. Figure 14.4 shows difference in image quality of the sample before and after substrate thinning, by using IR laser imaging and 50× magnification. We can clearly see the difference in image contrast, especially in the blocks in the top-right corner.

As mentioned before, it is possible to achieve thinner substrate but it is at the expenses of the device reliability. Indeed, it would induce higher mechanical constraints that can generate cracks. Such thickness is of interest for the use of high numerical aperture lenses or shorter operating wavelengths. With the current test setup, such objectives were not used and the laser wavelength was fixed to 1064 nm. As a conclusion, a thickness of 130 μm offered a good trade-off between energy maximization and keeping the device functional on the board.

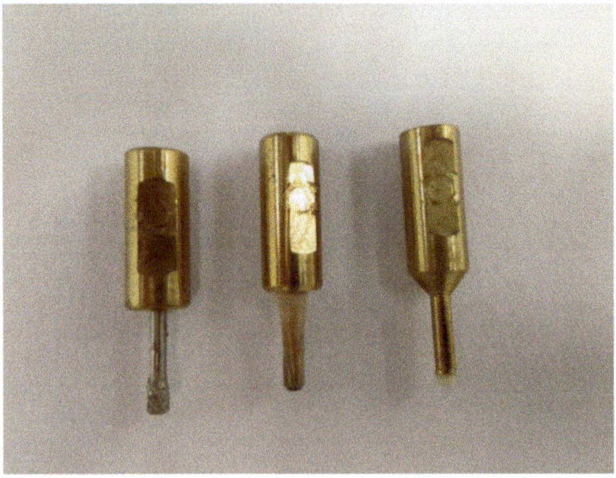

Fig. 14.3 ASAP-1 tools. From left to right: Diamond tool, Xylem tool, and Xybove tool

14.3.3 Device Under Test and Configuration

The target device, Virtex-5 FPGA (LX50T), consists of 12 metal layers, manufactured in 65-nm technology in a 1136-pin flip-chip BGA package. The device provides 3600 CLB (7200 slices) deployed in 12 clock regions. Each slice contains four 6-input look-up tables (LUTs) and four flip-flops. A number of BRAMs, digital clock mangers (DCMs), phase-locked loops (PLLs), and DSPs are located in columns of the logic resource array. A system monitor together with its temperature and power supply sensors are situated in the center of the die. Figure 14.5 (left) illustrates the basic architecture of the target device. The CLB structure in Xilinx FPGA contains two slices, together with the routing channel to a switch-box, as sketched in Fig. 14.5 (right).

The focal plane of the laser beam is critical for impacting the logic elements that are deployed under substrate. Due to the undisclosed bottom device information and the unknown dopant density in silicon that hinders the laser focalization, we had to empirically calibrate the focal plane to the active CLB layer relying on the number of generated faults, as an indicator, in a preliminary chip scan. As mentioned before, a diode pulse laser with a wavelength of 1064 nm was selected due to its superior penetration into silicon. The spot size of the chosen laser with a 5× lens was around $60 \times 14\,\mu m^2$. The output power of the laser could be adjusted with an embedded attenuator with 1% precision step from 0 to 100% of its full power strength (10 W). The entire setup for performing fault injection experiments is depicted in Fig. 14.6.

Importantly, our experiments show that only the very central part of the laser beam spot is powerful enough to trigger the faults (*"high-energy laser core"* illustrated in Fig. 14.7), which was empirically tested to be much smaller than the spot size at the substrate surface. This phenomenon is based on the nature of diode laser, and the *optical refraction* and *energy absorption* through the residual substrate ($\approx$130 µm).

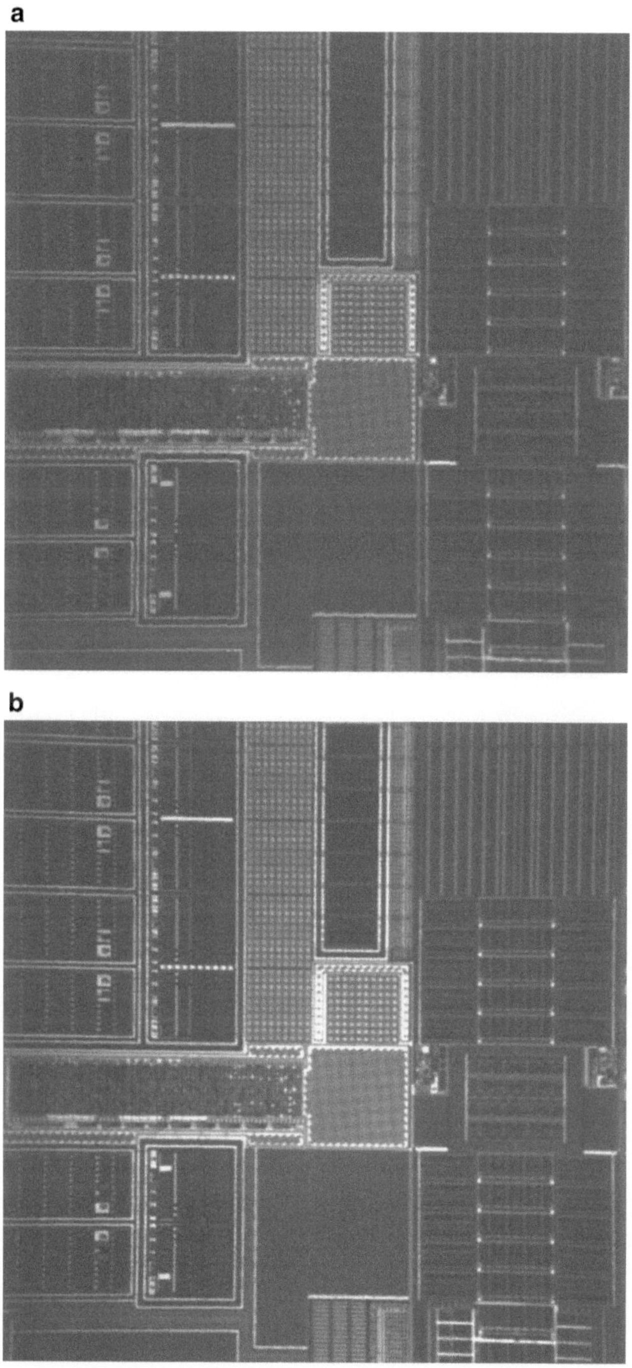

Fig. 14.4 Virtex-5 FPGA (a) before, and (b) after substrate thinning

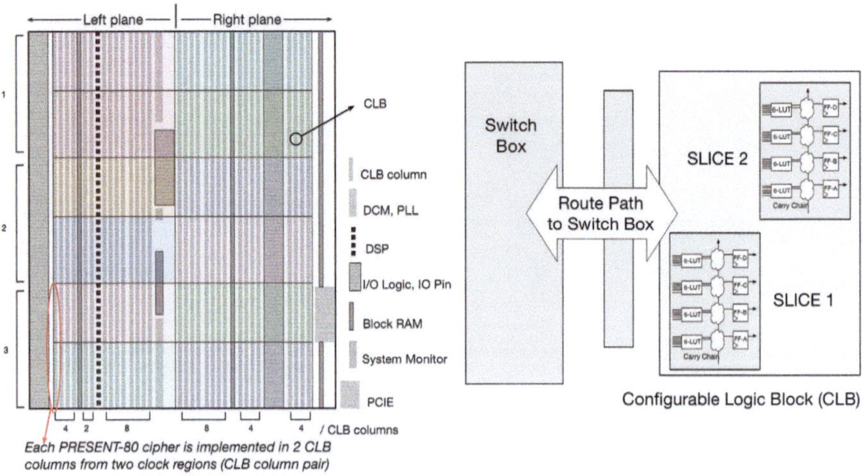

Fig. 14.5 Simplified architectural views of the target FPGA and CLB cell

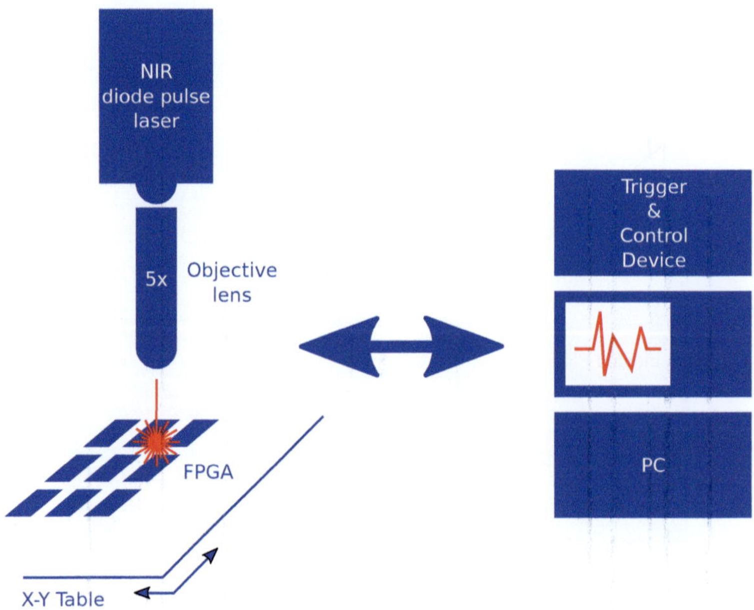

Fig. 14.6 Laser setup used for the experimental fault injection

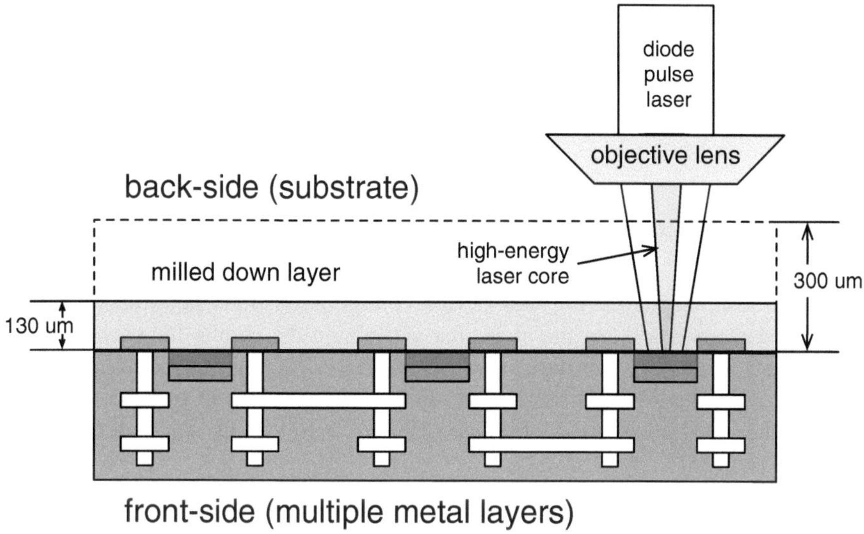

Fig. 14.7 Laser penetration through thinned silicon substrate to active transistor layer

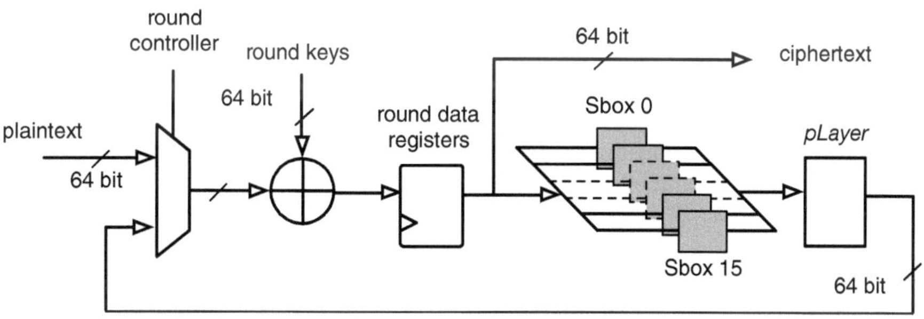

Fig. 14.8 Implemented PRESENT-80 cryptographic algorithm

Test Circuit

To allow automated profiling, it is necessary to select an adequate test circuit that allows precise localization and characterization of the injected fault. Since the aim of this chapter is to provide a profiling method for further cryptographic attacks, a cipher circuit was chosen for this purpose.

A lightweight block cipher PRESENT-80 [7] was used for profiling the logic array, which is a substitution–permutation network (SPN) with 64-bit block size, 80-bit key, and 31 computation rounds. Each round contains a key addition (`addRoundKey`), a substitution by an SBox (`sBoxLayer`), and a bit permutation (`pLayer`). Figure 14.8 illustrates the round-based architecture of the implemented cipher circuit. A single PRESENT can be tailored to be implemented in a CLB column pair. We define a CLB column pair as two adjacent CLB columns from two clock regions, as shown in Fig. 14.5 (left). We chose a CLB column pair as the cipher could not fit in a single CLB column. Moreover, the chosen CLB columns had to be vertically adjacent, as horizontally adjacent CLB columns would hinder establishment of column boundaries during the profiling.

14.4 Automated Profiling

After the sample preparation and device configuration, it is possible to use an automated method to scan the chip area and obtain the results for each position of the laser beam for given laser diode parameters. The analysis of unique faults resulting from disturbing a number of ciphers implemented in parallel allows to identify the laser sensitivity distribution of FPGA architecture—ultimately leading to device profiling.

14.4.1 Global Array Scan

We applied a strategy by implementing a large number of PRESENT-80 cipher primitives into logic resource array. Each core is restricted into a specific CLB column pair by applying the placement constraints at the implementation stage. It is remarked that other algorithms or even a simple cascaded logic chain could be used for this purpose as well. We have chosen a cryptographic algorithm in our work owing to the following advantages:

- PRESENT-80 occupies almost all the logic resources for each assigned CLB column pair, which provides a good coverage of resource occupation.
- The 31 encryption rounds provide a sufficiently large time window (31 clock cycles) to test the laser injection with varying glitch offsets.
- The exact logic points and affected timings could be simply determined by finding the collision round between the faulty ciphertext decryption and plaintext encryption.
- For the bit flips in the configuration memory of SRAM, the faults change the basic circuit configuration instead of the processed data, and it hence leads to permanent malfunction of the design [21]. The malfunction stays for the following encryptions until the FPGA is reconfigured with an uninfected bitstream. Therefore, a practical algorithm (e.g., a cipher) used here shows whether the faults are transient data bit upsets or permanent configuration bit flips in SRAM.

All the cores encrypt the same plaintext in parallel and all the output ciphertexts are compared at the output—a *tag* bit vector. The vector width is equal to the number of the implemented ciphers, and the value of each bit represents whether the corresponding cipher is correct or faulty (‘`0`’: correct; ‘`1`’: faulty). A fault in any

Algorithm 1: Automatic profiling of the device under test to laser fault injection

Input : P_S: step for increasing the laser power; D_S: step for increasing the laser pulse duration; T_S: step for increasing the trigger delay.

Output : R: table of resulting fault injections after successful fault occurred in one of the test vectors. Each entry in R contains the laser parameters, together with the coordinates and tag vector V.

```
Set laser power P := 0, laser pulse duration D := 0, and trigger delay T := 0;
do
    D := D + D_S;
    do
        P := P + P_S;
        do
            T : = T + T_S;
            for int x:= 0 to x_max do
                for int y:= 0 to y_max do
                    for int z:= 0 to z_max do
                        Run implemented test circuits in parallel;
                        Wait for the duration of T;
                        Inject the laser pulse of power P for the duration of D;
                        Obtain and analyze the tag vector V;
                        if V != 0 then
                            Store the measurement result to table R;
        while R is empty and T != T_max;
    while R is empty and P != P_max;
while R is empty and D != D_max;
return R;
```

of the PRESENT cores can be identified automatically, by checking the position of the exclusive tag bit. The scanning stage also records critical parameters, like scan coordinates, injection power, and timing. Hence, each fault can be associated to a particular cipher and specific location on chip.

The automated profiling method works according to Algorithm 1. First, the laser parameters are set to minimal values and in each iteration, one of them gets increased by a predefined step size. Then, the entire chip area is scanned with the set parameters, performing one laser injection at one spot. We note that in some cases, the z coordinate which specifies the distance between the laser source and the die surface is fixed. The algorithm analyzes the tag vector V, which contains information on whether there was any fault in one of the test circuits. In case there was a fault, it stores the measurement result into the result table R with all the important information—laser parameters, coordinates, and number of test circuit that was faulted. Increasing of the parameters is done until R is empty. In case R already contains some successful measurements after the area scan for given parameters is finished, the algorithm outputs it so that the evaluator can decide about the next steps.

Since the peripheral logic (e.g., the output comparison) also occupies some resources, it is beneficial to divide the complete die mapping into two parts: the left plane mapping and the right plane mapping. When the right part is scanned, peripheral logic can be deployed on the left side, and vice versa, to avoid control interruption. In that case, the coordinate system for the automated scanning should be adjusted accordingly.

In our case study, 48 PRESENT cores were implemented in the right region and 42 in the left side of the FPGA, corresponding to the device architecture. The results from the fault injection according to Algorithm 1 were then merged to construct the fault map of the entire chip. Relying on the recorded coordinates of each fault, we provide the 2D plot in Fig. 14.9. `X` and `Y` axes are the dimensions of the thinned chip, i.e., $12 \times 12\,\mathrm{mm}^2$. Blue dots represent the valid faults by laser injection (occurring in any single cipher). Red dots represent the unexpected invalid faults that simultaneously affected multiple ciphers.

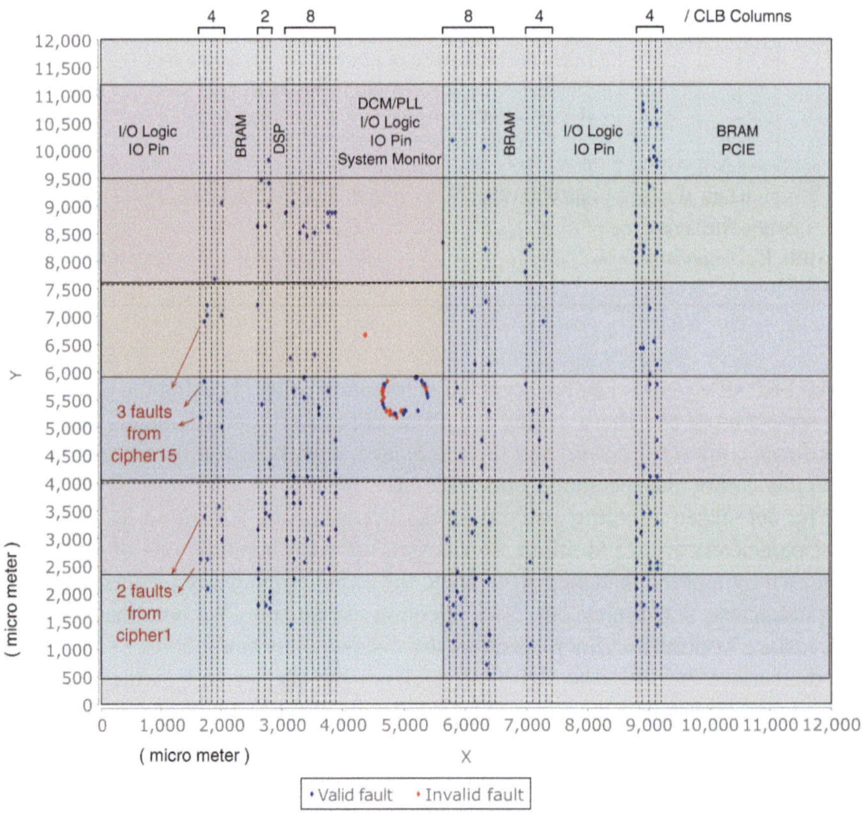

Fig. 14.9 Laser sensitivity properties of the device under test (DUT), profiled by the implemented algorithm. The plotted faults reveal the logic resource architecture of the DUT

According to our initial results, the faults from each cipher could be precisely mapped w.r.t. the chip area, as depicted in Fig. 14.9. The coordinates correspond to real dimensions of the FPGA chip in μm. Comparing the picture to the architectural view in Fig. 14.5, dimensions of other logic resources can be estimated. It is shown that the IO pad (IO Logic and IO Pin) and PCIE occupy a significant die space, and the width of BRAM and DSP are roughly equal to 4 and 2 CLB columns, respectively. Besides, there are no faults from the extreme top and bottom (gray) regions. This indicates that the active logic array does not extend to the very edge of the die. Due to the insufficient information, we could not determine the boundaries on the left IO pad region and the right BRAM&PCIE region. Nevertheless, we have clearly identified and mapped the CLB columns to the physical dimensions of the chip. Based on this mapping, we could further continue with a fine-grained scan within the CLB column to identify the laser sensitivity for slices.

14.4.2 Configurable Logic Block Column Scan

Laser fault experiments with a higher scan resolution were executed exclusively in the part of the CLB column where in total 10 CLBs (e.g., 20 slices) were occupied. In this case, the output of the FPGA was adjusted—only one PRESENT-80 was implemented in this area, and therefore the output was the ciphertext, instead of the tag vector. The `round data registers` of PRESENT-80 were implemented into the flip-flops of these CLBs. Algorithm 1 was adjusted in a way that the analysis part would backtrack the fault to determine where in the cipher the fault happened. The scan matrix for Algorithm 1 was 100 × 1400, and so totally 140,000 positions were evaluated in this CLB column, with one injection at each location. Note that either single-bit or multiple-bit faults from 4 flip-flops of each slice are tagged with the same color, which returns 20 different fault types, as plotted in Fig. 14.10. Hence, the fault sensitivity distribution of the ten CLBs can be distinctly identified, and a relative position of two slices inside each CLB can also be determined.

Figure 14.11 gives a closer view of the slice faults of CLB_6 from Fig. 14.10. The effective laser spot can impact flip-flops from both slices in this CLB, and therefore, Fig. 14.11 shows an overlapping region for this experiment. For most of the CLB regions, it was only possible to disturb the two slices from the CLB; however, the scanned regions had various sizes and different overlapping patterns. This phenomenon is mainly due to the uneven substrate layer because of manufacturing process variations, causing different energy levels of the laser beam at the logical layer. The thickness variation across the 12 × 12 mm die was within 15 μm.

Given the coordinates from both Figs. 14.10 and 14.11, the following important parameters can be estimated as follows:

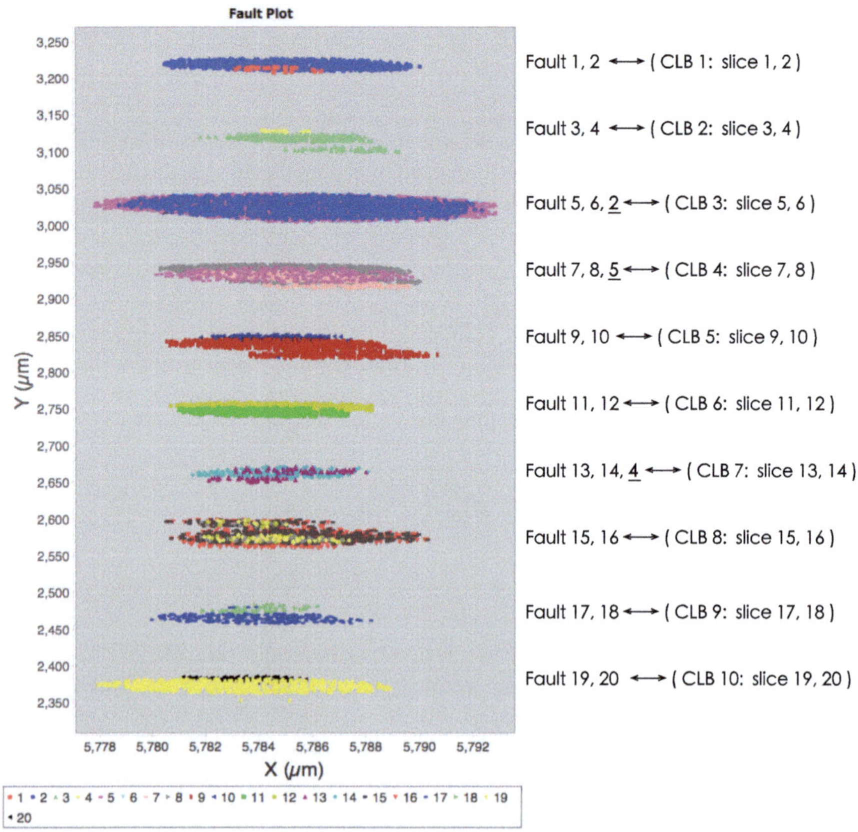

Fig. 14.10 2D laser sensitivity map from a CLB column (faults from different slices are colored differently)

- Distance between the neighboring CLBs: 60–80 μm;
- Width (X) of a CLB column: 7–15 μm;
- For this DUT, each clock region has 20 CLB rows. Regions are symmetrically divided by a global-clock routing channel. In Fig. 14.10, half of the clock region was measured, and the middle clock routing channel occupies around 700 μm. So, the height (Y) of a CLB column in a clock region (e.g., the height of the clock region) in this Virtex-5 FPGA is estimated as: $(3250 - 2350) * 2\,\mu\text{m} + 700\,\mu\text{m} \approx 2500\,\mu\text{m}$.

It should be noted that these dimensions are the laser fault sensitivity regions, instead of the precise component sizes. However, they show the critical areas that are sensitive to laser attacks. These parameters can help to efficiently navigate the laser to the POIs, for performing precise bit-level fault attacks.

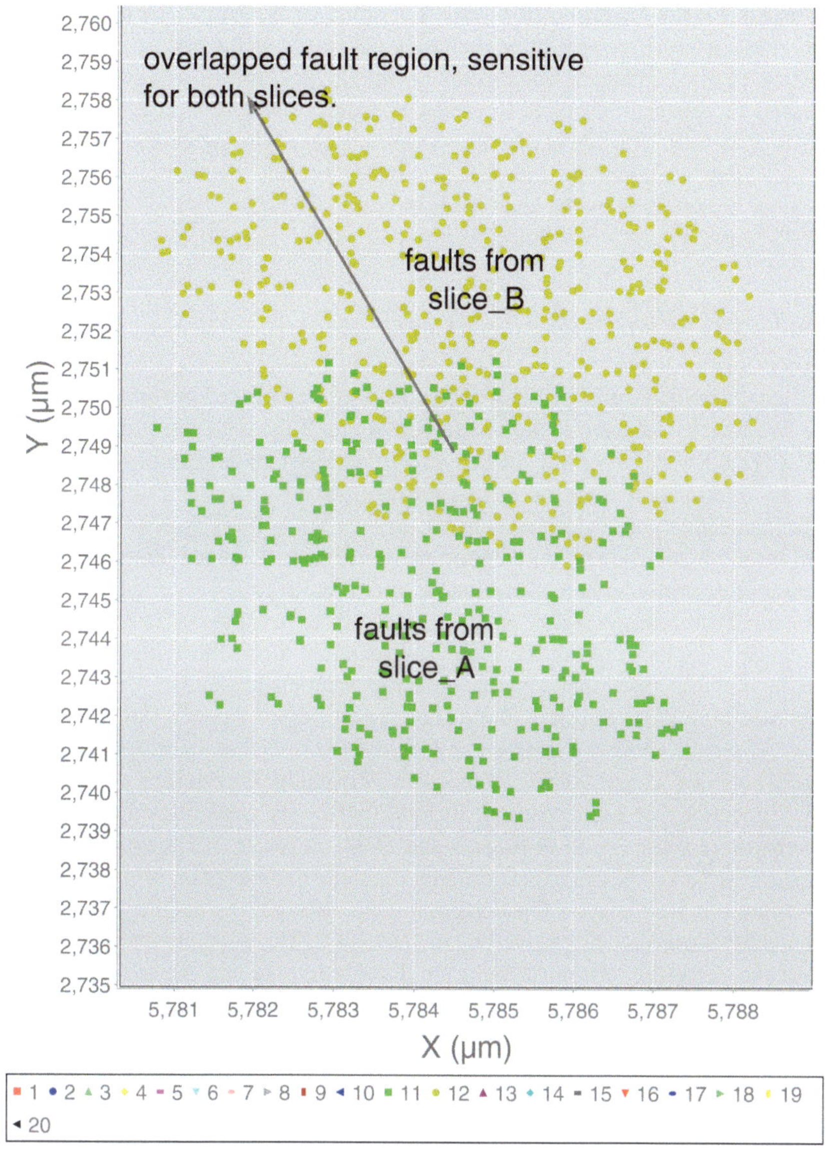

Fig. 14.11 Slice-exclusive faults for a single CLB

Discussion on the Unexpected Faults For some CLB regions, we could observe faults that showed a very different behavior compared to the rest of the faults that could be easily explained. For example, *fault_2* (denoted as a blue dot) is only supposed to appear in CLB_1. However, it occurred when the laser was targeted at CLB_3 as well. This phenomenon is mainly because the signal paths for register bits [4–7] that were deployed in slice_2 pass the routing channel close to CLB_3, and hence are affected by the laser while targeting CLB_3.

14.4.3 Flip-Flop Scan

After localizing particular CLBs, we could easily navigate the laser spot to a specific slice. We focused on a particular slice where 4 out of the total 64 round registers of PRESENT-80 were deployed. In this slice, the registers storing bits 0, 1, 2, and 3 of the intermediate state, were respectively placed in four flip-flops. The four LUTs inside this slice were left unused. In an FPGA, LUT is actually a 6-input ROM by nature, and any bit upset in this memory changes the implemented Boolean function (potentially leads to computation errors), until FPGA is refreshed by a new bitstream. Therefore, no matter whether the LUTs are used or not, it does not affect the registers implemented in the slice. The scanning algorithm remained the same as in the previous case, analyzing the ciphertext outputs for faults.

By scanning the interested single slice region ($6 \times 13\,\mu\text{m}^2$), we obtained the following results. With the laser glitch length fixed to 282 ns and the laser strength varying between 75 and 100%, we received 3918 faulty encryptions out of 10,000, with 1 injection per each position. In total, 6462 bits were flipped in the faulty ciphertexts, resulting in 3378 bit sets and 3084 bit resets. It shows that with the same laser settings, we can expect roughly the same number of bit sets and bit resets in flip-flops. If we focus on flip-flops that were affected, the majority of the faults changed the flip-flop A, as can be seen in Table 14.2. The other three flip-flops share almost the same proportion of faults. In Table 14.3, we can see the numbers for different fault models that were obtained. More than one half of all the faults were 1-bit flips, following by approximately one third of 2-bit flips. 3- and 4-bit flips were less likely to occur, however still possible to obtain. Moreover, with a high-precision scan, we could find the POIs affecting only one slice without accidentally injecting faults in neighboring slices.

Table 14.2 Percentages of faults for different registers (nonexclusive)

Register	% of faults
A	66.9
B	35.5
C	35.9
D	36.2

Table 14.3 Numbers of 1-, 2-, 3-, and 4-bit flips from the total 3918 faults

Fault model	# of faults
1-bit flip	2243
2-bit flip	947
3-bit flip	595
4-bit flip	135

Each slice in Xilinx FPGAs contains four flip-flops (FF-A, FF-B, FF-C, and FF-D). Therefore, each injection can in fact cause multiple bit flips if the laser spot is bigger than the flip-flop scale. We show the faults when two adjacent registers are flipped in Fig. 14.12. The red, green, and blue points represent 2-bit flips occurred on (FF-A, FF-B), (FF-B, FF-C), and (FF-C, FF-D), respectively, being caused by single injection. It is clearly shown that different regions overlap in X axis, caused by the effective laser spot size that covers two neighboring registers. More specifically, $X1$ and $X2$ constitute the middle lines of registers (C, D) and (A, B) in X axis ($X1 \approx 5782.4445\,\mu\text{m}$, $X2 \approx 5781.9900\,\mu\text{m}$). Due to the similarity of each register, $d/2 = (X2 - X1)/2 \approx 227\,\text{nm}$ should be roughly equal with the fault sensitive region of a single register. It is stressed that the register structure varies for devices manufactured with different technologies, and therefore this estimation is valid only for the tested Virtex-5 FPGA. However, the analysis method is applicable to other FPGA devices as well.

As mentioned before, none of the faults were found in the configuration memory. As our laser equipment was operating at its maximum capability, we could not find adequate parameters to inject configuration faults. This could be due to different structure and/or layer placement for flip-flops and configuration memory.

14.4.4 Impact of Substrate Thinning

To demonstrate the impact of thinning and polishing on laser fault injection, we repeated the experiments with another copy of the test board, where the FPGA substrate was not thinned down. Only the metal lid over the FPGA was removed. A global laser scan on the entire chip was repeated. The scan result has shown that faults only occur when conducting the laser injection in the central area of the chip, similar to the same area on a thinned sample in Fig. 14.9. The phenomenon demonstrates that only this area of the chip is sensitive to laser without any substrate thinning. We were not able to trigger any events in the active CLB logic array where the ciphers were implemented, even with the maximum laser power. Thus, we can conclude that substrate thinning is necessary in order to get exploitable transient faults with laser. The fault mechanism of the central area will be discussed in Sect. 14.5. Please note that the coordinates in all the following figures are preserved with respect to Fig. 14.9.

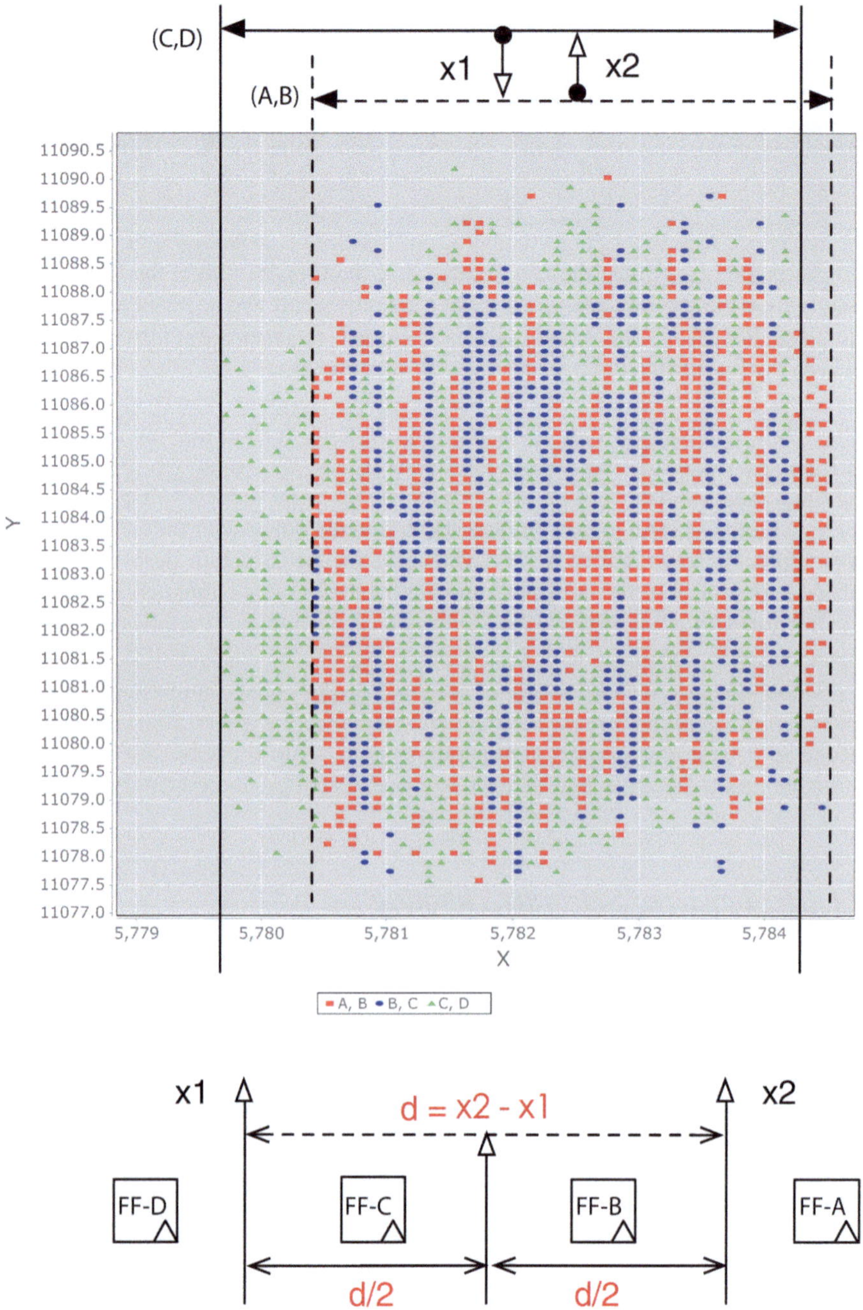

Fig. 14.12 Estimation of flip-flop laser sensitivity region based on 2-bit faults from adjacent flip-flops

14.5 Discussion

In this section, we first detail some other experiments to further analyze the fault topology and success probability. Next, we discuss the relevance of these fault models to fault attacks on cryptographic algorithms. Finally, we shed some light on the invalid faults found in the central region of the FPGA.

14.5.1 Success Rate

Apart from different types of faults, success rate is another important parameter. In this part, we determine the manipulating power of the attacker for a given target. It is important to know which laser settings are the most efficient for producing bit flips, random byte faults, etc. The objective is to ascertain the minimum power required for fault injection with each fault model.

The experiment was conducted by injecting laser with varying power in the range 0–100%. The injection campaign was performed on the POI of a slice region where 4-bit round data registers were implemented in the four flip-flops of this slice. 100 injections were performed per laser power, using PRESENT-80 encryption with random plaintext and fixed key. In Fig. 14.13, it can be observed that faults started appearing at 81% laser power. With >85% laser power, over 90% injections resulted in faults. The fault injection success went to 100%, when laser power was over 96%. These faults included both bit-flips and random byte/nibble.

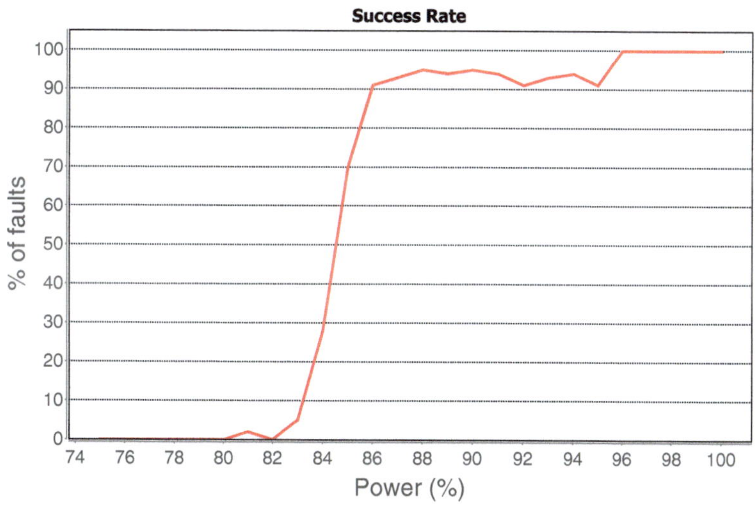

Fig. 14.13 Fault success rate for random byte flips

14.5.2 Compatibility with Cryptographic Fault Attacks

The observed fault models can now be easily translated in terms of fault-based cryptographic attacks. Proposed experiments reported laser fault injection in Virtex-5 FPGA with **single bit-flip** and **random byte fault** models. Scanning through the literature on differential [4, 5, 28] and algebraic [13, 30] fault attacks on cryptographic primitives (block ciphers, stream ciphers, hash functions, etc.), we found that majority of proposed attacks are based on these two fault models. This means that given a detailed profiling of the target device and the underlying algorithm, any cryptographic primitives can be exploited.

Dual-rail precharge logic (DPL) has been previously shown to be intrinsically resistive against most fault injections [24]. DPL generally employs complementary duplication encoding where each single logical bit is replaced by a complementary bit pair, e.g., 1 is $(0, 1)$ and 0 is $(1, 0)$. Moreover, it is a recommended practice in DPL to place complementary bit pairs in adjacent flip-flops of a slice [15] for achieving smaller silicon process variations in order to reducing the early propagation effect (EPE) [19]. Authors of [24] demonstrated that dual-rail logic resists all faults except symmetric faults which flip encoded $(0, 1)$ to $(1, 0)$ and vice versa. Faults which do not follow this pattern cannot be exploited for DFA or AFA, since they inevitably break the DPL and can be easily detected. As shown in previous experimental results, we found that 13% of random byte faults are actually symmetric, located in adjacent flip-flops. This fault pattern shows that various fault attacks can be practically realized in dual-rail protected cryptographic primitives, by stealthily injecting faults without breaking the dual-rail logic compensation. Similarly, the demonstrated fault model can also bypass error detection schemes, such as SBox parity in [29].

14.5.3 Discussion on Central Fault Region

A dense fault region appeared in the center of the FPGA die. This region was not an active CLB region and no user logic was implemented in this area. The nature of injected faults in this region was also very different from the `valid` faults, i.e., several cores were faulted by a single injection. Moreover, the faults started appearing at a much lower power (18% as compared to 81% for faults in CLB columns). To study this behavior, we have specially focused on this region with better scanning precision using a 20× laser lens. The size of the laser spot with this lens was $15 \times 3.5\,\mu m^2$. The energy density of the 20× lens was higher than that of the 5× lens. We varied the laser power from 17% to 25% of the full laser strength. Figure 14.14 gives the fault plot after the laser scan in this section. Points in different colors represent different laser strengths. Most faults were located in two regions, hereafter named "`Region A`" and "`Region B`", respectively. A very few number of faults were seen in some remote spots. A bitstream modification was never observed.

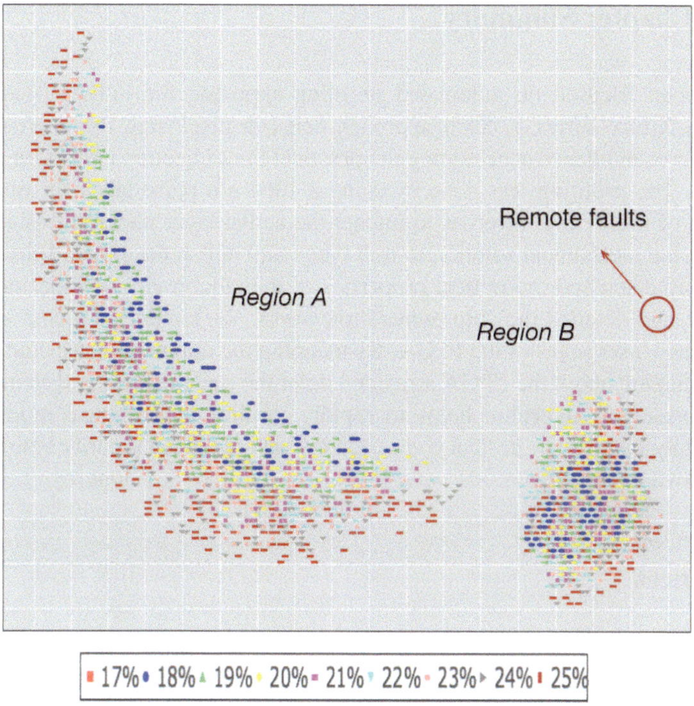

Fig. 14.14 Position and strength of faults in a laser scan focused on the center of FPGA

Due to undisclosed transistor-level device information, clarifying the internal mechanism of the faults here is challenging. Even when the cipher and its peripheral logics were placed in a distant FPGA corner, the fault characteristic of central region remained unchanged. Also, multiple ciphers could be faulted by a single injection, when targeting this region. Thus, laser injection in this region causes and propagates some global disturbance, which could affect multiple ciphers irrespective of the placement. Deeper analysis was conducted under two assumptions:

- The faults were triggered by the *global clock network*. Since the clock buffer that fans out the global clock is deployed in the die center in this FPGA, a fault on the buffer can spread to the whole chip. To validate, we removed the clock buffer and routed the clock system using the signal paths. However, the faults still persisted in the new experiment.
- The faults were triggered by the *system monitor*. System monitor is an environment sensor system (power supply, temperature, etc.), deployed near the center of the FPGA die. System monitor is activated by default and physically connected to the power network that can possibly propagate the voltage disturbance induced by laser impact. However, fresh experiments after disabling the system monitor, by connecting all of its IO pins to GND on board, still reported similar faults in central region.

14.6 Chapter Summary

This chapter focused on automated profiling approach for FPGAs, with a case study on Xilinx Virtex-5. Such approach helps in disclosing the internal device architecture, and hence accelerating the practical fault injection attacks on sensitive modules. The profiling was done by using a 1064-nm pulse laser, focused on the backside of the FPGA. In order to impact the active layer under chip surface, we relied on the mechanical solution to mill down and polish the silicon substrate. We thoroughly discussed the optical properties of the silicon circuit under laser fault injection, and detailed the chip preparation works. We conducted a chip-scale and fine-grained laser scans of the FPGA. By mapping the output data, we could restore the information about the FPGA array and defer the scale of the logic elements.

The presented algorithm helps to rapidly localize the sensitive modules and successfully identify the critical components of an embedded security system inside an unknown target chip.

References

1. M. Agoyan, J.-M. Dutertre, A.-P. Mirbaha, D. Naccache, A.-L. Ribotta, A. Tria, Single-bit DFA using multiple-byte laser fault injection, in *IEEE International Conference on Technologies for Homeland Security (HST), 2010* (IEEE, Piscataway, 2010), pp. 113–119
2. M. Alderighi, F. Casini, S. d'Angelo, M. Mancini, S. Pastore, G.R. Sechi, Evaluation of single event upset mitigation schemes for SRAM based FPGAs using the FLIPPER fault injection platform, in *22nd IEEE International Symposium on Defect and Fault-Tolerance in VLSI Systems, 2007. DFT'07* (IEEE, Piscataway, 2007), pp. 105–113
3. R.J. Anderson, *Security Engineering: A Guide to Building Dependable Distributed Systems* (Wiley, Hoboken, 2010)
4. N. Bagheri, R. Ebrahimpour, N. Ghaedi, New differential fault analysis on present. EURASIP J. Adv. Signal Process. **2013**(1), 145 (2013)
5. N. Bagheri, N. Ghaedi, S.K. Sanadhya, Differential fault analysis of SHA-3, in *Progress in Cryptology–INDOCRYPT 2015* (Springer, Cham, 2015), pp. 253–269
6. J. Beutler, Visible light LVP on bulk silicon devices, in *41st International Symposium for Testing and Failure Analysis (November 1–5, 2015)* (Asm, Novelty, 2015), pp. 1–8
7. A. Bogdanov, L.R. Knudsen, G. Leander, C. Paar, A. Poschmann, M.J. Robshaw, Y. Seurin, C. Vikkelsoe, PRESENT: an ultra-lightweight block cipher, in *Proceedings of the 9th International Workshop on Cryptographic Hardware and Embedded Systems, CHES '07* (Springer, Berlin, 2007), pp. 450–466
8. D. Boneh, R.A. DeMillo, R.J. Lipton, On the importance of eliminating errors in cryptographic computations. J. Cryptol. **14**(2), 101–119 (2001)
9. J. Breier, D. Jap, Testing feasibility of back-side laser fault injection on a microcontroller, in *Proceedings of the WESS'15: Workshop on Embedded Systems Security* (ACM, New York, 2015), p. 5
10. S.P. Buchner, F. Miller, V. Pouget, D.P. McMorrow, Pulsed-laser testing for single-event effect investigations. IEEE Trans. Nucl. Sci. **60**(3), 1852–1875 (2013)
11. G. Canivet, P. Maistri, R. Leveugle, J. Clédière, F. Valette, M. Renaudin, Glitch and laser fault attacks onto a secure AES implementation on a SRAM-based FPGA. J. Cryptol. **24**(2), 247–268 (2011)
12. F. Courbon, P. Loubet-Moundi, J.J.A. Fournier, A. Tria, Adjusting laser injections for fully controlled faults, in *International Workshop on Constructive Side-Channel Analysis and Secure Design* (Springer, Cham, 2014), pp. 229–242

13. N.T. Courtois, K. Jackson, D. Ware, Fault-algebraic attacks on inner rounds of DES, in *e-Smart'10 Proceedings: The Future of Digital Security Technologies* (Strategies Telecom and Multimedia, Montreuil, 2010)
14. J.-M. Dutertre, A.-P. Mirbaha, D. Naccache, A. Triaz, Reproducible single-byte laser fault injection, in *Conference on Ph.D. Research in Microelectronics and Electronics (PRIME), 2010* (IEEE, Piscataway, 2010), pp. 1–4
15. W. He, A. Otero, E. de la Torre, T. Riesgo, Customized and automated routing repair toolset towards side-channel analysis resistant dual rail logic. Microprocess. Microsyst. **38**(8), 899–910 (2014)
16. F.L. Kastensmidt, L. Tambara, D.V. Bobrovsky, A.A. Pechenkin, A.Y. Nikiforov, Laser testing methodology for diagnosing diverse soft errors in a nanoscale SRAM-based FPGA. IEEE Trans. Nucl. Sci. **61**(6), 3130–3137 (2014)
17. P. Kocher, J. Jaffe, B. Jun, Differential power analysis, in *Annual International Cryptology Conference* (Springer, Berlin, 1999), pp. 388–397
18. H. Lohrke, P. Scholz, C. Boit, S. Tajik, J.-P. Seifert, Automated detection of fault sensitive locations for reconfiguration attacks on programmable logic, in *Proceedings of the 42nd International Symposium for Testing and Failure Analysis* (ASM, Washington, 2016), pp. 1–6
19. A. Moradi, V. Immler, Early propagation and imbalanced routing, how to diminish in FPGAS, in *Cryptographic Hardware and Embedded Systems–CHES 2014* (Springer, Berlin, 2014), pp. 598–615
20. J.C.H. Phang, D.S.H. Chan, M. Palaniappan, J.M. Chin, B. Davis, M. Bruce, J. Wilcox, G. Gilfeather, C.M. Chua, L.S. Koh, H.Y. Ng, S.H. Tan, A review of laser induced techniques for microelectronic failure analysis, in *Proceedings of the 11th International Symposium on the Physical and Failure Analysis of Integrated Circuits. IPFA 2004*, July (2004), pp. 255–261
21. V. Pouget, A. Douin, D. Lewis, P. Fouillat, G. Foucard, P. Peronnard, V. Maingot, J. Ferron, L. Anghel, R. Leveugle, R. Velazco, Tools and methodology development for pulsed laser fault injection in SRAM-based FPGAs, in *8th LATW'07*, (IEEE Computer Society, Piscataway, 2007)
22. C. Roscian, J.-M. Dutertre, A. Tria, Frontside laser fault injection on cryptosystems-application to the AES'last round, in *IEEE International Symposium on Hardware-Oriented Security and Trust (HOST), 2013* (IEEE, Piscataway, 2013), pp. 119–124
23. C. Roscian, A. Sarafianos, J.-M. Dutertre, A. Tria, Fault model analysis of laser-induced faults in SRAM memory cells, in *Workshop on Fault Diagnosis and Tolerance in Cryptography (FDTC), 2013* (IEEE, Piscataway, 2013), pp. 89–98
24. N. Selmane, S. Bhasin, S. Guilley, T. Graba, J.-L. Danger, WDDL is protected against setup time violation attacks, in *Workshop on Fault Diagnosis and Tolerance in Cryptography (FDTC), 2009* (IEEE, Piscataway, 2009), pp. 73–83
25. B. Selmke, S. Brummer, J. Heyszl, G. Sigl, Precise laser fault injections into 90 nm and 45 nm SRAM-cells, in *International Conference on Smart Card Research and Advanced Applications* (Springer, Cham, 2015), pp. 193–205
26. P. Swierczynski, G.T. Becker, A. Moradi, C. Paar, Bitstream fault injections (BIFI) – automated fault attacks against SRAM-based FPGAS. IEEE Trans. Comput. **PP**(99), 1–14 (2017)
27. S. Tajik, H. Lohrke, F. Ganji, J.P. Seifert, C. Boit, Laser fault attack on physically unclonable functions, in *2015 Workshop on Fault Diagnosis and Tolerance in Cryptography (FDTC)*, September (2015), pp. 85–96
28. M. Tunstall, D. Mukhopadhyay, S. Ali, Differential fault analysis of the advanced encryption standard using a single fault, in *IFIP International Workshop on Information Security Theory and Practices* (Springer, Berlin, 2011), pp. 224–233
29. K. Wu, R. Karri, G. Kuznetsov, M. Goessel, Low cost concurrent error detection for the advanced encryption standard, in *Proceedings 2004 International Test Conference, ITC 2004* (IEEE, Piscataway, 2004), pp. 1242–1248
30. F. Zhang, S. Guo, X. Zhao, T. Wang, J. Yang, F.-X. Standaert, D. Gu, A framework for the analysis and evaluation of algebraic fault attacks on lightweight block ciphers. IEEE Trans. Inf. Forensics Secur. **11**(5), 1039–1054 (2016)

Index

J. Breier et al. (eds.), *Automated Methods in Cryptographic Fault Analysis*,
https://doi.org/10.1007/978-3-030-11333-9

MIX
Papier aus verantwortungsvollen Quellen
Paper from responsible sources
FSC® C105338

If you have any concerns about our products,
you can contact us on
ProductSafety@springernature.com

In case Publisher is established outside the EU,
the EU authorized representative is:
Springer Nature Customer Service Center GmbH
Europaplatz 3, 69115 Heidelberg, Germany

Printed by Libri Plureos GmbH
in Hamburg, Germany